Davor Margetić

High Pressure Organic Synthesis

Also of interest

Organic Chemistry.
Fundamentals and Concepts
McIntosh, 2018
ISBN 978-3-11-056512-6, e-ISBN 978-3-11-056514-0

Organophosphorus Chemistry.
Novel Developments
Keglevich (Ed.), 2018
ISBN 978-3-11-053453-5, e-ISBN 978-3-11-053583-9

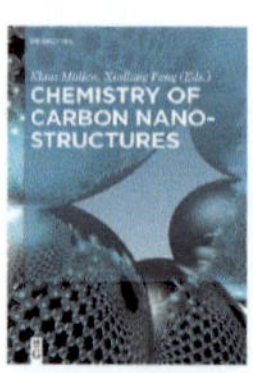

Chemistry of Carbon Nanostructures.
Muellen, Feng (Eds.), 2017
ISBN 978-3-11-028450-8, e-ISBN 978-3-11-028464-5

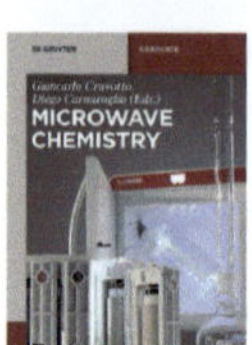

Microwave Chemistry.
Cravotto, Carnaroglio, 2017
ISBN 978-3-11-047992-8, e-ISBN 978-3-11-047993-5

Pharmaceutical Chemistry.
Volume 1: Drug Design and Action
Volume 2: Drugs and Their Biological Targets
Campos Rosa, Camacho Quesada, 2017
Set-ISBN 978-3-11-056680-2

Davor Margetić

High Pressure Organic Synthesis

DE GRUYTER

Author
Dr. Davor Margetić
Ruđer Bošković Institute
Division of Organic Chemistry and Biochemistry
Bijenička 54
HR-10000 Zagreb
Croatia
Davor.Margetic@irb.hr

ISBN 978-3-11-055595-0
e-ISBN (PDF) 978-3-11-055684-1
e-ISBN (EPUB) 978-3-11-055602-5

Library of Congress Control Number: 2019934660

Bibliographic information published by the Deutsche Nationalbibliothek
The Deutsche Nationalbibliothek lists this publication in the Deutsche Nationalbibliografie; detailed bibliographic data are available on the Internet at http://dnb.dnb.de.

Typesetting: Integra Software Services Pvt. Ltd.
Printing and binding: CPI books GmbH, Leck
Cover image: moolyboo/iStock/getty images

www.degruyter.com

Dedicated to Professor Ron Warrener on the occasion of his 85th birthday

Preface

Today, the extremely high pressure (HP) is an experimental method used widely in synthetic organic chemistry. Whereas in earlier times, studies of reactions in hyperbaric conditions were more focused on the investigation of the mechanistic and physical–organic aspects. Nowadays, as the physicochemical effects of pressure and the synthetic utility of the technique are recognized, HP becomes an important tool for synthesis. Many examples have been given for the reactions promoted by pressure, which otherwise could not be accomplished. This is the third monograph sparked by my continuing interest in "green chemistry" methodologies, and follows the books on microwave reactions (*Microwave Assisted Cycloaddition Reactions*, Nova Science Publishers, New York 2011) and solid state mechanochemistry (*Mechanochemical Organic Synthesis*, Elsevier, Amsterdam 2016).

This monograph covers the vast number of publications up to 2018 dealing with high-pressure organic reactions. The book is divided into nine chapters covering different chemical transformations, which are preceded by an introductory chapter on the experimental aspects of the technique.

The cover of this book depicts a macadamia nut cracker, which is used to open the toughest nuts in the world, a tree species indigenous to Australia also known as Queensland nut. These nuts are hard to crack – similar to many of the reactions described in this book. A hard hammer blow will crush the nuts, whereas the cracker slowly imposes pressure until shell fractures. Macadamias were introduced to me at the same time as the HP technique by Professor Ron Warrener, Central Queensland University, to whom I am very indebted. Also I would like to thank Walter de Gruyter GmbH publishing house for accepting this book and my family for their understanding and support during the manuscript preparation.

https://doi.org/10.1515/9783110556841-201

Contents

Abbreviations

Δ	conventional (classical) heating
Ac	acetyl
AcOH	acetic acid
AcOEt	ethylacetate
AIBN	azobisisobutyronitrile
All	allyl
Ar	aryl
BINAP	2,2′-bis(diphenylphosphino)-1,1′-binaphthyl
Bn	benzyl
Boc	*t*-butoxycarbonyl
Bu	*n*-butyl
t-Bu	*tert*-butyl
Bz	benzoyl
CAAC	copper catalyzed alkyne-azide 1,3-dipolar cycloaddition
Cbz or Z	benzyloxycarbonyl
CD	cyclodextrin
Chex	cyclohexyl
CPD	cyclopentadiene
DA	Diels–Alder reaction
DABCO	1,4-diazabicyclo[2.2.2]octane
DBN	1,5-diazabicyclo[4.3.0]non-5-ene
DBU	1,8-diazabicyclo[5.4.0]undec-7-ene
DCE	1,2-dichloroethane
DCC	dicyclohexylcarbodiimide
DCM	dichloromethane
DEM	diethyl mesoxalate
DDQ	1,2-dichloro-4,5-dicyano-1,4-benzoquinone
DIPEA	diisopropylethyl amine
DMA	*N,N*-dimethylacetamide
DMAD	dimethylacetylene dicarboxylate
DMAP	4-(dimethylamino)pyridine
DMC	dimethylcarbonate
DME	1,2-dimethoxyethane
DMF	*N,N*-dimethylformamide
dmpe	1,2-bis(dimethylphosphino)ethane
DMSO	dimethyl sulfoxide
DNA	deoxyribonucleic acid
dppp	1,3-bis(diphenylphosphino)propane
dppe	1,2-bis(diphenylphosphino)ethane
dppy	2-(diphenylphosphino)pyridine
ee	enantiomeric excess
Et	ethyl
EtOH	ethanol
Et_2O	diethyl ether
EWG	electron withdrawing group

https://doi.org/10.1515/9783110556841-202

FG	functional group
Fmoc	9-fluorenylmethoxycarbonyl
h	hour
HDA	hetero Diels–Alder reaction
HFIP	1,1,1,3,3,3-hexafluoro-2-propanol, hexafluoroisopropanol
HMPT	hexamethylphosphortriamide
HP	high pressure
h*ν*	photochemical irradiation
IL	ionic liquid
IMDA	intramolecular Diels–Alder reaction
IPB	isopropyl benzene (cumene)
mCPBA	*m*-chloroperoxybenzoic acid
MCH	methylcyclohexane
Me	methyl
MeCN	acetonitrile
MEK	methyl ethyl ketone (butanone)
MeOH	methanol
min	minute
MTBE	2-methoxy-2-methylpropane
MVK	methyl vinyl ketone
MW or μW	microwave
NMM	*N*-methylmorpholine
NMP	*N*-methylpyrrolidone
NMR	nuclear magnetic resonance
ODCB	orthodichlorobenzene (1,2-dichlorobenzene)
PBH	pinacol borane
PEG	polyethylene glycol
Ph	phenyl
Phth	*N*-phthalimido
PMB or MPM	*p*-methoxybenzyl
Pr	propyl
PTAD	4-phenyl-1,2,4-triazolin-3,5-dione
Py	pyridine
QD	quinidine
QN	quinine
RT	room temperature
TBAB	tetrabutylammonium bromide
TBAF	tetrabutylammonium fluoride
TBDMS	*t*-butyldimethylsilyl
TBDPS	*t*-butyldiphenylsilyl
TBS	*t*-butyldimethylsilyl
TCE	2,2,2-trichloroethanol
TCNE	tetracyanoethylene
Tf	trifluoromethanesulfonyl
TFE	2,2,2-trifluoroethanol
THF	tetrahydrofuran
TMS	trimethylsilyl

TMSCN	trimethylsilyl cyanide
Tol	tolyl
Tos	tosyl
Trt	triphenylmethyl, trityl
UV	ultraviolet

1 Introduction

1.1 Historical background

Studies of the influence of high pressure (HP) on chemical compounds and reactions have started at the end of nineteenth [1] and at the beginning of the twentieth century [2]. The investigations of chemical reactions under HP were closely linked by the advancement of HP technology [3], for which Percy Bridgman have received the Nobel Prize in physics in 1946. The early investigations of organic chemical reactions in the liquid phase at extremely HPs have been carried by Cohen, Bridgman, Conant and Tammann. The hydrolysis of esters was investigated by Cohen [4] as well as the inversion of cane sugar [5, 6]. Polymerizations of conjugated dienes were the topic of work by Bridgman [7], Conant [8, 9] and Tammann [10]. Two seminal books by Cohen have greatly influenced the field: *Piezochemie kondensierter Systeme* (1919) [11] and *Physico-chemical Metamorphosis and Some Problems in Piezochemistry* (1926) [12]. The research on organic reactions under HP was intensified in the golden age of physical organic chemistry, resulting in many books and reviews. Today, the HP technique is an indispensable tool in organic synthesis and complements other methods.

1.2 Basic physicochemical principles

The rate enhancement of reactions under HP stems from the change of the rate constants as a function of pressure defined by the thermodynamic relationship given by eq. (1.1). Here k stands for rate constant, p the pressure and $\Delta V^{\neq}$ the activation volume [13]. In addition to the acceleration of reaction rates, pressure could have effect on the physicochemical properties of organic molecules, such as acido-basic properties [14, 15]. By using activation volume value of $\Delta V^{\neq} = -20\ \mathrm{cm^3\ mol^{-1}}$ and eq. (1.1), the influence of pressure on rate constants could be estimated (at $T = 25\ °C$), where k_p is the rate constant at a given pressure and k_{1bar} is the rate constant at ambient pressure (Table 1 1) [16]. At 15 kbar, the enormous rate increase by five orders of magnitude is predicted. Similarly, large negative $\Delta V^{\neq}$ of $-33\ \mathrm{cm^3\ mol^{-1}}$ obtained experimentally

https://doi.org/10.1515/9783110556841-001

Table 1.1: Rate constants change by an increase in pressure.

Pressure /kbar	k_p/k_{1bar}
1	2.2
5	5.7×10^1
10	3.2×10^3
15	1.8×10^5

Reproduced with permission from Synthesis, 1986, Copyright Georg Thieme Verlag KG

translates into a rate enhancement of approximately 10^3 at pressures between 10 and 20 kbar [17, 18].

$$\delta \ln k/\delta p = -\Delta V^{\neq}/RT \tag{1.1}$$

Beneficial effect of HP on the rates of reactions of hydrolysis of epoxides was established by le Noble. A detailed kinetic study of acid-catalyzed hydrolysis of trimethylene oxide was carried out in water over a pressure range of 6.8 kbar, and the rate increase is illustrated in Figure 1.1 [19]. Another illustration of the correlation of kinetics versus pressure in the homo Diels–Alder reaction under pressure from 1 to 2166 bar is given by Kiselev [20].

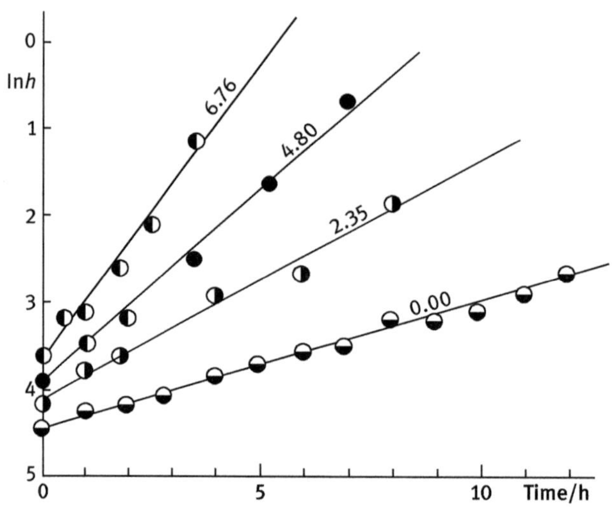

Figure 1.1: A first-order plot of the hydrolysis of trimethylene oxide catalyzed by 0.003 M perchloric acid at 25 C at various pressures (*h*, peak height in gas chromatogram). Adapted with permission from W. J. le Noble, M. Duffy The Journal of Physical Chemistry A 1964, 68, 619. Copyright (1964) American Chemical Society.

The activation volume ($\Delta V^{\neq}$) is defined as the difference in molar volumes between the transition state and the reactants (eq. (1.2)). It is obvious from eq. (1.2) that reactions with negative activation volumes will be accelerated by pressure. Many bond-forming (condensation) reactions such as cycloadditions have negative activation volumes, and some characteristic values for various chemical reactions are listed in Tables 1.2–1.4. Their absolute values are influenced by the solvents and temperature. On the other hand, some processes with a positive reaction volumes such as E1cB eliminations are decelerated by pressure [21].

$$\Delta V^{\neq} = V^{\neq}_{(TS)} - \sum V_{(reactants)} \tag{1.2}$$

Table 1.2: Experimentally determined activation ($\Delta V^{\neq}$) and reaction volumes (ΔV)[a] for Diels–Alder reactions.

Reactions	$\Delta V^{\neq}$	ΔV	Reference
Diels–Alder	-37.4_{EtOAc}		Int J Chem Kinet 2013 613
	-39.8_{DCM}		
	-32.5_{MeNO2}		
	-39.3_{DMC}		
	-37.5_{MeCN}		
	$-38.5_{DiiPrether}$		
	$-38.0_{Clbutane}$		
	$-39.0_{Acetone}$		
	-37.0_{DCE}		
	-38.1 to -32.2_{EtOH}		JPOC 2004 221
	-37.2 to -35.3_{CHCl3}		
	$-32.8_{DCM.\ 60\ °C}$		Liebigs 1995 1
	$-34.2_{DCM.\ 100\ °C}$		
	-38.1 to -31.5_{BrBu}		TH 1972 3113
	-36.5_{neat}		JACS 1962 152
	-21.4_{IPB}	-25.5 to -32.1_{DMF}	BCSJ 1985 2490
	-30.0_{IPB}	-33.5_{IPB}	
	-16.8_{cumen}	-25.5_{cumen}	Chem Lett 1984 1855
	-24.8 to -36.6_{Tol}		Perkin 2 1975 1555
HDA–O	-20.1 to -26.0_{DCM}		JACS 1988 4065
	-23.4 to -24.2_{DCM}		Ber 1989 1179
	-24.4 to -35.0_{DCM}		Ber 1992 2249
	-28.5 to -32.5_{DCM}		HP Res 1990 638
Cyclic dienes	-37.4_{EtOAc}		Aiche J 1970 766
IMHDA–O	-33.1_{DCM}		Ber 1994 2241
	-34.2_{THF}		
	-13.4_{Tol}		
	-17.0_{MeCN}		

[a] in $cm^3\ mol^{-1}$

Table 1.3: Experimentally determined activation ($\Delta V^{\neq}$) and reaction volumes (ΔV)[a] for various cycloaddition reactions.

Reaction	$\Delta V^{\neq}$	ΔV	Reference
1,3-Dipolar	-17.8 to -23.6_{ClBz}	-25.1 to -27_{ClBz}	JOC 1983 1035
	-15.3 to -23.6_{MeCN}	-27.8 to -16.7_{MeCN}	
	-24_{Hex}	-34.9_{Hex}	
	-23.2_{Tol}	-26.8_{Tol}	
	$-30.0,-45.9_{Tol}$		Perkin 2 1975 1555
[2 + 2]	-29.5 to -45.3_{Tol}		Perkin 2 1975 1555
	28.0_{neat}		BCSJ 1996 1667
	-35 to -37_{DCM}		JACS 1979 151
	-20.9_{DCM}		Rec Trav Chim 2010 267
	-24.5_{MeCN}		
	-14.9 to -17.8_{neat}		OM 1992 490
	-44_{Bz}		Perkin 2 1983 37
	-52_{Tol}		
	-9_{Tol}		Aust JC 1980 1419
	-10_{MeOH}		
	-50_{CCl4}		THL 1973 3773
	-43_{Bz}		
	-37_{DCM}		
	$-35_{Acetone}$		
	-29_{MeCN}		
[2 + 2] Cycloreversion	-16.7_{MeOH}		JACS 1975 5398
	-2 to -3_{DCM}		TH 1998 2771
[4 + 4]	-15.5 to -15.8_{Tol}		JOC 1989 5016
[6 + 4]	-7.5 to -30.5_{Dioxan}		Perkin 2 1986 1491
	-30.1 to -32.6_{DMF}		BCSJ 1987 977
	-33.1 to -37.6_{Cumene}		JACS 1975 5398

[a] in $cm^3\ mol^{-1}$

The decrease in the volume at the transition state is due to the volume shrinkage due to the bond-forming processes. Important contribution to the extent of the volume decrease is a result of the increasing solvation of the formed charges. This effect is generally referred to as electrostriction and represents the volume effect generated from changes in solute–solvent interactions during the process [22]. Particularly important electrostriction contributions are obtained in reactions involving ionic species such as zwitterions. These species interact with each other as well as with the medium, exerting the volume effects by orientation, polarization and attraction, which compress the overall volume. For ionogenic reactions, the volume of activation is composed of two terms (eq. (1.3)):

Table 1.4: Experimentally determined activation ($\Delta V^{\neq}$) and reaction volumes (ΔV)[a] for various reactions.

Reaction	$\Delta V^{\neq}$	**ΔV**	**Reference**
Menshutkin	-21.4 to -50_{MeCN}		JOC 1989 570
			JACS 1975 1778
	-26 to -47_{MeOH}		JCS 1961 146
	-21.9 to $-35_{acetone}$		TH 1970 4119
Cope	-8.8 to -13.3_{Tol}		HP Res 1994 7
Michael	-19.7_{THF}		NJC 1999 525
	-31.5_{H2O}		
	-46.5_{CHCl3}		
aza-Michael	-60 to -65_{MeCN}		TH 1996 13557
	-65_{C6F14}		JPOC 2003 265
	-55_{Et2O}		JPOC 1999 619
	-54_{CHCl3}		
	-56_{MeCN}		
	-35_{MeOH}		
	$-33_{ethylene\ glycol}$		
	-21_{DMF}		
	-25_{H2O}		
Baylis–Hillman	-50 to $-80_{Acetone}$		THL 1986 5007
Heck	-5 to -37_{MeCN}		EJOC 2003 2375
Wittig	-19_{DCM}		THL 1986 995
	-21_{Dioxan}		
	-29_{MeCN}		
Henry nitroaldol	-20.5 to -25.2_{neat}		NJC 1999 525
Ketone reduction	-12.2 to -20_{H2O}		Perkin 2 1997 1465
[1, 7] Shift	$-5.45_{Bz/Tol}$		THL 1988 3021
	-5.70_{EtOH}		
	$-5.15_{EtOH/H2O}$		
[1, 5] Shift	$+10$ to $-30_{diglyme}$		JACS 1978 5961
	-26.5_{CDCl3}		JACS 1983 3988
	-4 to12.5_{Bz}		Chem Lett 1986 1203
	-2.2_{Bu2O}		
Valence isomers	$+5_{Pyr}$		JACS 1982 3150
	-12_{BrBz}		
E/*Z* isomerization	-6.9 to -12.7_{pent}		TH 1978 887

Table 1.4 (continued)

Reaction	$\Delta V^{\neq}$ [a]	ΔV	Reference
Ar electrophilic nitration	-10 to -23.5_{AcOH}		Trans Farad Soc 1961 2231
S_NAr–N	-8 to -64_{H2O}		JACS 1959 3504
Ene	-30.9 to -42.1_{DCM}		Perkin 2 1992 137
Polymerization	-16.5 to -55_{CHCl3}		Chem Comm 1979 851
Allylation	-30.3 to -32.1_{DCM}		TH 2010 1570
H-transfer	-13 to -35_{MeCN} -25.9 to -38.6_{MTBE}		Perkin 2 1978 709[24, 25] CEJ 2004 2707
Ester aminolysis	-19.3_{MeOH}		J Prakt C 1997 359
C–C homolytic cleavage	$+11.4$		HP Res 1991 133
C–N cleavage	-2_{H2O} -4_{MeCN} -15_{CHCl3}		JACS 1972 5402 J Prakt C 1996 691
Maillard	-12_{H2O}		Acta C Scand 1979 275
Amadori	-14_{H2O} formation $+17_{H2O}$ decomp.		JPOC 1996 639

[a] in cm^3mol^{-1}

$$\Delta V^{\neq} = \Delta V^{\neq}{}_{S} + \Delta V^{\neq}{}_{\varepsilon} \tag{1.3}$$

In this equation, $\Delta V^{\neq}{}_{S}$ represents the standard structural activation volume change caused by alterations in the nuclear positions of two reactants during the development of the transition state. $\Delta V^{\neq}{}_{\varepsilon}$ is the volume change caused by electrostriction interactions. Since the structural activation volume is affected by steric effects, it could be divided into two terms that are given in eq. (1.4): $\Delta V^{\neq}{}_{\sigma}$ (steric hindrance contribution to activation volume) and $\Delta V^{\neq}{}_{o}$ (the activation volume of the unhindered reference reaction) [23].

$$\Delta V^{\neq}{}_{S} = \Delta V^{\neq}{}_{o} + \Delta V^{\neq}{}_{\sigma} \tag{1.4}$$

1.3 Technical details (equipment)

1.3.1 Piston-cylinder apparatus

The basic equipment employed in HP organic synthesis is the piston-cylinder apparatus, which could generate HPs (commonly up to 30 kbar) and at the same time could work at elevated temperatures (up to 200 °C). Besides the home-built HP systems, several of designs are commercially available and some of the technical designs are illustrated in Figures 1.2 and 1.3. All these apparatuses operate in the pressure region above 1 kbar (sometimes called kilo bar region) [26], where 1 kbar = 0.1 GPa = 100 MPa.

Figure 1.2: Hofer HP apparatus at Central Queensland University (CQU), Rockhampton, Australia.

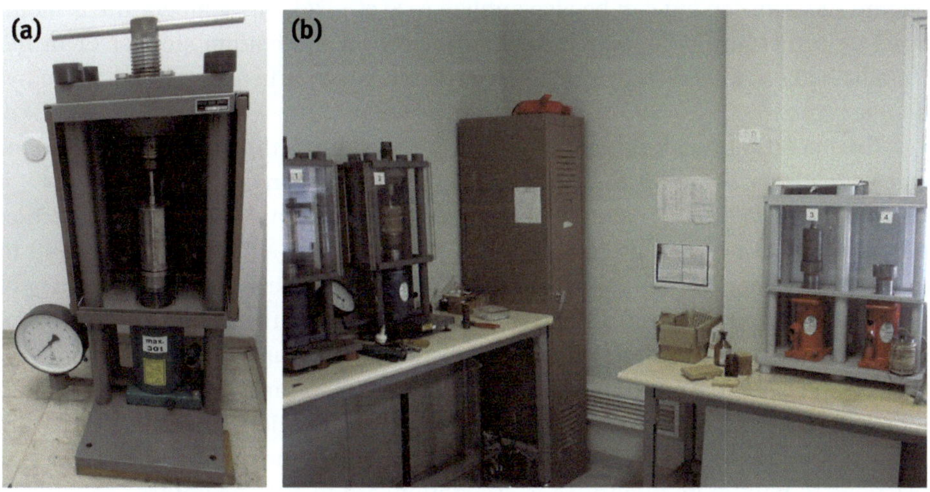

Figure 1.3: HP piston apparatuses (a) at the Ruđer Bošković Institute, Zagreb, and (b) at CQU, Rockhampton.

The main construction details of piston-cylinder HP-type apparatus are outlined in Figure 1.4a [28]. The HP vessel is made of two external steel rings with an internal conical steel vessel. The vessel is closed from the bottom with a steel stopper and from above by a mobile piston. The mobile piston and the stopper below are sealed using rubber O-rings and brass sealing rings or equipped with Bridgman triple seals (rubber, Teflon and bronze). For the reactions performed at higher temperatures, an external heating jacket was used. The whole reaction chamber is being placed in a press frame and the apparatus was raised to the desired pressure by hydraulic press by manual application of pressure by a hand pump [29] or by automatic pressurizing device. To avoid damaging the apparatus, it is important to follow the guidelines for the maximum pressure/temperature conditions that are given by the manufacturer.

Similar general design is applied in the apparatus for reactions with gaseous reactants, which has a gas inlet (Figure 1.4b) [30]. Here, the internal reactor part is a beryllium brass vessel (E) which is supported by two steel rings (C and D). External jacket (G) is available for circulating the heating medium. Thermocouple in the bottom stopper H serves for temperature monitoring. For the sealing (F) of the stopper (H) and the upper mobile piston (A), the O-ring in conjunction with Bridgeman-type metallic sealing was used. Gas could be introduced through the inlet B. The other design of HP instrument features Kennedy-type double action hydraulic press (with upper and lower presses) [31, 32].

Rarely, organic reactions are carried out in a diamond anvil cell, where a minute quantities of material are used, just hundreds of nanoliters [33].

Heating of the HP system could be achieved either by an in-built heating mantle with circulating heating fluid or simply by an external flexible heating tape (Figure 1.5). The cooling can be also applied (−20 °C) [34]. The temperature is monitored by a thermocouple, which is either placed in the bottom stopper or in the metallic body of the reactor chamber. A pressure gauge is usually placed at the hydraulic press.

The operating (working) volume greatly depends on the design of HP reactor and varies from about 10 to 100 mL. Although the settings shown in Figure 1.4 could be used to directly insert reaction mixture, it is more common that reaction mixtures are inserted in the HP reactor in small reaction ampules and the void in the reaction chamber is filled with the pressure-transmitting medium. A variety of piezotransmitter liquids was used such as castor oil, pentane, hexane, ligroin, kerosene, gasoline, ethanol, dioctyl adipate, silicone oil, mixture of decaline and isooctane (1:1). The actual reaction volume of HP reactor is smaller than the reported total working volume and depends on the type of reaction vessel used and the space in the reactor bore, which is filled with pressure transfer fluid. The scaling up reactions to multigram quantities is straightforward in HP apparatuses capable of holding a 50 mL or larger reaction volumes. The engineering problems associated

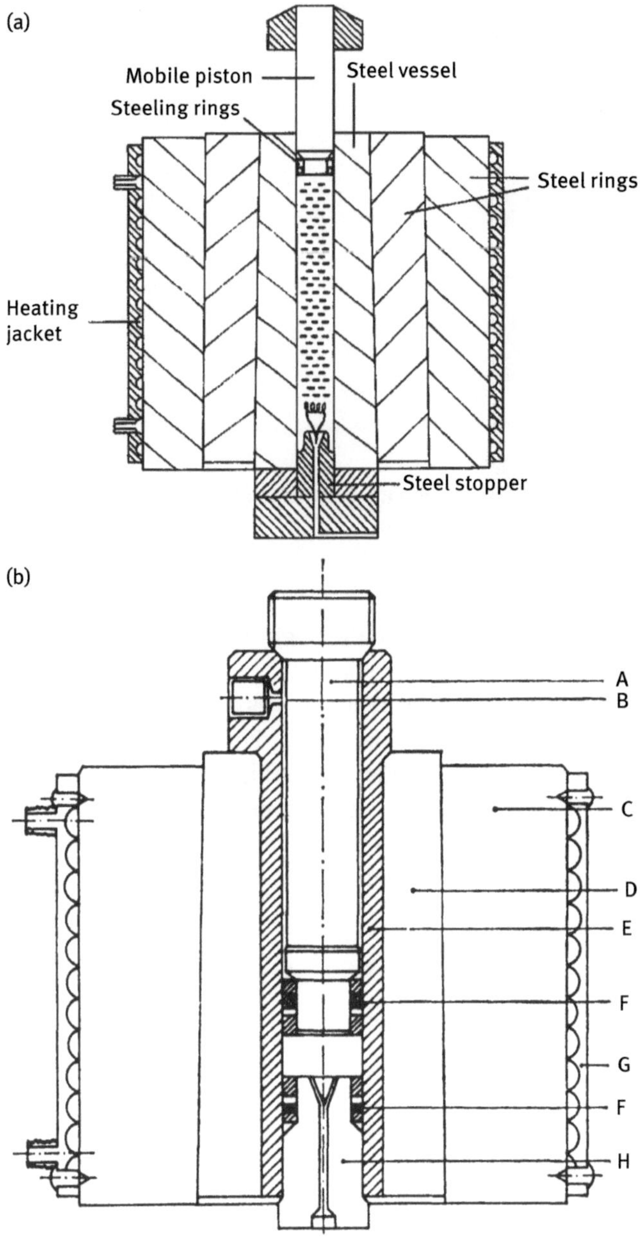

Figure 1.4: (a) Reaction chamber of piston-cylinder HP apparatus. Reproduced with permission from Jurczak J, Chmielewski M, Filipek S. Synthesis, 1979, 41–42. Copyright Georg Thieme Verlag KG; (b) Apparatus with gas inlet. Reproduced with permission from Jurczak J. Bull Chem Soc Jpn 1979, 52, 3438–42. Copyright Japan Chemical Society.

Figure 1.5: External heating of HP reactor chamber.

with the scaling-up of the pressure apparatus limit the operation of large-scale reactions (100 mL or more) at pressures greater than 7 kbar [35].

Many types of ampules are employed as reaction vessels. The majority of them are made of Teflon (polytetrafluoroethylene, PTFE). For instance, these are cylindrical cells closed with a sliding stopper (Figure 1.6a) [36] or a screw-capped Teflon cylinder. A variety of flexible PTFE vessels are used: collapsible small bags sealed by welding, by wire (Figure 1.6b) or that had been clamped using a brass clamp, as well as flexible vessels sealed at both ends with brass screw

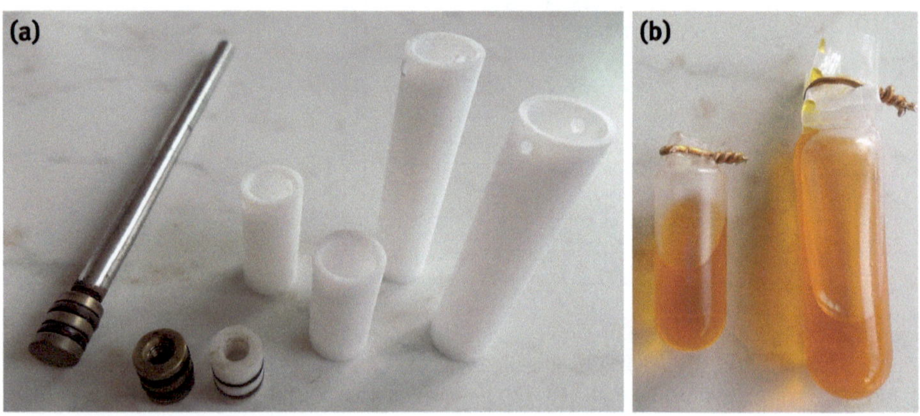

Figure 1.6: PTFE vessels for HP reactions a) PTFE vials and lid extractor tool; b) sealed collapsible PTFE bag.

clamps or by stainless steel plugs. Reaction mixtures are also sometimes sealed in a plastic syringe. Compressible metallic bellowed tubes made of copper or copper-beryllium are also available for HP synthesis. Small metallic cylinders and glass cells are rarely used. Larger apparatuses allow the placement of several smaller reaction vessels at the same time [37]. This feature is important for better efficiency, screening of reaction conditions and safety.

During the operation of HP apparatus, safety precautions have to be made, as there are reports on relatively violent reactions that occurred upon depressurization [14, 38, 39]. We had similar experience with benzyl azide reactions. Very exothermic, explosive decomposition or carbonization of reaction mixture under pressure [40] could be a result of the generation of reactive species such as radicals or carbenes. Hence, it is highly recommended that during the most critical manipulations with HP samples (the decompression and the manual removal of the piston from HP chamber) safety mask is used, and operator or experimentator should stay behind the safety shield. Measures that could diminish the occurrence of out-of-control reactions include slow elevation of pressure to the final value, with short intervals between two steps [41]. A fast increase in pressure for reactive compounds could be associated with exothermic reactions, leading to carbonization and generation of gaseous products with pressure built-up. From the safety standpoint, it is also advisable to perform HP reactions on the smaller scale, especially with reactive reagents and simultaneously pressurize several experiments [42]. The reactions that are carried out in HP for the first time should be done on a small scale.

Specialized modifications of HP apparatuses were made to provide means of kinetic measurements [43], in situ UV-spectrophotometric [44, 45], infrared [46] and EPR [47] measurements, as well as photochemical reactions [48]. For these purposes, HP optical cells with sapphire windows [49], sampling devices [50] or probes were built in. Solid-state reactions are rarely carried out in the HP equipment for which a hydraulic press designed for the preparation of everyday IR pellets was used [51]. Current development of HP technology is the translation from batch to flow reactors. Until now, pressures from 100 to 300 bar were applied in organic synthesis in continuous flow HP reactors [52].

1.3.2 HP solvents

Ample variety of solvents was employed for HP reactions. The choice of solvent to be used in HP experiments is determined by the solubility of the substrate and the solvent freezing points (m.p./ °C at given pressures p/kbar), which could be found in the literature [53]. Working conditions applied vary in the temperature range from −9 to 197 °C and pressures from 0.39 to 40 kbar. In general, polar solvents such alcohols could withstand higher pressures before reaching solidification.

Table 1.5: Maximum pressures used experimentally for halogenated solvents.

Solvent	Pressure/kbar	Temp, time	Reference
DCM, CH_2Cl_2	10	RT, 48 h	EJOC 2015 4457
	10.5	100 °C, 24 h	HP Res 1988 67
	12	−20 °C, 24 h	TH 2000 873
	15	150 °C, 3 h	JOC 1978 1471
	16	28 °C, 20 h	J Het Chem 2001 645
	17	80 °C, 168 h	HP Res 2008 675
	19	25 °C, 24 h	THL 2000 2723
	20	55 °C, 20 h	JOC 1985 3963
	20	100 °C, 3 h	BCSJ 1979 544
DCM/DMF 4:1	8	RT, 20 h	Heterocycles 2010 799
DCM/MeCN 3:1	17	65 °C, 48 h	THL 1986 3729
DCM/iPrOH 4:1	8	RT, 20 h	Heterocycles 2010 799
DCM/t-BuOH 9:1	11	RT, 96 h	JPOC 2002 590
DCM/toluene 1:1	10	RT, 10 h	Synlett 2009 2346
DCM/hexane	7	RT, 65 h	TH 1995 10033
CD_2Cl_2	11	25 °C, 1 h	JOC 2008 1099
$CHCl_3$ chloroform	3	30 °C, 24 h	HP Res 1995 321
	8	RT, 12 h	Angew 2002 1031
	8	100 °C, 80 h	Arkivoc 2004 70
	9.5	110 °C, 24 h	HP Res 1995 321
	10	60 °C, 100 h	BCSJ 1986 3197
	12	50 °C, 15 h	THL 1983 4613
	15	23 °C, 20 h	THL 2000 515
	20	100 °C, 3 h	BCSJ 1979 544
$CHCl_3/CH_2Cl_2$ 5:2	15	65 °C, 16 h	Synth Commun 1992 2965
$CHCl_3$/THF 1:1	16	20 °C, 69 h	TH 1996 8307
$CHCl_3$/2-propanol 10:1	15	50 °C, 19 h	TH 1996 8307
$CHCl_3/CH_3CN$ 3:2	15	50 °C, 19 h	TH 1996 8307
$CDCl_3$	12	100 °C, 22 h	EJOC 1999 2757
$CDCl_3/CD_3CN$ 1:1	3	60 °C, 2 h	JOC 2010 4950
	5	RT, 25 h	JOC 2010 4950
$CDCl_3$/DMSO-d_6 1:1	3	60 °C, 1 h	JOC 2010 4950
$CDCl_3$/DMSO 2:8	2	28 °C	Perkin 2 1987 1477
CCl_4	10	35 °C, 66 h	TH 1995 8953
	20	100 °C, 3 h	BCSJ 1979 544
$Cl_2CHCHCl_2$	8	100 °C, 2 d	Angew 2004 2105
	10	20 °C, 24 h	Synlett 2005 227
	15	100 °C, 3 h	JOC 1978 1471
$CF_3(CF_2)_4CF_3$	8	100 °C, 2 days	Angew 2004 2105
C_6F_{14} perfluorohexane	6	22 °C, 6 h	JOC 2011 5392
CH_3I iodomethane	8	100 °C, 2 days	Heterocycles 2007 187
	9	22 °C, 6 h	JOC 2011 5392
Ethyl iodide	8	100 °C, 2 days	Heterocycles 2007 187
n-Propyl iodide	8	100 °C, 2 days	Heterocycles 2007 187
1,2-Dichloroethane DCE	2.6	35 °C	Int J Chem Kinet 2013 613

Table 1.6: Maximum pressures used experimentally for alcohols.

Solvent	Pressure/kbar	Temp, time	Reference
MeOH	8	100 °C, 24 h	Synlett 2005 2254
	12	90 °C, 24 h	Dalton 1990 1115
	15	60 °C, 5 d	JOM 2006 349
	16	RT, 24 h	EJOC 2010 6423
	18	40 °C, 20 h	J Het Chem 2001 645
	22	60 °C, 8 h	Helv 1983 222
EtOH	7	60 °C, 24 h	THL 2001 2493
	11	55 °C, 16 h	JOC 2009 4311
	12	RT, 24 h	EJOC 2010 6423
	12	90 °C, 24 h	Dalton 1990 1115
	16	28 °C, 20 h	J Het Chem 2001 645
EtOH aq. 80%	1	60 °C, 20 h	JACS 1976 920
EtOH aq.	2.8	4–12 h	Bunseki Kagaku 2000 551
EtOH/H_2O 9:1	15	20 °C, 11.5 h	THL 1988 3021
EtOH/H_2O 83%	8	107 °C, 24 h	Perkin 1 1977 1200
EtOH/H_2O 1.3:1	9.1	75 °C, 5 min	JACS 1959 2151
propanol	10	RT, 2 h	THL 2012 5287
iPrOH	8	RT, 20 h	Heterocycles 2010 799
	9	65 °C, 2 h	Dalton 1990 1115
	10	RT, 24 h	JOC 2015 10375
n-Butanol	8	RT, 24 h	THL 2001 4807
	16	110 °C	Chemosphere 1973 31
t-Butanol	10	RT, 2 h	THL 2012 5287
Pentanol	8	RT, 24 h	THL 2001 4807
TCE trichloroethanol	10	RT, 24 h	JOC 2015 10375
TFE trifluoroethanol	10	RT, 24 h	JOC 2015 10375
HFIP hexafluoroisopropanol	14	RT, 17 h	JOC 2015 10375
HFIP/DCM	10	RT, 24 h	JOC 2015 10375
Ethylene glycol 50% H_2O	1.8	30 °C, 3 h	JACS 1983 1745
Ethylene glycol	3	30 °C, 16 h	JPOC 2003 265
	3	80 °C, 24 h	NJC 2000 203
	3	90 °C, 38 h	Synth Comm 1997 1475

The increase in working temperature results in a possibility to operate at higher pressures before achieving solvent freezing. The mixing with higher freezing point solvent also increases the pressure limit. Tables 1.5–1.11 give a comprehensive overview of solvents and most extreme conditions reported in literature, for reactions that are covered in the subsequent chapters. In addition to literature freezing point values, these data could quickly guide synthetic chemists for the choice of the appropriate solvent and reaction conditions.

Table 1.7: Maximum pressures used experimentally for aromatic solvents.

Solvent	Pressure/kbar	Temp, time	Reference
Benzene	7	170 °C, 7 h	Chem Comm 1992 100
	10	160 °C, 7 h	Phosph Silicon 2010 1558
	11	25 °C, 60 h	Perkin 1 1985 1277
	15	100 °C, 3 h	JOC 1978 1471
Toluene	3	100 °C, 36 h	Fuller Sci Technol 1995 45
	9	83 °C, 17 h	CEJ 1999 2119
	10	125 °C, 4 d	THL 1994 73
	15	100 °C, 3 h	BCSJ 1979 544
	16	28 °C, 20 h	J Het Chem 2001 645
	19	50 °C, 20 h	Helv 1983 218
	22	25 °C, 6 h	Helv 1982 1021
Toluene/MeCN 4:0.8	12	100 °C, 24 h	EJOC 2004 1405
Toluene/MeCN 8:2	10	125 °C, 2.5 d	THL 1994 73
Toluene/Benzene 7:3	12.4	20 °C, 20 h	JOC 1979 3347
Toluene/Benzene 3:7	15	40 °C, 36 h	JOC 1984 4293
Br-benzene	1.4	140 °C	JACS 1982 3150
	2	42 °C	
	15	100 °C, 3 h	BCSJ 1979 544
Cl-benzene	3	130 °C, 15 h	Perkin 1 1998 3219
	15	100 °C, 3 h	BCSJ 1979 544
F-benzene	1	34 °C, 4 h	JPOC 2005 268
Cumene = iPr-benzene	2	135 °C	BCSJ 1985 2490
isopropylbenzene IPB			
	3	130 °C, 10 h	Chem Lett 1986 1315
Mesitylene	1.4	140 °C	JACS 1982 3150
	2	42 °C	
Isodurene/heptane	3	110 °C	Ber 1989 1179
Isodurene 1,2,3,5-triMebenzene	3	75 °C	Ber 1988 781
1,2-diClbenzene ODCB	5	120 °C, 5 h	Synth Comm 1997 1475
Nitrobenzene	4.8	25 °C, 6 h	Russ Chem Bull 2004 45
Benzonitrile	5	25 °C	Perkin 2 1983 37
	15	100 °C, 3 h	BCSJ 1979 544
Furan	3	20 °C, 24 h	HP Res 1999 243
	7	20 °C, 24 h	JOC 1985 2576
	8	160 °C, 20 h	Rev High Press Sci Technol 1998
	8.5	65 °C, 24 h	HP Res 2002 511
	9	30 °C, 14 d	Main Gr Met Chem 1997 581
	10	45 °C, 48 h	Perkin 1 1999 2255
	11	20 °C, 48 h	Chem Comm 2000 1191
	15	RT, 72 h	TH 1997 11843
	15	60 °C, 16 h	BCSJ 1982 496
Thiophene	8	100 °C, 2 d	Angew 2004 2105

Table 1.8: Maximum pressures used experimentally for ketones.

Solvent	Pressure/kbar	Temp, time	Reference
Acetone	10	60 °C, 48 h	THL 1993 4031
	11	55 °C, 16 h	JOC 2009 4311
	15	RT	EJOC 2001 553
	15	100 °C, 3 h	BCSJ 1979 544
Acetone-d_6	40		JOC 1977 4170 DA Gladysz
Cyclohexanone	2	RT, 72 h	Synlett 2003 1655
Cyclohexenone	6	60 °C, 20 h	Synlett 2008 1402
Methyl ethyl ketone	2	RT, 96 h	Synlett 2003 1655
	7	70 °C, 45 h	JOC 1976 2495
Pinacolone	3	30 °C, 16 h	JPOC 2003 265
Ketones various	3	30 °C, 16 h	THL 2003 447

Table 1.9: Maximum pressures used experimentally for water and aqueous mixtures.

Solvent	Pressure/kbar	Temp, time	Reference
H_2O	2.1	−22 °C	JOC 2006 3126
	2.8	40 °C, 24 h	OBC 2012 7321
	6	60 °C, 20 h	Synlett 2008 1402
	8	80 °C, 60 h	Synlett 2008 1402
	10	RT, 12 h	BBRC 2002 145
H_2O/MeOH/dioxane 9:7:1	8	25 °C, 14 h	Heterocycles 2015 1164
H_2O/EtOH 1:1	9	22 °C, 14 h	JOC 2011 5392
H_2O/EtOH 3:1	9	22 °C, 14 h	JOC 2011 5392
tris-HCl pH 7.7 buffer	8	RT, 18 h	Biochimie 1996 862
D_2O tris-HCl pH 7.7 buffer	10	RT	Chem Pyhs Chem 2014 138
0.1 M imidazole-HCl pH 6.6 buffer	8	18 h	Int J Biomol Macromol 1994 153
20% aq NaOH/dioxane	8	100 °C, 36 h	Synlett 2000 116
20% aq NaOH/THF	8	70 °C, 20 h	Synlett 2000 116
20% aq NaOH/EtOH	8	70 °C, 20 h	Synlett 2000 116

Table 1.10: Maximum pressures used experimentally for ethers.

Solvent	Pressure/kbar	Temp, time	Reference
Diethyl ether, Et_2O	5	150 °C, 6 h	Russ JOC 2008 411
	8	60 °C, 8 h	Synlett 2006 1968
	10	25 °C, 7 d	Chem Comm 1978 543
	10	130 °C, 20 h	TH 1996 12185
	15	80 °C, 240 h	HP Res 2008 675
	15	100 °C, 3 h	BCSJ 1979 544

Table 1.10 (continued)

Solvent	Pressure/kbar	Temp, time	Reference
	20	65 °C, 5 h	JOC 1981 2230
	22	50 °C	Chem Comm 1983 540
	25	20 °C, 20 h	Synthesis 1979 41
Methyl phenyl ether	15	100 °C, 3 h	BCSJ 1979 544
Methyl *t*-butyl ether	3.5	40 °C	CEJ 2004 2707
	12	20 °C, 24 h	JOC 1987 365
MTBE/MeCN 9:1	4	106 °C	EJOC 1999 2757
Dibutyl ether	1.4	130 °C	Chem Lett 1986 1203
Amyl ether	3	40 °C, 5 h	Proc RSC 1936 684
Diisopropyl ether	2.6	35 °C	Int J Chem Kinet 2013 613
THF	6	130 °C, 4 h	JOC 1977 3101
	8	100 °C, 72 h	THL 2006 587
	9	60 °C, 7 d	BCSJ 1989 3138
	14	64 °C, 2 d	THL 1997 5631
	12	90 °C, 12 h	HP Res 2004 149
	15	100 °C, 36 h	THL 2008 5620
	16	RT, 24 h	EJOC 2005 5296
	16	60 °C, 5 min	Macromol Chem 1976 3265
	16	50 °C, 60 h	OL 2007 4159
THF/H_2O 9:1	9	60 °C, 15 h	JOC 2006 70
THF/MeOH 5:1	10	30 °C, 23 h	TH Asym 1995 1241
THF/MeCN 1:1	10	70 °C, 48 h	TH 1996 12185
1,2-Dimethoxyethane (DME)	8.41	50 °C, 24 h	JPC 1964 2361
Diglyme	4	100 °C	JACS 1978 5961
Dioxane 1,4-dioxane	3	120 °C, 20 h	BCSJ 1989 1567
	5.5	100 °C, 15 h	JACS 1968 6145
	8	70 °C, 20 h	Synlett 2000 116
	9	40 °C, 24 h	HP Res 2005 1

Table 1.11: Maximum pressures used experimentally for other solvents.

Solvent	Pressure/kbar	Temp, time	Reference
MeCN acetonitrile	6	60 °C, 20 h	Synlett 2008 1402
	8	70 °C, 5 d	Synlett 1999 650
	10	RT, 48 h	EJOC 2015 4457
	10	125 °C, 4 d	Liebigs 1997 501
	13	RT, 82 h	Heterocycles 2006 2006 589
	13	40 °C, 50 h	J Het Chem 2001 645
	15	50 °C, 48 h	EJOC 2013 6955

Table 1.11 (continued)

Solvent	Pressure/kbar	Temp, time	Reference
	15	100 °C, 3 h	BCSJ 1979 544
MeCN-d_3	6	60 °C, 72 h	TH 1987 4555
MeCN/H_2O 60:1	10	30 °C, 4 d	JOC 1991 5737
MeCN/DMA 2:1	10	60 °C, 48 h	Liebigs 1997 887
MeCN/DCM 2:1	15	65 °C, 15 h	JOC 1982 5042
DMF	2	135 °C	BCSJ 1985 2490
	8	60 °C, 8 h	Synlett 2006 1968
	10	140 °C, 4 h	THL 1995 5547
	13	RT, 5 d	THL 1997 5949
	15	40 °C, 7 d	Bioorg Med Chem Lett 2001 191
$HCONH_2$ formamide	3	30 °C, 16 h	JPOC 2003 265
NMP *N*-methyl-2-pyrrolidone	1.8	90 °C, 24 h	ChemSusChem 2015 504
Ethyl acetate	11	55 °C, 16 h	JOC 2009 4311
	12	RT, 24 h	JOC 1991 6946
	14	34 °C, 90 h	Perkin 1 1985 1277
	15	100 °C, 3 h	JOC 1978 1471
Butyl acetate	1	45 °C	Russ Chem Bull 2009 21
Pentane	7.5	25 °C	Chem Comm 2002 1272
	10	100 °C, 17 h	TH 1996 12185
Hexane	4.6	153 °C	Angew 1994 1079
	10.2	20 °C, 20 h	JOC 1979 3347
	12	53 °C, 20 h	CEJ 1999 2119
n-Heptane	2.5	172 °C	JACS 1998 6212
n-Heptane/*t*-butylmethyl ether 9:1	3	50 °C	JACS 2005 18107
Cyclohexane	2.98	25 °C	Russ Chem Bull 2004 45
Methylcyclohexane	3	25 °C, 15 min	OBC 2004 1295
1-Bromobutane	8.14	70 °C	TH 1974 3081
1-Chlorobutane, *n*-butylchloride	4	110 °C	Ber 1994 2241
Nitromethane	6.96	25 °C, 6 h	Russ Chem Bull 2004 45
	8	RT, 12 h	Angew 2002 1031
	10	60 °C, 42 h	THL 1996 3727
i-$PrNO_2$	10	20 °C, 24 h	Synlett 2005 227
Acetic acid	8.87	25 °C, 6 h	Russ Chem Bull 2004 45
Ionic liquids (ILs)	1.5	5 °C	JPOC 2007 109
Benzoyl chloride	1	34 °C, 4 h	JPOC 2005 268
Dimethyl carbonate	2.6	35 °C	Int J Chem Kinet 2013 613
DMAD	15	60 °C, 7 d	Chem Comm 1979 1091
DMSO	8	60 °C, 8 h	Synlett 2006 1968
	10	25 °C, 10 h	JPOC 1991 639
Sulfolan	8–10	200 °C, 15 h	TH 1975 619
CS_2	5.5	100 °C, 26 h	JOC 1981 446

1.4 Quantum–chemical calculations

Predictions of the reactivities of molecules in HP conditions and the experimental findings are sometimes rationalized by quantum–chemical calculations and molecular modeling. Basically, computational methodologies employed could be grouped as the ones that are carried out in standard conditions ($p = 1$ bar, gas phase) and the ones in which the pressure is explicitly taken into account [54]. Although theoretical calculations simulate the reaction in the gas phase at a very low pressure, the results could be in accord to experiments [55], for instance, for regioselectivity prediction, often the stereochemical predictions are not satisfactory (activation enthalpies obtained by the semiempirical AM1 method) [56]. The inclusion of solvent in calculations could improve the accuracy, such for the calculations of $\Delta V^{\neq}$ and ΔV in different solvents to obtain preference of one of two possible isomers by estimating $\Delta\Delta V^{\neq}$ [57]. Theoretical calculations were performed using hybrid functional B3LYP, and different solvents were simulated by polarized continuum model.

Custom-built programs for the calculation of volumes of activation and reaction were developed. Klärner employed program MOLVOL that uses the molecular structures that were obtained from force field and quantum mechanical ab initio calculations, and $\Delta V^{\neq}$ are calculated by a Monte-Carlo computer simulations [58, 59]. The method of calculation of molecular volumes employing the program MOLSV by Firestone [60] led to the estimation of the intrinsic volumes of the molecules before and after the reaction.

Molecular dynamics simulations allow direct calculations of the effect of external pressure. Mukherjee [61] used the GROMACS program package with AMBER99sb force field with parmbsc0 modifications. Water was modeled using the TIP3P parameters. Whereas in the HP study of host-guest equilibria molecular dynamics (MD) simulations and density functional theory (DFT) calculations at the B3LYP/6-31G(d) level of theory with explicit solvent with Amber force field parameter GAFF [62].

SIBFA (sum of interactions between fragments *ab initio* computed), an anisotropic polarizable molecular mechanics procedure, was developed by Dumas [63]. The modification of MM was made to take into account the effect of pressure on the molecular conformations at 1 bar and 15 kbar.

The use of quantum–chemical electronic structure calculations to study pressure effects is not straightforward and several ways of estimation of the effect of pressure on molecules by quantum chemical calculations were put forward. Mukherjee has developed a way of estimating the effect of pressure on distortion of molecular structure using quantum chemical calculations [61]. Rubtsov has predicted the activation volumes with good accuracy by devising an equation in the form of a linear relationship between free energies and describing the effect of the pressure on the rate constants [64]. The polarizable continuum model for extreme pressure was developed by Cammi to estimate the effects of HP on the potential energy surface of organic reactions in solution. DFT B3LYP [65] and M062X [66] functionals with 6-31G(d) were used to describe molecules.

References

[1] Rothmund, V. Ueber den Einfluss des Druckes auf die Reaktionsgeschwindigkeit. Zeit Phys Chem 1896, 20, 168–179.
[2] Demazeau, G. Un siècle de Hautes Pressions: Développements technologiques et scientifiques. C R Chimie 2009, 12, 933–942.
[3] Cohen, E., & de Boer, RBB. Piezochemische Studien. IX. Ein Druckautomat für hohe drucke. Zeit Phys Chem 1913, 84, 32–40.
[4] Cohen, E., & Kaiser, HFG. Piezochemische studien. XIII. Der einfluss des drucks auf die reaktionsgeschwindigkeit in kondensierten systemen. II. Zeit Phys Chem 1915, 89, 338–364.
[5] Cohen, E., & de Boer, RB. Piezochemische Studien. X. Einfluss des Druckes auf die Reaktionsgeschwindigkeit in kondensierten Systemen. I. Zeit Phys Chem 1913, 84, 41–82.
[6] Cohen, E., & Valeton, AM. Piezochemische Studien. XIV. Der Einfluss des Druckes auf die Reaktionsgeschwindigkeit in kondensierten Systemen. III. Zeit Phys Chem 1918, 92, 433–470.
[7] Bridgman, PW., & Conant, JB. Irreversible transformations of organic compounds under high pressures. Proc Natl Acad Sci USA 1929, 15, 680–683.
[8] Conant, JB., & Tongberg, CO. Polymerization reactions under high pressure. I. some experiments with isoprene and butyraldehyde. J Am Chem Soc 1930, 52, 1659–1669.
[9] Conant, JB., & Peterson, WR. Polymerization reactions under high pressure. II. The mechanism of the reaction. J Am Chem Soc 1932, 54, 628–635.
[10] Tamman, G., & Pape, A. Der Einfluß des Druckes auf die polymerisationsgeschwindigkeit. Zeitschr Anorg Chem 1931, 200, 113–132.
[11] Cohen, E., & Schut, W. Piezochemie kondensierter Systeme. Akademische verlags, Leipzig, 1919.
[12] Cohen, E. Physico-chemical metamorphosis and some problems in piezochemistry. McGraw Hill, N. York, 1926.
[13] le Noble, WJ. Kinetics of Reactions in Solutions under Pressure. in Progress in Physical Organic Chemistry, Vol. 5, A Streitwieser, Jr. & RW Taft (Eds.), Wiley, N. York 1967, pp. 207–330.
[14] Dauben, WG., & Bunce, RA. Organic reactions at high pressure. A Robinson annulation sequence initiated by Michael addition of activated cycloalkanones with hindered enones. J Org Chem 1983, 48, 4642–4648.
[15] Holan, G., Evans, JJ., & Linton, M. Formation of benzimidazoles at high pressure. J Chem Soc Perkin Trans 1 1977, 1200–1203.
[16] Dauben, WG., Gerdes, JM., & Look, GC. Organic reactions at high pressure: The effect of pressure change upon reaction rates of bimolecular processes. Synthesis 1986, 532–535.
[17] Isaacs, NS., Maksimovic, Lj., Rintoul, GB., & Young, DJ. The high pressure addition of aldehydes to allylstannanes: a 'Diels-Alder Like' transition state. J Chem Soc Chem Commun 1992, 1749–1750.
[18] Isaacs, NS. The role of high pressure methods in organic chemistry. Tetrahedron, 1991, 47, 8463–8497.

[19] le Noble, WJ., & Duffy, M. The effect of pressure on the rates of Acid-Catalyzed hydrolysis of trimethylene oxide and of epichlorohydrin. J Phys Chem 1964, 68, 619–622.
[20] Kiselev, VD., Shakirova, II., Kornilov, DA., Kashaeva, HA., Potapova, LN., & Konovalov, AI. Homo-Diels-Alder reaction of a very inactive diene, bicyclo[2.2.1]hepta-2,5-diene, with the most active dienophile, 4-phenyl-1,2,4-triazolin-3,5-dione. Solvent, temperature, and high pressure influence on the reaction rate. J Phys Org Chem 2012, 26, 47–53.
[21] Brower, KR., Muhsin, M., & Brower, HE. Activation volume as a measure of E2-E1cB character in elimination reactions. J Am Chem Soc 1976, 98, 779–782.
[22] Okamoto, Y., & Shimizu, H. Steric hindrance under high pressure. Reactions of 2,4,6-Tri-*t*-butyl-*N*-methylaniline with alkyl iodides. J Am Chem Soc 1968, 90, 6145–6148.
[23] Jenner, G. High pressure relief of steric inhibition in ionogenic organic reactions. High Press Res 2004, 24, 163–168.
[24] Hubbard, CD., Wilson, CJ., & Caldin, EF. Activation volumes and volumes of reaction for the proton-transfer reaction between 4-nitrophenylnitromethane and 1,1′,3,3′-tetramethylguanidine in some aprotic solvents. J Am Chem Soc 1976, 98, 1870–1875.
[25] Palmer, DA., & Kelm, H. Pressure effects on the rates of some hydrogen/deuterium transfer reactions between 2,2-Diphenyl-1-picrylhydrazyl and substituted phenols. Austral J Chem 1977, 30, 1229–1239.
[26] Matsumoto, K., Hashimoto, S., Otani, S., Atnita, F., & Osugi, J. Mannich reaction under high pressure. Dimethylamino-methylation of ketones with bis(Dimethylamino) methane under mild conditions. Synth Commun 1984, 14, 585–590.
[27] Rogachev, VO., & Metz, P. Thermal and high pressure intramolecular Diels-Alder reaction of vinylsulfonamides. Nature Protocols 2006, 1 3076–3087.
[28] Jurczak, J., Chmielewski, M., & Filipek, S. Organic syntheses under high pressure; I. High-Pressure Diels-Alder reactions between 1-methoxybuta-1,3-diene and carbonyl compounds. Synthesis 1979, 41–42.
[29] Deshong, P., Dicken, CM., Perez, JJ., & Shoff, RM. A simple, inexpensive system for performing high pressure reactions. Org Prep Proced Int 1982, 14, 369–372.
[30] Jurczak, J. Stereochemistry of Diels-Alder reactions at high pressure II. Influence of high pressure on asymmetric induction in condensation of (-)-Di-(*R*)-menthyl fumarate with butadiene and isoprene. Bull Chem Soc Jpn 1979, 52, 3438–3442.
[31] Matsumoto, K., Uchida, T., & Sera, A. Organic synthesis under high pressure; I. Synthesis 1985, 1–26.
[32] Brown, CL., Muderawan, IW., & Young, DJ. High-Pressure synthesis of oligoanilines. Synthesis 2003, 16, 2511–2517.
[33] Melai, V., Brillante, A., & Zanirato, P. High pressure assisted 1,3-dipolar cycloadditions: formation of 1,2,3-triazoles from aryl azides and (trimethylsilyl)acetylene. J Chem Soc Perkin Trans 2 1998, 2447–2449.
[34] Plietker, B., Seng, D., Fröhlich, R., & Metz, P. High Pressure Intramolecular Diels-Alder Reactions of Vinylsulfonates and Vinylsulfonamides. Tetrahedron 2000, 56, 873–879.
[35] Dauben, WG., Gerdes, JM., & Smith, DB. Organic reactions at high pressure. The preparative scale synthesis of cantharidin. J Org Chem 1985, 50, 2576–2578.
[36] Margetić, D., Butler, DN., & Warrener, RN. Sulfur-bridged molecular racks: *O, S*-sesquinorbornadienes, *CNS*-[3] and *CNOS*-[4]polynorbornanes. Arkivoc 2002, vi, 234–236.
[37] Kocój, M., Jarosińska, E., & Chaładaj, W. High-Pressure Au-Catalyzed cycloisomerization of internal 1,6-and 1,7-Enynes. Synlett 2018, 29, 1319–1323.
[38] Bengelsdorf, IS. High pressure-high temperature reactions. I. The trimerization of aromatic nitriles. J Am Chem Soc 1958, 80, 1442–1444.

[39] Raistrick, B., Sapiro, RH., & Newitt, DM. Liquid-phase reactions at high pressures. Part VI. The explosive decomposition of cyclopentadiene. J Chem Soc 1939, 1770–1773.

[40] Almarzoqi, A., George, AV., & Isaacs, NS. The quarternisation of tertiary amines with dihalomethane. Tetrahedron 1986, 42, 601–607.

[41] Golebiowski, A., & Jurczak, J. High-pressure [4+2] cycloaddition of 1-methoxy-1,3-butadiene to *N,O*-protected D-threoninals and D-*allo*tThreoninals. Tetrahedron 1991, 47, 1037–1044.

[42] Dauben, WG., & Bunce, RA. Organic reactions at high pressure. dipolar cycloaddition-ring contraction reactions of hindered silyl enol ethers and arylsulfonyl azides. J Org Chem 1982, 47, 5042–5044.

[43] van Eldik, R., Gaede, W., Wieland, S., Kraft, J., Spitzer, M., & Palmer, DA. Spectrophotometric stopped flow apparatus suitable for high-pressure experiments to 200 MPa. Rev Sci Instrum 1993, 64, 1355–1357.

[44] Fleischmann, FK., Conze, EG., Stranks, DR., & Kelm, H. High pressure cells for the study of optical properties of liquid systems. Rev Sci Instrum 1974, 45, 1427–1429.

[45] le Noble, WJ., & Schlott, R. All quartz optical cell of constant diameter for use in high pressure studies. Rev Sci Instrum 1976, 47, 770–771.

[46] Buback, M., Tost, W., Tietze, LF., & Voβ, E. *hetero* Diels-Alder reactions of enamine carbaldehydes at high pressure. Chem Ber 1988, 121, 781–785.

[47] Sueishi, Y., Tobisako, H., & Kotake, Y. Distinctive pressure effects on the association equilibrium in cyclodextrin group-inclusion complex as studied with electron paramagnetic resonance. J Phys Chem B 2004, 108, 12623–12627.

[48] Hara, K., & Morishima, I. High-pressure cell and its inner capsule for optical studies of liquids. Rev Sci Instrum 1988, 59, 2397–2398.

[49] Isaacs, NS., & Rannala, E. Reaction studies at high pressure. Part I. activation volumes of some [2+2] and dipolar cycloadditions. J Chem Soc Perkin Trans 2 1975, 1555–1560.

[50] Grieger, RA., & Eckert, CA. A new technique for chemical kinetics at high pressures. AiChE J 1970, 16, 766–770.

[51] Chen, H., Ni, B-B., Gao, F., & Ma, Y. Pressure-accelerated copper-free cycloaddition of azide and alkyne groups pre-organized in the crystalline state at room temperature. Green Chem 2012, 14, 2703–2705.

[52] Razzaq, T., & Kappe, CO. Continuous flow organic synthesis under high-temperature/pressure conditions. Chem Asian J 2010, 5, 1274–1289.

[53] Hubbard, CD., & van Eldik, R. Effect of pressure on organic reactions, in: R van Eldik & CD Hubbard eds., Chemistry under extreme or non-classical conditions, Wiley, New York, 1996, pg 103–161.

[54] Weinberg, N. Theoretical models in high pressure reaction kinetics: from empirical correlations to molecular dynamics. Rev High Press Sci Technol 1998, 8, 86–95.

[55] Margetić, D., Eckert-Maksić, M., Trošelj, P., & Marinić, Ž. Reaction of 2,5-bis-trifluoromethyl -1,3,4-oxadiazole with 7-oxanorbornenes revisited: experimental and quantum-chemical study of reaction stereospecifity. J Fluor Chem 2010, 131, 408–416.

[56] Branchadell, V., Sodupe, M., Ortuño, RM., & Oliva, A. Diels-Alder cycloadditions of electron-rich, electron-deficient, and push-pull dienes with cyclic dienophiles: high-pressure-induced reactions and theoretical calculations. J Org Chem 1991, 56, 4135–4141.

[57] Tsypysheva, IP., Borisevich, SS., Lobov, AN., Kovalskaya, AV., Shamukaev, VV., Safiullin, RL., & Khursan, SL. Inversion of diastereoselectivity under high pressure conditions: Diels-Alder

reactions of 12-*N*-substituted derivatives of (-)-cytisine with *N*-phenylmaleimide. Tetrahedron Asymm 2015, 26, 73–77.
[58] Klärner, F-G., Breitkopf, V. The effect of pressure on retro diels-alder reactions. Eur J Org Chem 1999, 2757–2762.
[59] Klärner, F-G., Krawczyk, B., Ruster, V., & Deiters, UK. The pressure-dependence of the dimerization of 1,3-butadiene: Experimental and theoretical volumes of activation and reaction. High Press Res 1994, 13. 1–5.
[60] Firestone, RA., & Smith, GM. The roles of changes in bonding vs. packing fraction in the pressure-induced acceleration of the diels-alder reaction. Chem Ber 1989, 122, 1089–1094.
[61] Kumar, PS., Mukherjee, A., & Hazra, A. Theoretical study of structural changes in DNA under high external hydrostatic pressure. J Phys Chem B 2015, 119, 3348–3355.
[62] Vazdar, M., Mitchell, AS., Warrener, RN., & Margetić, D. High pressure promoted exchange of guests from hemicarceplexes. Tetrahedron 2015, 71, 550–553.
[63] Dumas, F., Fressigné, C., Langlet, J., & Giessner-Prettre, C. Theoretical investigations of the influence of pressure on the selectivity of the michael addition of diphenylmethaneamine to stereogenic crotonates. J Org Chem 1999, 64, 4725–4732.
[64] El'yanov, BS., Shakhova, S., & Rubtsov, GA. Pressure dependence of the rate constants for Diels-Alder reactions and calculations of activation volumes. Bull Russ Acad Sci 1975, 12, 2563–2569.
[65] Cammi, R. A new extension of the polarizable continuum model: Toward a quantum chemical description of chemical reactions at extreme high pressure. J Comput Chem 2015, 36, 2246–2259.
[66] Yang, T., Fukuda, R., Cammi, R., & Ehara, M. Diels-Alder cycloaddition of cyclopentadiene and C_{60} at the extreme high pressure. J Phys Chem A 2017, 121, 4363–4371.

2 Diels–Alder cycloaddition reactions

Introduction

The formation of six-member rings by the Diels–Alder (DA) cycloaddition reactions is of great synthetic importance and represents the most often performed reaction in high-pressure (HP) conditions. From way back in 1939, it is known that DA reaction is accelerated by pressure [1]. The applications of HP-DA reactions vary from medicinal chemistry, natural product synthesis [2] to material science. There is an enormous volume of literature concerned with cycloadditions under pressure. Literature shows that even the highly unreactive aromatics such as naphthalene [3], anthracene [4], benzene[5, 6] and [2,2]paracyclophane could react as dienes at elevated pressures [7].

DA reactions are characterized by very large negative activation volumes. This suggests that [4 + 2] cycloadditions are highly favored by the application of pressure and can be interpreted as the following one-step concerted mechanism, which is characterized by a "quasi-cyclic" transition state with simultaneous formation of two covalent bonds [8]. For example, $\Delta V^{\neq}$ for isoprene reaction with quinones range from −38.1 to −32.2 $cm^3\ mol^{-1}$ in EtOH and −37.2 to −35.3 $cm^3\ mol^{-1}$ in $CHCl_3$ [9]. In a mechanistic study of DA reactions under HP conditions (6.1 kbar) by Grieger and Eckert [10], $\Delta V^{\neq}$ were obtained as follows: for the addition of 1,3-cyclohexadiene with maleic anhydride, it is −37.2 $cm^3\ mol^{-1}$ in DCM; for addition of cyclopentadiene (CPD) with dimethyl acetylenedicarboxylate, it is −30.2 $cm^3\ mol^{-1}$ in EtOAc; for isoprene with maleic anhydride, it is −39.0 $cm^3\ mol^{-1}$ in acetone; and for *trans*-1-methoxy-1,3-butadiene-maleic anhydride, it is −45.4 $cm^3\ mol^{-1}$ in *n*-butyl chloride. Activation volumes could also be affected by the solvent. For the reaction of 9-anthracenemethanol and *N*-ethylmaleimide, the following activation volumes

https://doi.org/10.1515/9783110556841-002

are obtained: $\Delta V^{\neq} = -36.0$, -31.4 and $-22.4\ cm^3\ mol^{-1}$ in water, butan-1-ol and heptane, respectively [11]. The largest $\Delta V^{\neq}$ for water may reflect a rate contribution from pressure favoring increased solvation of the transition state.

To further promote the reactivity of dienes and dienophiles, a variety of catalysts were used in HP-DA reactions: $EtAlCl_2$ [12], $ZnCl_2$ [13], $TiCl_2(OiPr)_2$ [14], $AlCl_3$ [15], $HfCl_4$.2THF, $AlCl_3$.2THF, $Sc(OTf)_3$ [16], and lanthanide catalysts such as $Eu(fod)_3$, $Eu(tfc)_3$, $Pr(tfc)_3$, $Yb(tfc)_3$ [17] and $Yb(NO_3)_3.5H_2O$ [18]. Besides Lewis acid catalysts, other catalysts were employed in hyperbaric conditions, such as thiourea [19] or tRNA [20]. Larger activation volumes were determined in catalyzed DA reactions. For instance, Isaacs reported that for reaction between isoprene and 9-anthracene methanol uncatalyzed and catalyzed by aluminum chloride and lithium perchlorate, $\Delta V^{\neq} = -36.1$, -41.7 and $-45.4\ cm^3\ mol^{-1}$ diethyl ether, respectively [11]. This finding is associated with ion-pair separation by pressure, which enhances the potency of the catalyst, as well as by pressure enhancement of lithium coordination to diene. Larger $\Delta V^{\neq}$ are also reported for catalyzed hetero DA (HDA) reactions of acrolein and crotonaldehyde with ethyl vinyl ether in the presence of $Eu(fod)_3$ or its absence: $\Delta V^{\neq} = -31.7/-29.6$ and $-31.9/-27.7\ cm^3\ mol^{-1}$ in chloroform [21]. Likewise, activation volume of ZnI_2 catalyzed methyl acrylate with CPD at 1.5 kbar in ionic liquids ($\Delta V^{\neq} = -27.1\ cm^{-3}\ mol^{-1}$) is $4.5\ cm^3\ mol^{-1}$ more negative than the uncatalyzed reaction [22].

Besides catalysts, different solvents or neat conditions were investigated, as well as the solid phase support [23]. Hydroquinone is sometimes added as radical scavenger to preclude polymerization [24].

2.1 DA cycloaddition reactions

The power of HP conditions is well illustrated by the DA reaction of naphthalene with maleic anhydride, which led to the formation of a mixture of the *endo*-**3** and *exo*-**3** cycloadducts only at pressures higher than 7 kbar in relatively concentrated solutions (Scheme 2.1) [25, 26]. In more dilute solutions, the cycloaddition stops at a relatively low conversion. The *endo/exo*-ratio represents the equilibrium mixture. The solubility of products is pressure dependent and products precipitate from solution at HP, which shifts the equilibrium toward the cycloadducts. In optimal reaction conditions (reactants c = 1.1 M, $CHCl_3$, 100 °C, 48 h, 12 kbar), adducts are obtained in

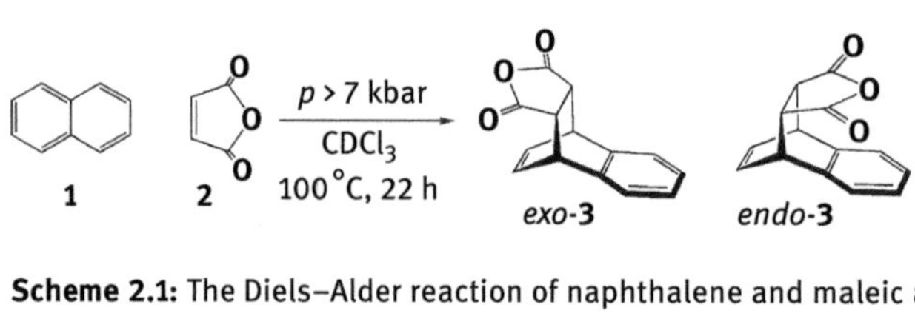

Scheme 2.1: The Diels–Alder reaction of naphthalene and maleic anhydride.

94% yield. In lower concentrations, at 7 kbar (reactants 0.5 M, 22 h), *exo*-**3** and *endo*-**3** were obtained in 2.1% and 1.1%, respectively (ratio 64:36), whereas yield at 11 kbar was significantly increased to 16.5% and 8.5% (without the change in ratio 66:34). At 12 kbar, yields are further increased to 17.4% and 15.9%, with partial precipitation from the solution.

Hyperbaric conditions often impede multiple DA reactions. These are either domino DA reactions of excess of diene, or multiple DA cycloadditions of bis- or tris-dienes that take place simultaneously. A classical example of this type of reactions in which complex cyclic structures are formed was published by Stoddart. Stereoregular oligomerization by repetitive DA reactions was achieved by the employment of proper cycloaddends that direct the diastereoselective preparation of the complex structures. When *syn*-bisdiene **4** was pressurized at 10 kbar in the presence of *syn*-bisdienophile **5**, a mixture of 2:1 and 3:2 adducts **6** and **8** was obtained possessing all-*syn*-configuration (Scheme 2.2) [27]. In addition, small amounts of the *syn, anti*-2:1 adduct **7** were isolated, which is formed by the minor reaction pathway, where both diene and dienophile units approach each other from the *exo*-side. By the employment of the same synthetic strategy, that is, triple stereoregular DA reaction of the U-shaped tris-diene **9** with tris-dienophile **10**, a caged molecule trinacrene **11** was formed with 4% yield (Scheme 2.3) [28]. In this synthesis, the first intermolecular DA reaction is followed by two intermolecular cycloadditions.

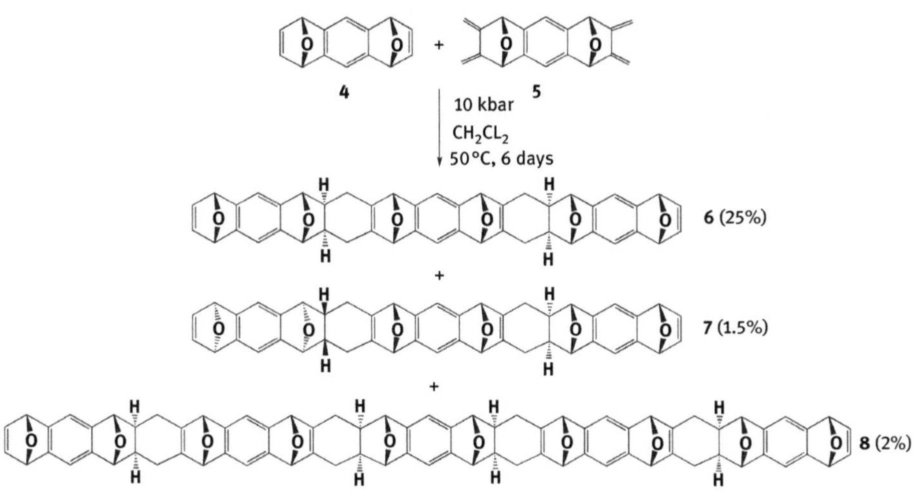

Scheme 2.2: Multiple Diels–Alder reactions.

With a proper choice of bis-diene and bis-dienophile initial geometries, single cyclic structures were obtained by Klärner (Scheme 2.4) [29, 30]. In HP conditions, stereoregular cycloaddition of bisdienophile **12** with the bisdiene **13** afforded cyclic

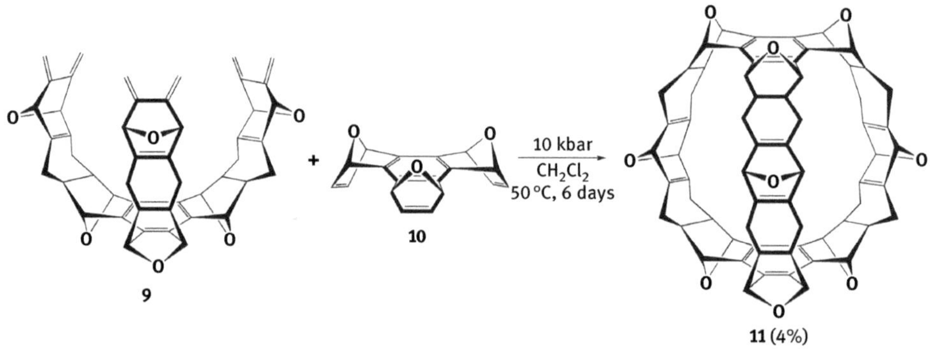

Scheme 2.3: Multiple Diels–Alder reactions leading to a cyclic structure.

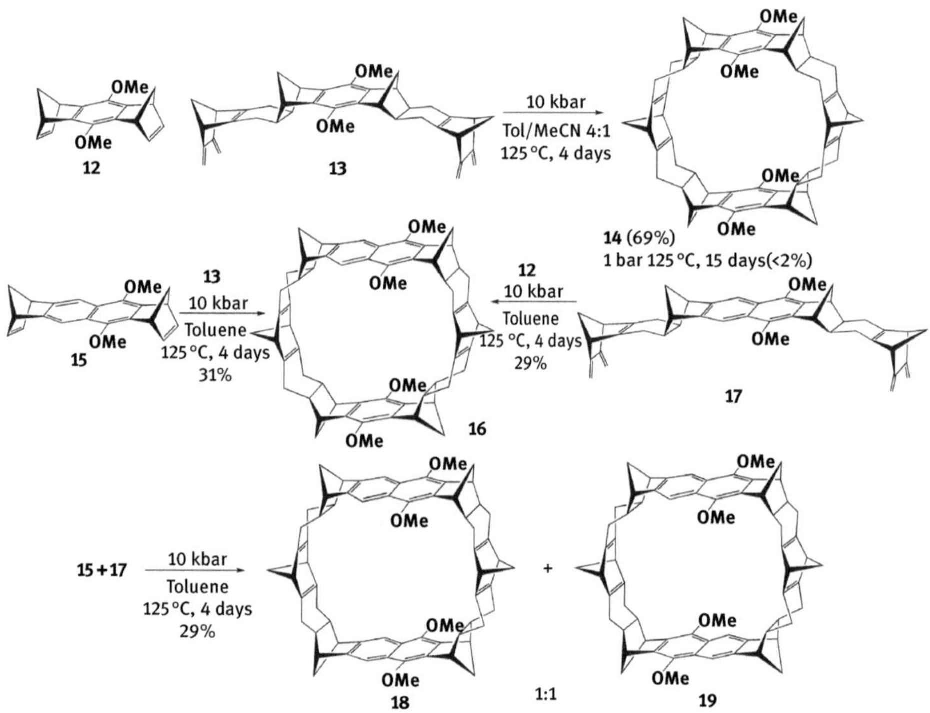

Scheme 2.4: Bis-Diels–Alder reactions.

product **14** in high yield. On the other hand, by reacting two components at ambient pressure only traces of products were obtained. Larger ring of **16** with one naphthalene moiety was obtained by the cycloaddition of either **15** with **13** or **17** with **12.** Finally, when **15** was reacted with **18,** a mixture of two regiomeric bisnaphthalene products **18** and **19** was obtained. A similar bis-DA reaction was reported by Klärner for the derivative of bis-dienophile **12** with benzonorbornane dimethoxydiene **13** (12 kbar, 24 h, NEt_3, toluene/MeCN) [31].

Cycloadditions involving CPD pose a potential problem of the competitive self-dimerization, which is also accelerated by pressure. This side-reaction consumes the amount of CPD monomer, an active diene reagent, and decreases the conversion, as shown in the reaction of CPD with methyl-*p*-benzoquinone (to 20%) (Scheme 2.5) [32]. Because of such problem, CPD is added usually in great excess. To minimize the effects on the competing diene dimerization, CPD was added in several batches (each time reaction was decompressed) to afford a mixture of *endo-anti-endo-* and *endo-anti-exo-*2:1 cycloadducts **22** and **23** with conversion of about 60%. On the other hand, cycloaddition reaction of dihydropyranones with CPD using sevenfold excess of CPD (15 kbar, 48 h, 25 °C, DCM) was reported to reach the full conversion to products [33].

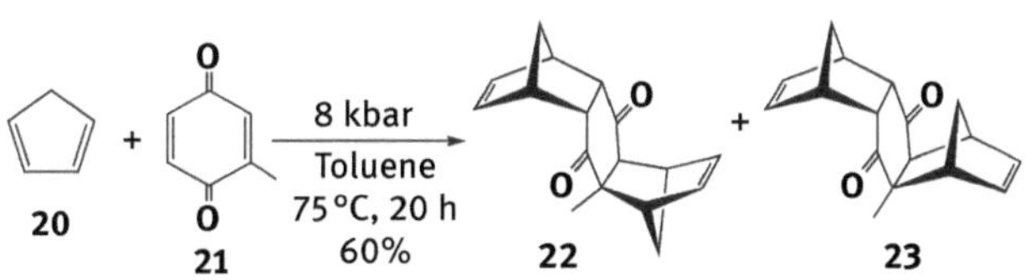

Scheme 2.5: Bis-Diels–Alder reactions of cyclopentadiene.

The application of HP is an extremely valuable synthetic method in the cases where thermal instability of products causes further transformation after the initial DA reactions. Takeshita has found that the thermal reaction of 2-chlorotropone **24** with 2,3-bis(methoxycarbonyl)-7-oxabicyclo[2.2.1]heptadiene **25** is rather complicated, affording four 1:1 cycloadducts (**26–29**) at 130 °C, mixed with cycloreversed homobarrelenones **31** and **33**, indanone **30** and furan **32** (Scheme 2.6) [34]. When

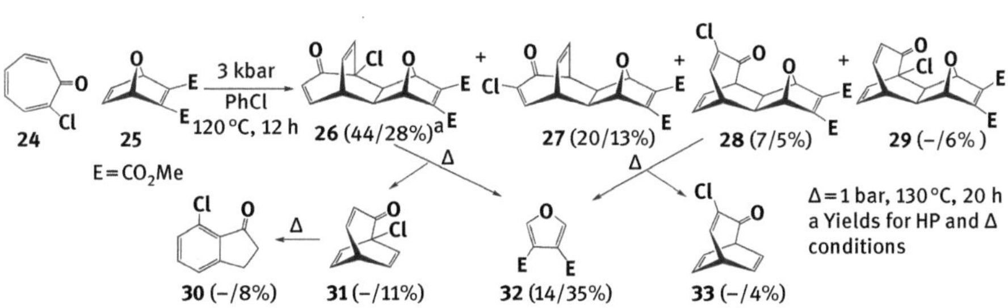

Scheme 2.6: The Diels–Alder reactions involving 2-chlorotropone.

reactants were subjected to 3 kbar, higher yields and much cleaner mixture of products were obtained. The *endo*/*exo*-ratio in HP conditions was 9.1, whereas in thermal reaction it is 3.7 (6.9 when the yield of **32** was added to the *exo*-adduct), showing that the *endo*/*exo*-ratio is slightly affected by pressure.

The employment of HP can cause a change in the course of a reaction as well as in the product specificity. Different types of products were obtained in the DA reactions of isophorone dienamine **34** with methyl acrylate and with acrylonitrile in mild conditions. Under pressures of 8.7–15.4 kbar at room temperature, only DA products **36** were isolated, which were formed in a regiospecific and stereospecific manner (Scheme 2.7) [35]. In thermal conditions, the very same reagents have provided different product distributions, the mixtures of ketones **37** and **38**, whose formation is associated with thermal decomposition of substrate or products (loss of morpholino and piperidino substituents).

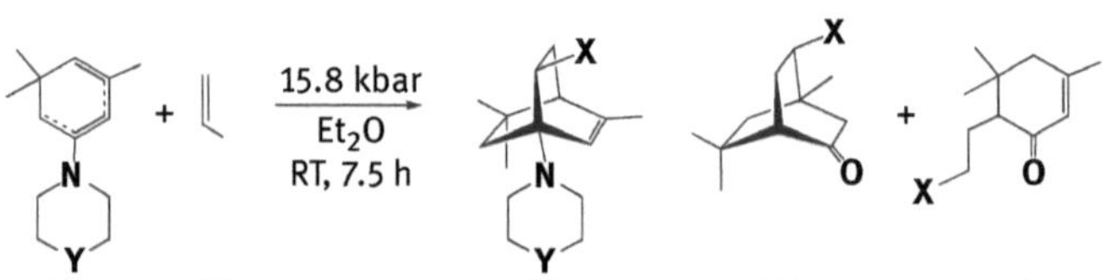

34 35 36 37 38

		36	37	38
Y = O, X = CO_2Me	HP	53%	–	–
	Δ[a]	–	29%	23%
Y = CH_2, X = CN	HP[a]	90%	–	–
	Δ	3%	10%	20%

[a]Δ = 100 °C, 24 h [b]8.4 kbar, 13 h

Scheme 2.7: The Diels–Alder reaction of isophorone dienamine with methyl acrylate.

Sometimes, unusual and unexpected reactions take place in hyperbaric conditions. Thermal reaction of cyanoacetylene **40** with [(2.2](2,5)furanoparacyclophane **41** at 160 °C provides a myriad of products **41–46** (Scheme 2.8) [36]. These are formed by 1:2 and 1:3 cycloadditions and subsequent thermal transformations, starting by the [2 + 2] cyclodimerization of cyanoethylene followed by the DA addition of furan. Klärner has found that different products could be obtained in hyperbaric conditions. In these conditions, the major products were 1:2 adducts **47** and **48** alongside small amount of 1:2 adduct **41**, which was the main product at 160 °C.

Unexpected products are sometimes the consequence of the transformation of the starting diene before the cycloaddition step, such as the diene rearrangements by acid generated from maleic anhydride of oxa[4.4.4]propelladiene [37] or methyl palustrate **49** to methyl levopimarate **53** (Scheme 2.9) [38]. Whereas *N*-phenylmaleimide provides a single cycloadduct **52**, reaction with maleic anhydride led to the

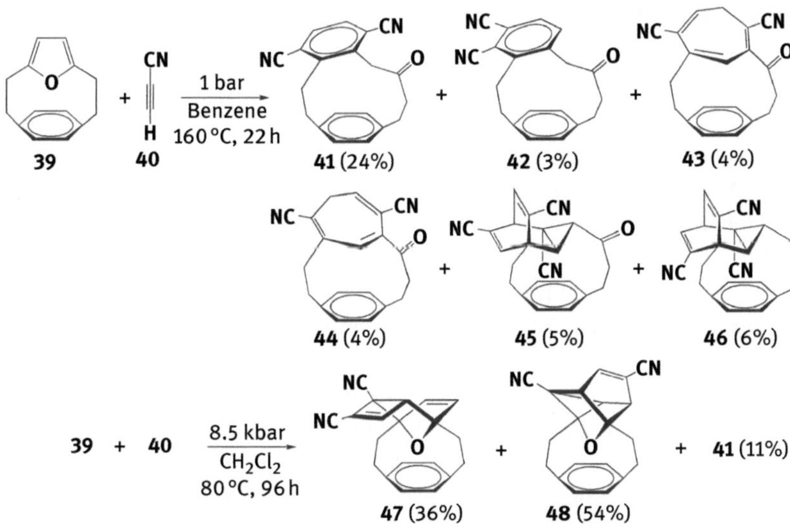

Scheme 2.8: The Diels–Alder reactions of cyanoacetylene with [(2.2](2,5) furanoparacyclophane.

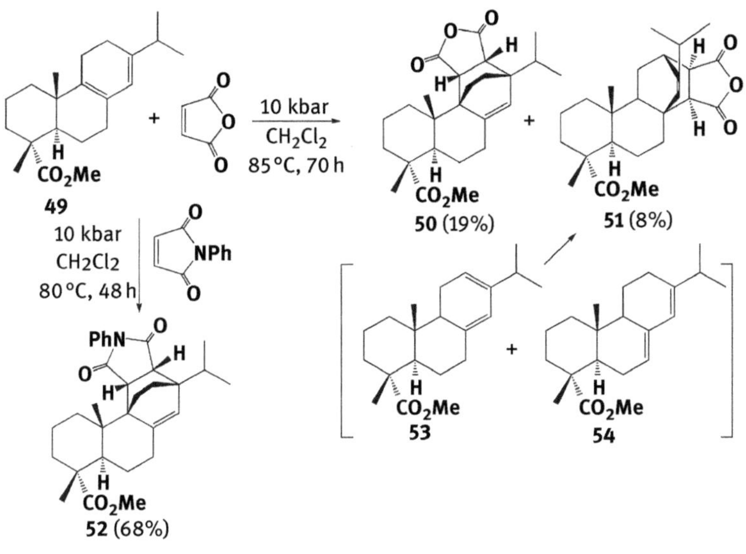

Scheme 2.9: Rearrangements in the Diels–Alder reaction of methyl palustrate.

formation of cycloadducts **50** and **51**. Low yield could be ascribed to facile acid-catalyzed isomerization to the more stable methyl abietate **54** and steric hindrance, which prevented cycloadditions at ambient pressure.

2.2 Heterocyclic dienes

2.2.1 Five-membered dienes

2.2.1.1 Furan

Furan is a reactive diene, and many accounts on its utilization to provide 7-oxabicyclo[2.2.1]hepta-2,5-diene skeleton were published. Under HP conditions, with an excess of furan often multiple DA reactions take place affording the mixture of products [39]. For instance, neat reaction of furan with diethyl (acetoxymethylene)malonate **56** (6.9:1 ratio) under 11 kbar gives the mixture of several cycloadducts (Scheme 2.10) [40]. The major product is the 1:1 *endo*-**57**, with much smaller amount of the *exo*-**57** cycloadduct. In addition, four bis-cycloadducts (**58** and **59**), trisadduct **60** and tetrakisadduct **61** were obtained in small yields. When furan:**56** ratio was lowered to 2.8, cycloadducts **57** were prepared in better yield (56%). Stereospecificities of the addition of the second molecule of furan affording both 2:1 isomers **58** and **59** differ from the Okamoto's report that reaction of furan with dichloromaleic anhydride (5 kbar, 50 °C, 15 h, neat) led exclusively to 2:1 adduct corresponding to **59** [41]. The nonbonding interactions between the two oxygen bridges were suggested to cause steric disadvantage for this mode of furan approach [42]; however, it may be also the consequence of the thermal instability of the linear product. Since both the linear and bent type 2:1 furan adducts were obtained in reaction with 1,4-epoxynaphthalene at 14 kbar [43], and formed furan adducts are often thermally unstable [44] the second explanation is more probable. The observed *endo*/*exo*-ratio in **57** is similar to the *endo*-preference obtained for furan–acrylic ester additions, which are not affected by pressure (10–20 kbar) or solvent [45].

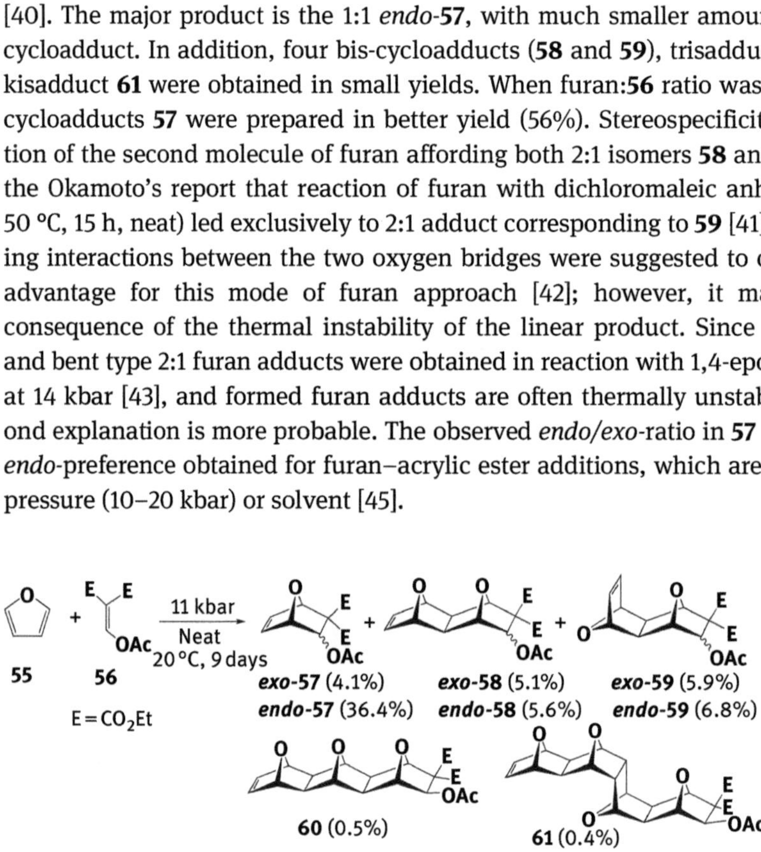

Scheme 2.10: The Diels–Alder reaction of furan with diethyl (acetoxymethylene) malonate.

Sometimes, HP cycloadditions of furan derivatives afford unexpected products [46]. For instance, in the DA reaction with tropone, furans act as dienophilic component. Under 3 kbar pressure, furan produced a mixture of two 1:1 adducts, *endo*-**64a** and *exo*-**65a** (Scheme 2.11) [47]. When 1-methoxyfuran **63b** was subjected to HP in the

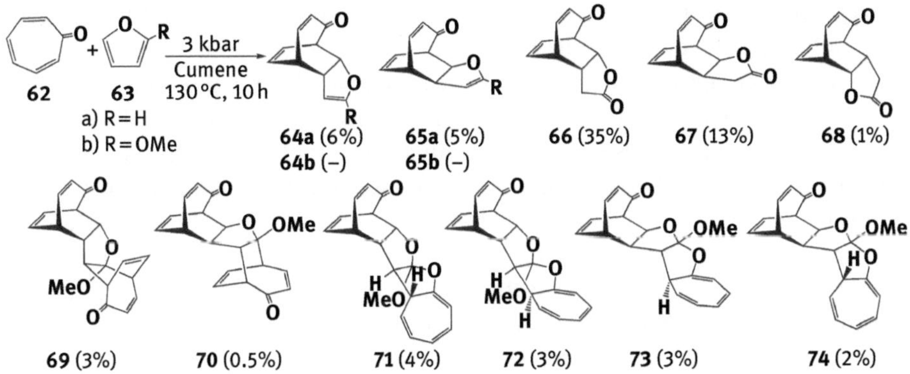

Scheme 2.11: The Diels–Alder reaction of furan with tropone.

presence of tropone, a wealth of cycloadducts was obtained. In addition to three 1:1 adducts **66–68**, six 2:1 adducts **69–74** are isolated, which arise from the second tropone cycloaddition to initial cycloadducts **64b** and **65b**. These two initial adducts were not detected, but further transformed in the reaction conditions. Heating of **63b** with tropone in sealed tube at 130 °C provides much less complicated reaction mixture containing of just two products (**66** and **67**).

Synthetic work toward CD-ring fragment precursors of paclitaxel employing HP cycloadditions of furans and citraconic anhydride was carried out by Scheeren and coworkers (Scheme 2.12) [48]. The DA reaction of 2-methyl furan with citraconic anhydride at 15 kbar provided a 1:1 mixture of two regioisomeric *exo*-adducts **79**, which were due to their cycloreversibility, without the isolation hydrogenated to **80**.

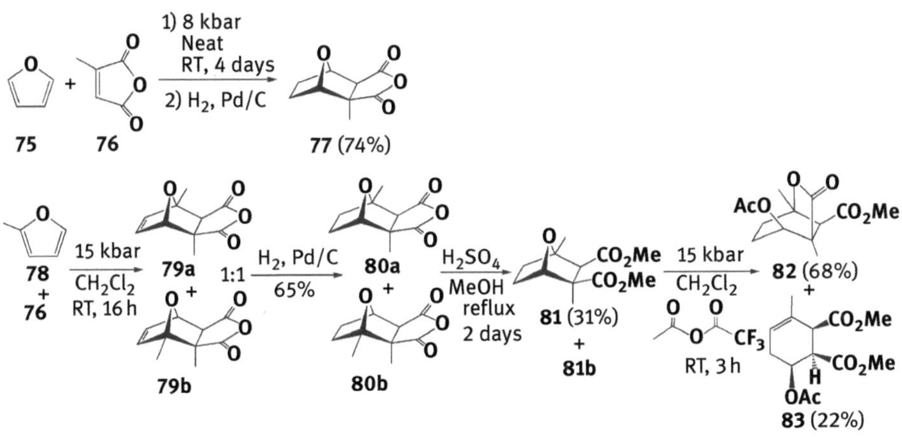

Scheme 2.12: The Diels–Alder reaction of furans and citraconic anhydride.

2-Alkoxyfurans ($CH_2OC(O)C(CH_3)_3$, CH_2OAc and CH_2OBn) reacted in an analogous manner affording the corresponding *exo*-cycloadducts in 54%, 59% and 31% yields, respectively. Their lower reactivity required four times longer time (60 h) than 2-methyl furan reaction; however, the regiomeric ratio was preserved (1:1, 5:6 and 7:5). The reported *exo*-selectivities of citraconic anhydride addition to furans are in accord to earlier studies on synthesis of palasonin **77** in neat furan [49, 50]. Further synthetic transformations toward CD-ring fragment of paclitaxel were carried out by esterification of regioisomeric mixture **80a/b**, in which **80a** is chemoselectively esterified to **81**. Ether bridge cleavage of product **81** was also affected under HP conditions with acetic trifluoroacetic anhydride. In this process, two products with complete regioselectivity were obtained: cyclohexene **83** and lactone **82**.

2.2.1.2 Pyrrole

Takayama and coworkers have shown that hyperbaric conditions could complement classical cycloaddition reactions to obtain different products than in thermal conditions (Scheme 2.13) [51]. The DA reaction of 3,5-dihydro-l*H*-thieno[3,4-*c*]pyrrole-2,2--dioxide **84** with dimethyl acetylenedicarboxylate (DMAD) at 12 kbar provides 2:1 adduct **88** (by the DA reaction followed by double Michael-type addition), which could not be obtained at lower pressure or by heating at atmospheric pressure. Instead, 1:1 and 3:1 adducts **85** and **87** were obtained. Interestingly, SO_2 elimination takes place at 4 kbar, and HP conditions are required to suppress desulfonylation process. In the intramolecular DA reaction (IMDA) of the *N*-derivative of **84**, desulfonylation could not be suppressed even at 12 kbar and room temperature [52]. HP conditions facilitate $ZnCl_2$ catalyzed reactions of 1-triflylpyrroles bearing electron-withdrawing groups that participate in the DA reactions as efficient dienophiles (12 kbar) [53]. In this system, product yield increased, without the change in the mechanism.

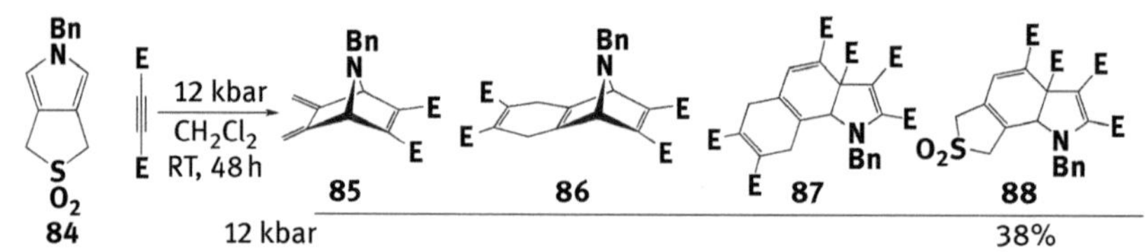

E = CO_2Me

	85	86	87	88
12 kbar				38%
4 kbar			62%	
100 °C 4 h	28%	0.1%	47%	(1 bar, benzene)
140 °C 46 h			97%	(1 bar, benzene)

Scheme 2.13: The Diels–Alder reaction of pyrrole derivative with DMAD.

Hyperbaric DA reaction of pyrroles was also employed in natural product synthesis, for the construction of 7-azabicyclo[2.2.1]heptane system of epibatidine by reaction of *N*-acyl pyrrole with *p*-toluenesulfonylacetylene (12 kbar) [54]. Similarly, HP has played an important role in the synthesis of hetero-substituted sesquinorbornadienes via the DA cycloadditions of *N*-substituted pyrroles (Scheme 2.14) [55]. The reaction of pyrrole **89** with the azanorbornadiene **90** requires drastic conditions (140 °C) to proceed. In these conditions, the *exo,exo*-product **92** is formed by the diene approach at the unsubstituted π bond of dienophile. Under HP, reaction occurs at considerably lower temperature forming the isomeric adduct **91**. The adventitious use of HP is demonstrated for the addition of **89** to 7-oxanorbornadiene **92**, by obtaining better yields, the change in regioselectivity and stereospecificity. Products **96** and **97** are obtained by kinetic, whereas **94** and **95** are products of the thermodynamic control.

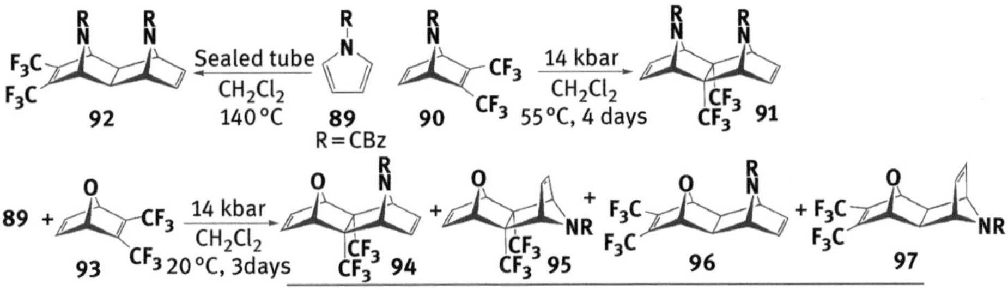

RT 5 days	100%	0	0	0
55 °C 4 days	94%	0	0	4%
90 °C 2 days	56%	0	7%	27%
130 °C 3 days	0	0	3%	62% (1 bar)

Scheme 2.14: The Diels–Alder reaction of pyrrole with azanorbornadiene.

Benzannulation of heterodienes to isobenzo species increases their reactivity [56]. To facilitate the participation of less reactive *N*-benzyl-4,5,6,7-tetrafluoroisoindole **98** in [4 + 2] cycloaddition reaction with *N*-methyl-3,4-dibromomaleimide, forcing conditions (14 kbar) were required. The *exo*- and *endo*-adducts **99** and **100** were formed in a 10:1 ratio (Scheme 2.15) [57, 58]. Isoindoles **102** exhibit exclusive regioselectivity in reactions with diene **103** adding to the cyclobutene C=C bond; however, stereospecificity depends on the indole substitution as well as on reaction conditions [59]. At ambient pressure, reactive *N*-benzyloxyisoindole provides the bent-frame (*endo, endo*)-structure **105** exclusively. On the other hand, cycloaddition of less reactive *N*-methyl-4,5,6,7-tetrafluoroisoindole at 14 kbar pressure led to exclusive formation of the extended frame (*endo,exo*)-isomer **104**, whereas *N*-benzyl-4,5,6,7-tetrafluoroisoindole provides the mixture of both isomers, in which **104** is the major product.

2.2.1.3 Thiophene

Thiophene is notoriously unreactive diene due to its high aromaticity and does not react with maleic anhydride even at 15 kbar at room temperature. Only the

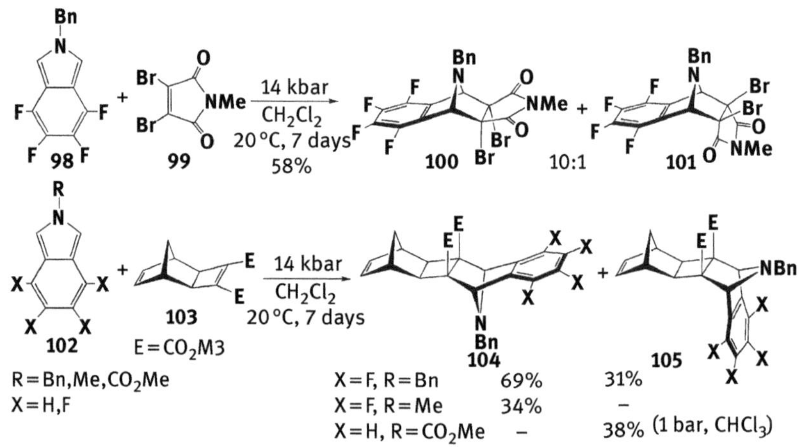

Scheme 2.15: The Diels–Alder reaction of isoindoles.

combination of HP (15 kbar) and heating at 100 °C was successful to overcome a striking absence of diene character in thiophene (Scheme 2.16) [60]. By this protocol, 7-thiabicyclo[2.2.1]hept-2-ene skeleton of **108** was obtained in satisfactory yield. The highest yield (47%) was obtained at 20 kbar [61]. Reaction yields are diminished when other solvents were employed, with the exception of perfluorohexane where the yield was significantly improved [62]. Remarkable yield was obtained by Kotsuki in neat conditions at 100 °C, where thiophene was used as solvent, even the pressure was only 8 kbar. Only the *exo*-adduct was formed in maleic anhydride case, which is in contrast to maleimide or acrylate solvent-free cycloadditions affording mixtures of *endo*- and *exo*-cycloadducts.

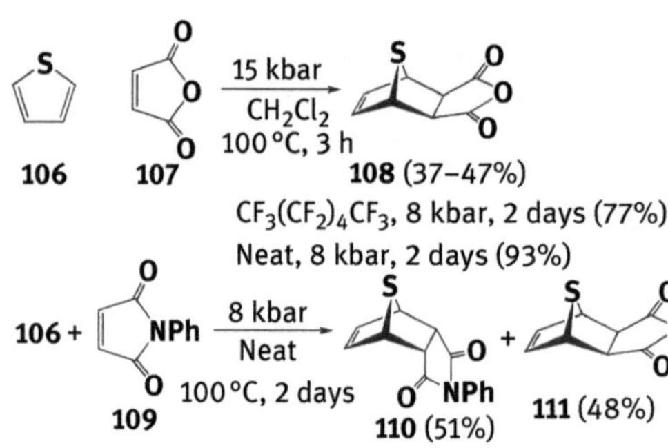

Scheme 2.16: The Diels–Alder reactions of thiophene.

Similar conditions (10 kbar, 100 °C, 1 day, CH_2Cl_2) were employed in reactions with maleimides and 2,5-dimethyl thiophene [63]. Addition of thiophene to maleic anhydride, maleimide (60% yield) or dimethyl maleate was also carried out at low temperature by the increase in reaction time and pressure (40 °C, 71 h, 17 kbar, CH_2Cl_2) [64]. Reactivity of thiophenes could be increased by oxidation to sulfoxides to allow cycloadditions at moderate temperatures (10 kbar, 60 °C, 1–5 days, MeCN) [65] or by the employment of highly reactive dienophiles such as **112** (Scheme 2.17) [66].

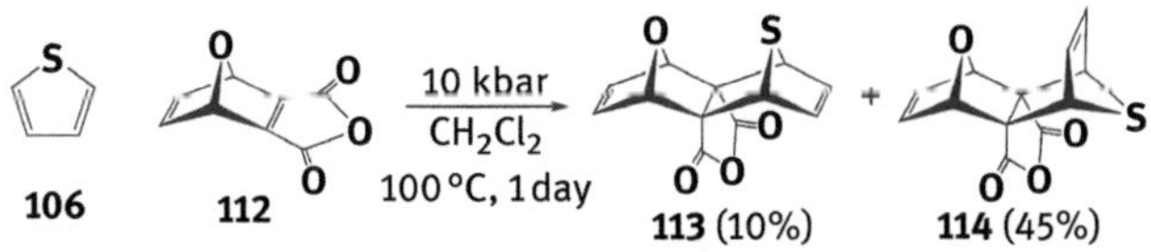

Scheme 2.17: The Diels–Alder reaction of thiophene.

2.2.1.4 Siloles

The DA cycloadditions of 1-sila-2,3,4,5-tetraphenyl-1,1-dimethyl-cyclopenta-2,4-diene (silole) to 7-oxanorbornadienes are found to be stereospecific, providing bent (*exo, endo*)-adducts [67–69]. Similar *exo,endo*-stereospecificity was observed for the germole cycloadditions in which 1-germa-2,3,4,5-tetraphenyl-1,1-dimethyl-cyclopenta-2,4-diene provides the corresponding 7-germabicyclo[2.2.1]heptene skeleton [70]. Experimentally observed stereospecificities could be rationalized by the steric hindrance of the oxygen atom and methyl substituents. For an illustration, when two equivalents of silole **115a** were reacted with dienophile **116**, only the 2:1 adduct **118** was obtained. HP reactions of equimolar amounts of **116** and silole **115a** or germole **115b** afforded only 1:1 adducts **117a** and **117b** (Scheme 2.18) [71]. Silole **115a** and germole **115b** also readily undergo the DA reaction with maleimides and maleic anhydride in hyperbaric conditions (8 kbar, 70 °C, 3 days) to stereospecifically afford *endo*-adducts [72].

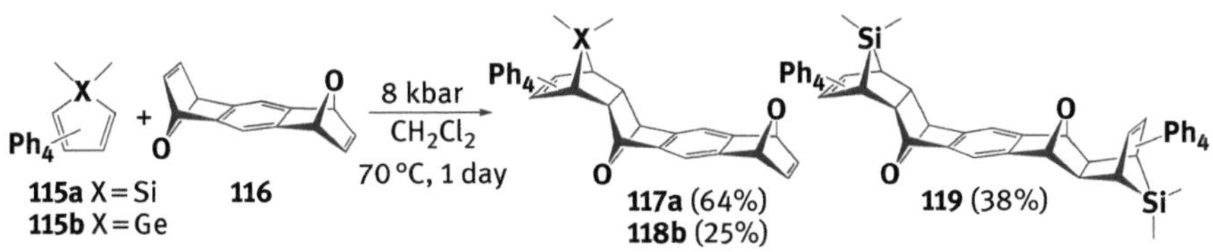

Scheme 2.18: The Diels–Alder reaction of siloles and germoles.

The 1,2-disilacyclohexa-3,5-diene reaction with phenyl vinyl sulfoxide under 10 kbar provides 7,8-disilabicyclo[2.2.2]octa-2,5-diene moiety **121** (Scheme 2.19) [73]. The elimination of benzenesulfonic acid from the DA adduct takes place under ultra-HP. Dimer **122** and 1,2-bis(phenylsulfinyl)ethane **123** are also obtained as side products. The use of HP conditions is a prerequisite of this cycloaddition, which does not take place at ambient pressure.

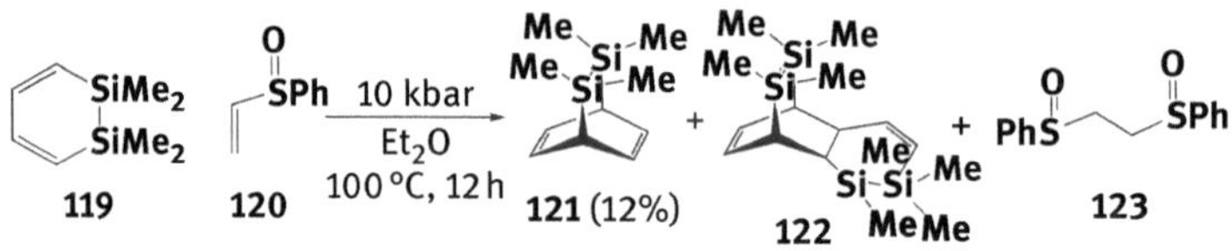

Scheme 2.19: The Diels–Alder reaction of disilacyclohexadiene.

2.2.1.5 Oxazoles

Cycloadditions of oxazoles, similarly to furan, provide 7-oxabicyclo [2.2.1] heptene moiety. Besides the rate acceleration by pressure, Matsumoto discovered that the nature of solvent has a profound effect on the composition of the reaction mixture (Scheme 2.20) [74]. Hyperbaric reaction of 5-methoxy-2-methyl-4-(*p*-nitrophenyl) oxazole **124** with *N*-phenylmaleimide **125** in benzene affords the mixture of *endo*- and *exo*-adducts **126** and **127**. The yield of the *exo*-adduct **127** increases with time, which was ascribed to more facile retro DA reaction (RDA) of the *endo*-adduct **126**, and further cycloaddition accumulating adduct **127**. Interestingly, when this reaction was carried out in dichloromethane or acetonitrile, the elimination from **126** and **127** provides pyridine derivative **128** and the reaction in methanol gives **128** as the sole product. This may be the consequence of the catalytic action of protic species such as water contained in the solvents.

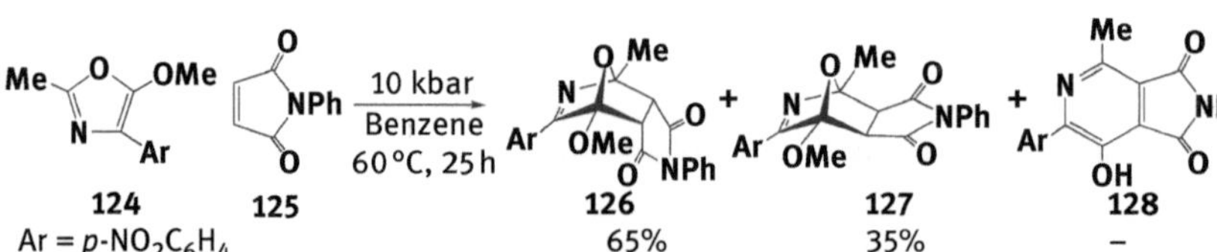

	126	127	128
	65%	35%	–
EtOAc	65	35	–
Et_2O	63	19	18
CH_2Cl_2	24.2	55.5	18.5
MeCN	17%	–	73
MeOH	–	–	100
1 bar	41	18	–

Scheme 2.20: The Diels–Alder reaction of oxazole with *N*-phenylimide.

2.2.2 Six-membered dienes

2.2.2.1 Pyrones

The [4+2] cycloadditions of *α*-pyrones (2*H*-pyran-2-on) lead to the formation of 3-oxo-2-oxabicyclo[2.2.2] skeleton. Unsubstituted *α*-pyrones react with strong dienophiles at 5–8 kbar in toluene providing the 1:1 cycloadducts in high yields

(60–92%) [75]. Maleic anhydride and fumaroyl chloride reactions proceed in higher yields than at atmospheric pressure (maleic anhydride 4.5 kbar, 55 °C 92% vs. 1 bar 50%; fumaroyl chloride 5.5 kbar, 85 °C, 85% vs. 1 bar 110 °C, 55%). Furthermore, in the synthesis of (±)-sativene, Takeshita used 5-methoxycarbonyl-2-pyrone with 2,3-dimethyl-1,3-cyclohexadiene as the starting reaction. At 50 °C under 3 kbar pressure the *exo*-adduct was isolated in 34% yield [76]. Some by-products were also obtained, however, in lesser amounts than the reaction at ambient pressure (100 °C), which led to the complicated mixture of products due to thermal isomerization of 2,3-dimethyl-1,3-cyclohexadiene. Similarly, reaction of electron-deficient Eschemoser's *α*-pyrone **129** with cyclopropenone ketal **130** provided the *exo*-cycloadduct **131** exclusively (Scheme 2.21) [77]. Boger has demonstrated that this [4+2] cycloaddition was proved to be ineffective at atmospheric conditions (even in the presence of Lewis acid catalyst), as a result of unfavorable steric interactions present in the transition state of the DA reaction.

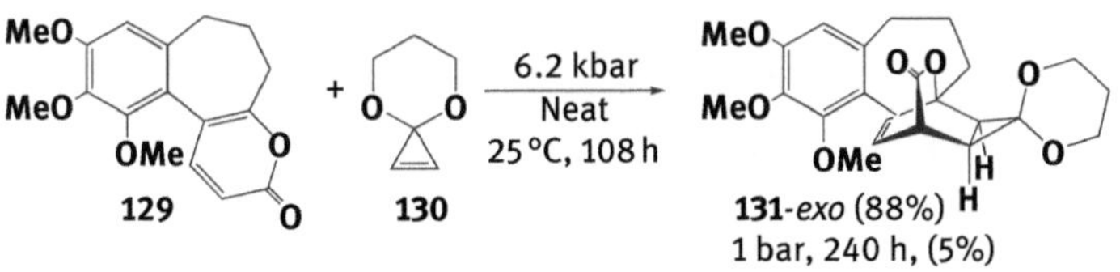

Scheme 2.21: The Diels–Alder reaction of Eschemoser's pyrone with cyclopropenone ketal.

A total synthesis of transtaganolide E and F employs the IMDA reaction of geranylated 2-pyrone derivative **331**, which is assumed to be an intermediate in the biosynthesis of transtaganolides. Classical thermal and Lewis acid conditions failed to convert **132** to **133** through the IMDA route. On the other hand, Scheeren has found that this cyclization proceeds in HP conditions (15 kbar), providing tricyclic lactone **133** as the 2:1 diastereomeric mixture (Scheme 2.22) [78]. Cycloaddition occurs in the

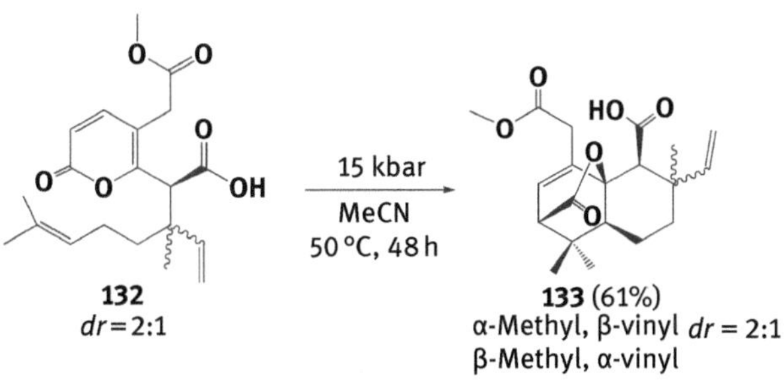

Scheme 2.22: Intramolecular Diels–Alder reaction of geranylated 2-pyrone.

exo-fashion, due to repulsive pseudoaxial 1,3-interactions in the transition state between the methyl ester and the C-9 carboxylic acid substituents.

2.2.2.2 Pyridone

Kotsuki reported an interesting example of products obtained from the unusual reactions that can take place only in hyperbaric conditions. Unexpected 1:3 adduct **136** from l-(5′,5′-dimethylcyclohex-2′-en-l'-on-3′-yl)-2 (1*H*)-pyridone **134** and DMAD was obtained after 15 kbar pressure was applied (Scheme 2.23) [79]. About the same ratio of 1:1 and 1:3 adducts **135** and **136** were formed. The expected DA reaction led to the formation of bicyclo[2.2.2] system of bicyclic lactam **135**, whereas mechanistically complex reaction sequence is proposed for the formation of adduct **136**, which involves multiple Michael additions to DMAD and presumably 1,3-dipolar cycloaddition. Another type of 1:3 adduct **137** (as the *exo,endo*-mixture) was obtained when *N*-methyl pyridone was pressurized with DMAD, but in minor proportion (4%) [80, 81].

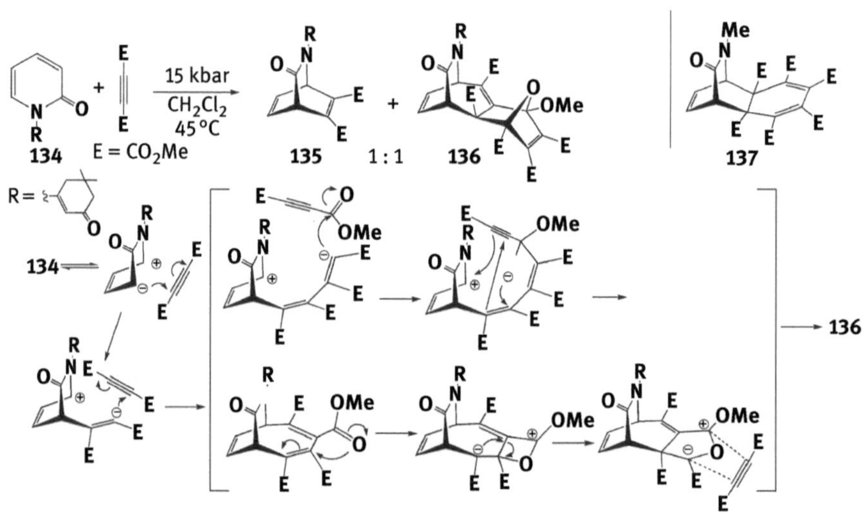

Scheme 2.23: The Diels–Alder reaction of pyridone with DMAD.

2.2.2.3 Tetrazines

Synthetic utility of tandem DA reactions of *sym*-tetrazines in construction of larger polycyclic molecules was realized by Warrener. It relies on HP due to the irreversible prototropic shift of intermediate to inactive 1,4-diene **143** in thermal conditions. Tandem DA reactions of *sym*-tetrazines such as 3,6-di(2′-pyridyl)-*s*-tetrazine **139** with norbornenes take place via initial [4+2] cycloaddition, which is followed by nitrogen elimination from adduct **141** and formation of the new 1,3-diene system of 4,5-dihydropyridazine **142**. In the closing mechanistic step, **142** reacts with another molecule of dienophile to obtain final diazabicyclo[2.2.2] structure **140** in a stereospecific manner (Scheme 2.24) [82].

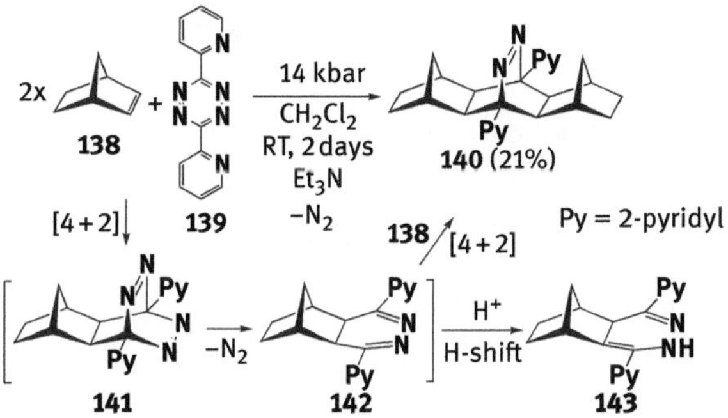

Scheme 2.24: The Diels–Alder reaction of *s*-tetrazine with norbornene.

Hyperbaric conditions allow the synthesis of a number of complex polycyclic systems [83, 84] such as the bisporphyrin **145** (Scheme 2.25) [85].

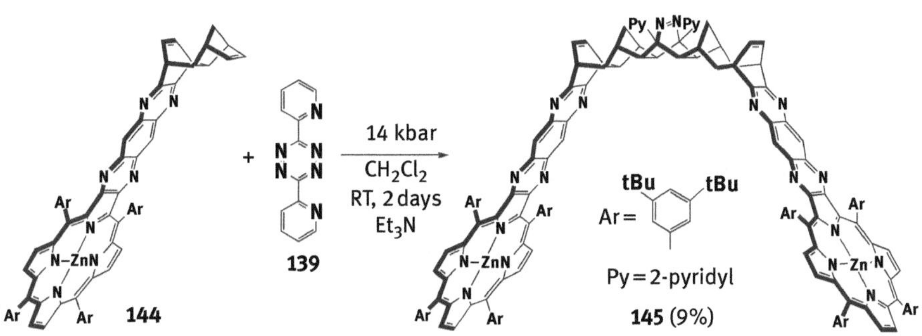

Scheme 2.25: The Diels–Alder reaction of *s*-tetrazine with norbornene porphyrin.

Pressure-promoted inverse-electron demand cycloaddition reactions of 1,2,4-triazine **147** with electron-rich morpholino and pyrrolidine enamines **146** were employed in the pyridyl CD ring construction of streptonigrin (Scheme 2.26) [86]. Analogously to tetrazine cycloaddition mechanism, initial DA adducts loose dinitrogen which was then followed by RN elimination to afford pyridine ring. The issues of low reactivity of **147**, which demanded the vigorous reaction conditions with loss of the regioselectivity and the thermal instability of the enamine **146b**, were circumvented by the application of pressure at ambient temperature. Similarly, dinitrogen is eliminated under HP (15 kbar, 55 °C) in other triazine cycloadditions [87].

In comparison with tetrazines, diazines are less reactive and higher temperatures are employed in combination with pressure to promote reaction (Scheme 2.27)

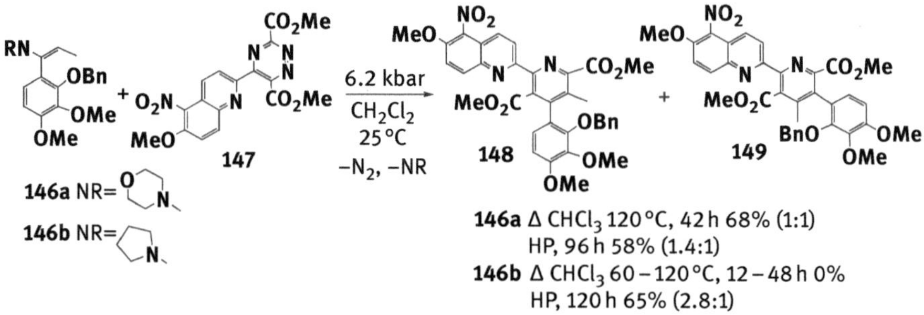

Scheme 2.26: The Diels–Alder reaction of triazine with enamines.

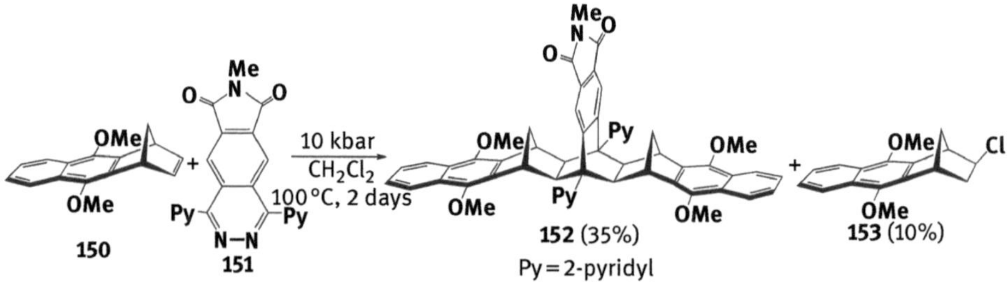

Scheme 2.27: The Diels–Alder reaction of phthalazine with naphthonorbornadiene.

[88]. Under 10 kbar, reaction of phthalazine **151** with naphthonorbornadiene **150** provided 2:1 cycloaddition product **152**. In these conditions, the chlorinated side product **153** was formed in significant amount. It probably arises from a Cl radical formed from dichloromethane and is not observed at ambient pressure. The addition of a radical scavenger (benzoquinone) suppressed the formation of **153** and increased the yield to 48%.

2.3 HDA reactions

The carbon–oxygen bond formation processes are the most often employed in HDA reactions. HDA reaction involving the oxygen atom results in the formation of 3,4-dihydro-2*H*-pyran ring. In these reactions, dienes such as α,β-unsaturated aldehydes and ketones, electron-deficient heterodienes bearing a terminal oxygen atom were employed. For instance, reactions of *trans*-4-methoxy-2-oxo-3-butenoate **154** with electron-rich ethyl vinyl ether proceed predominantly through an *endo*-transition state (either by the thermal or pressure-promoted cycloaddition reaction) (Scheme 2.28) [89]. The selectivity toward the *endo*-product increased as the temperature of the reaction was decreased and the pressure increased (6–13 kbar). On

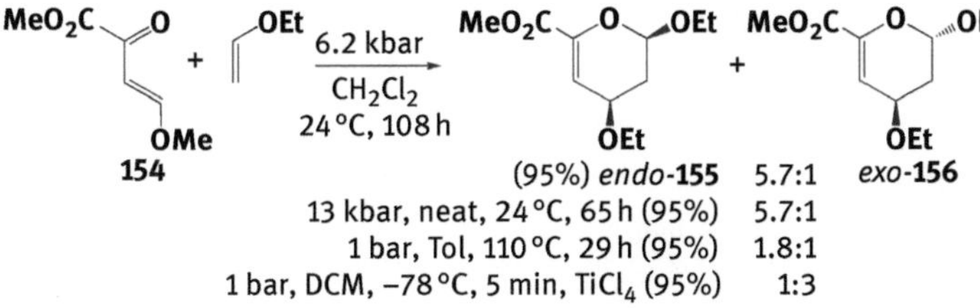

Scheme 2.28: The Diels–Alder reaction of butenoate with ethyl vinyl ether.

the other hand, an employment of Lewis acid catalysts led to the preferred formation of the *exo*-adducts.

The [4+2] cycloaddition of phosphonates such as **157** with ethyl vinyl ether, under different conditions of temperature and pressure, afforded new 5-dialkoxyphosphoryl-3,4-dihydro-2*H*-pyrans **158** (Scheme 2.29) [90]. Maddaluno have isolated these adducts as a mixture of *trans*- and *cis*-diastereomers, where their ratio depends on the reaction conditions. The increase in pressure and the addition of equimolar amount of *tert*-butanol led to higher *trans*-diastereoselectivity. In addition to HP favoring the more compact transition state, the presence of ^{t}BuOH also supports the *endo*-approach by simultaneous hydrogen bonding to carbonyl and ethereal oxygen atoms.

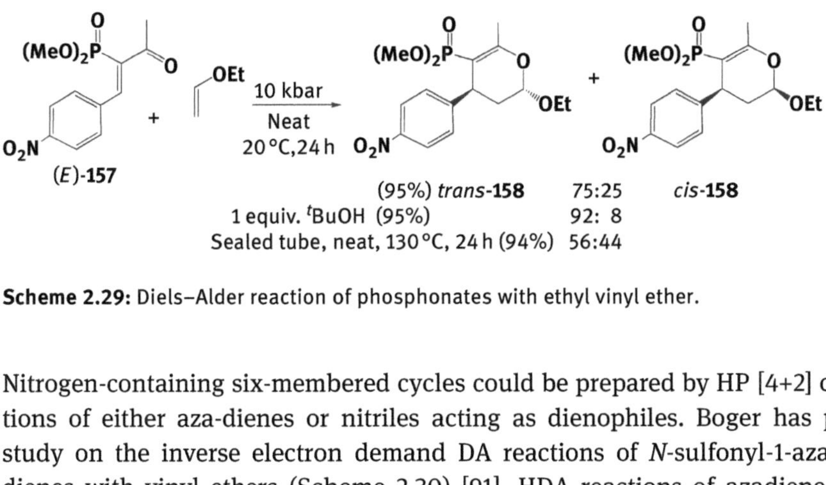

Scheme 2.29: Diels–Alder reaction of phosphonates with ethyl vinyl ether.

Nitrogen-containing six-membered cycles could be prepared by HP [4+2] cycloadditions of either aza-dienes or nitriles acting as dienophiles. Boger has published study on the inverse electron demand DA reactions of *N*-sulfonyl-1-aza-1,3-butadienes with vinyl ethers (Scheme 2.30) [91]. HDA reactions of azadiene **159** with vinyl ether are accelerated by pressure and proceed with high *endo*-selectivity to octahydroisoquinoline **162**. As an explanation, authors postulated that the transition state for the *endo*-approach benefits from the n–σ* stabilization in a manner analogous to the ground-state anomeric effect.

Cycloaddition of naphthoquinone **164** with crotonaldehyde *N,N*-dimethylhydrazone **165** under HP (10 kbar) in short reaction time (6 h) provided only cycloadduct

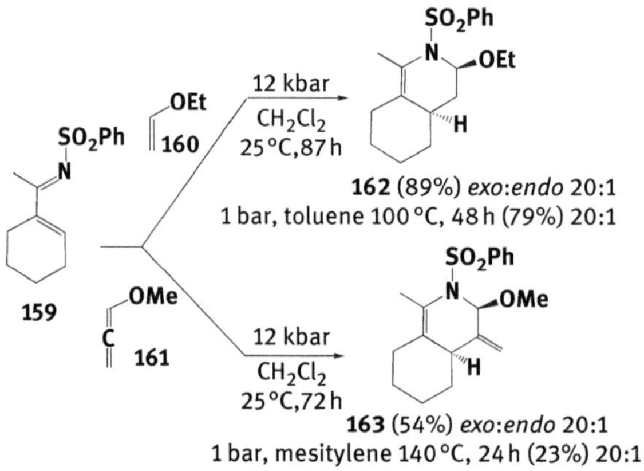

Scheme 2.30: Diels–Alder reactions of azadiene.

167, while upon longer pressurization (24 h), the yield of dihydro azaanthraquinone **167** has diminished to 24%, with the parallel formation of quinones **168** (15%) and **169** (15%) (Scheme 2.31) [92]. The formation of these side products is by the addition of dimethylamine obtained from the elimination of dimethylamine from transient cycloadduct **166.** Alternative synthetic conditions, sonication or stirring at room temperature gave product **167** in significantly lower yields, in which sonication also enhanced the formation of products **168** and **169.**

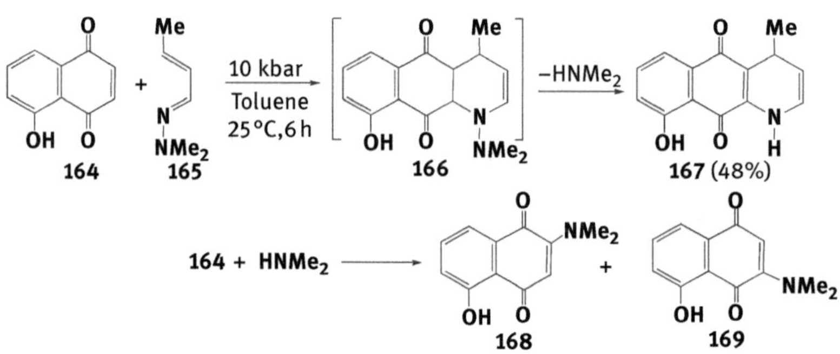

Scheme 2.31: The Diels–Alder reaction of naphthoquinone with crotonaldehyde *N,N*-dimethylhydrazone.

The addition of 2,3-dimethyl-1,3-butadiene to perfluorooctanonitrile affords the 3,4-dimethyl-6-perfluoroheptyl-2,5-dihydropyridine **173** as the final product (Scheme 2.32) [93]. Reaction time defines the composition of the reaction mixture; as the reaction is prolonged, more pyridine product **173** is formed, by the oxidative aromatization of the initial DA adduct **172.** The exclusive formation of product **172** is achieved after short time (3.9 h), albeit in very low yield.

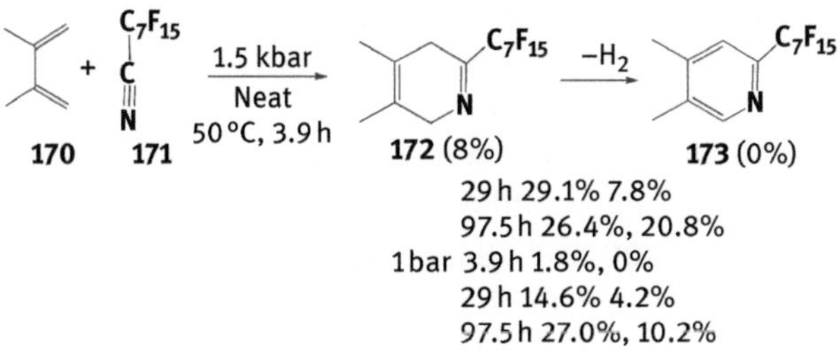

Scheme 2.32: The Diels–Alder reaction of dimethylbutadiene with perfluorooctanonitrile.

The *α*-phosphono *α*,*β*-unsaturated dithioesters are sulfur analogs of the hetero [4+2] cycloaddition of phosphonates described earlier in this chapter. Their HP reactions (11 kbar, RT) provide new C–S and C–C bond of phosphono 3,4-dihydro 2*H*-thiopyrans. High yields are obtained (76–90%) with dominant *trans*-selectivity in hyperbaric conditions and with *cis*-selectivity favored at atmospheric pressure, similar to phosphonate reactions [94].

Weinreb has shown that *N*-alkyl-*N*-sulfinyl compounds possessing low reactivity readily engage in the HDA reactions with 1,3-dienes at 12 kbar pressure (Scheme 2.33) [95]. Cycloadditions of **175** with *E*,*E*-2,4-hexadiene provided dihydrothiazine oxides with *syn*-selectivity as an epimeric mixture of **176** and **177**. Varying mixtures of sulfur epimers are obtained in dependence of the reaction conditions, and yields obtained in HP conditions are much better than catalytic runs at atmospheric pressure. These HDA reactions were also affected in an intramolecular manner, by reaction of *N*-

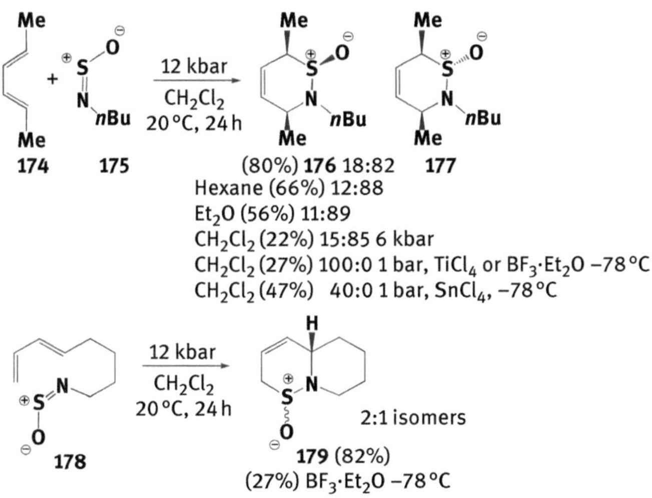

Scheme 2.33: The Diels–Alder reactions of *N*-alkyl-*N*-sulfinyl compounds.

sulfinyl compound **178**, which did not show affinity to undergo thermal cycloaddition. Adduct **179** was obtained as a 2:1 mixture of isomers, both at 12 kbar or in BF_3–Et_2O-catalyzed reactions.

Simultaneous formation of C–S and C–N bonds could also be achieved when *N*-sulfinylamines were employed as dienophilic partner in HDA reaction. Reactions of *N*-sulfinylazulenylamines **180** and **183** with 1,3-butadiene provided under 10 kbar pressure azulene-substituted 6*H*-2,3-dihydro-1,2-thiazine-1-oxides **182** and **184** in high yields (Scheme 2.34) [96]. These heterocumulenes did not easily react under atmospheric pressure, and the yield of product **184** was rather poor (29%).

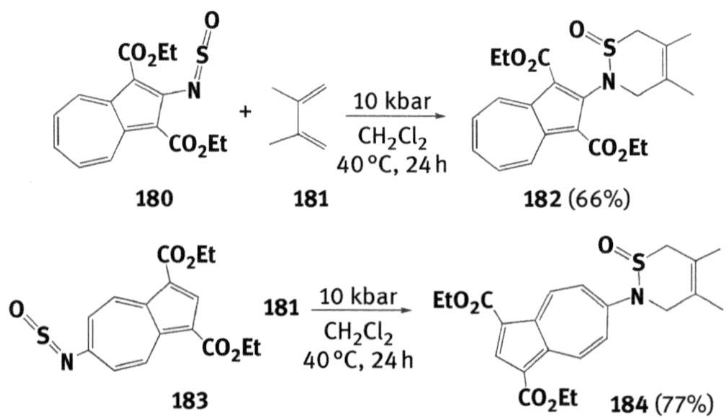

Scheme 2.34: The Diels–Alder reactions of *N*-sulfinylamines.

2.4 Intramolecular DA reactions

Kinetic studies of IMDA reactions with furan (IMDAF) by Isaacs indicate that intramolecular reaction proceeds with a smaller diminution of volume than intermolecular ($\Delta V^{\neq} = -25\ cm^3\ mol^{-1}$) [97]. This finding could be rationalized by two contributions to the volume contraction during the activation process: the first one because of the formation of two covalent bonds (about $-10\ cm^3\ mol^{-1}$ per bond) and the other resulting from the close approach of the reagents (amounting to $-10\ cm^3\ mol^{-1}$). The IMDA reactions lack the second contribution (or it is diminished), which is reflected in the less negative $\Delta V^{\neq}$. The activation volumes for IMDA reactions of (*E*)-1,3,8-nonatriene were determined to be $\Delta V^{\neq} = -24.8$ and $-24.8\ cm^3\ mol^{-1}$ (in hexane) for *cis*- and *trans*-addition, and for (*E*)-1,3,9-decatriene reaction $\Delta V^{\neq} = -37.6$ and $35.0\ cm^3\ mol^{-1}$ (in heptane) indicate the ring-size dependence [98].

As an illustration of often used IMDAF reactions is the cyclization of furyl amides **185** at 14–19 kbar, which afforded the corresponding *exo*-adducts **186** in moderate yields (Scheme 2.35) [99].

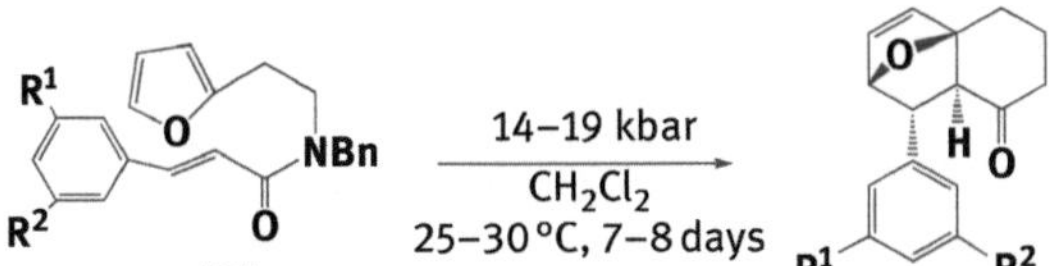

185
R^1,R^2 = H,OMe

186 (48–55%)

Scheme 2.35: Intramolecular Diels–Alder reaction of furyl amide **185**.

HP IMDA reactions of vinylsulfonic acid derivatives, vinylsulfonates and vinylsulfonamides **187** are readily converted to two diastereomeric cycloadducts **188** and **189** at 13 kbar (Scheme 2.36) [100]. The *endo/exo*-selectivity in this case is influenced by pressure. The more compact *endo*-transition state forming *cis*-fused products **189** is favored at HP, whereas there is a very small *exo*-preference at ambient pressure. In variance to acyclic dienes, in IMDA reactions of cyclic dienes **190**, only *endo*-cycloadducts are formed. This result is explained by nonbonding interactions of H-atom on the saturated C_2-bridge, which is in *syn* position in relation to the six-membered heterocycle in the *exo*-transition states. The HP greatly prefers the adduct **191** with methyl group in the equatorial position, while in thermal conditions this preference is diminished.

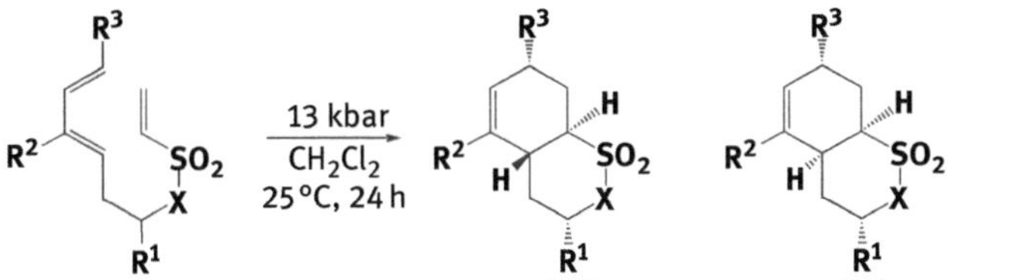

187 **188** **189**

	HP	Δ[a]	HP	Δ
X = O, R^1 = R^2 = R^3 = H	1:2:3	1:1	88%	76%
X = O, R^1 = R^2 = Me R^3 = H	1:2	1:4	78%	64%
X = NBn, R^1 = Me R^2 = H R^3 = Me	1:1.9	1.6:9	81%	61%

[a]Toluene reflux, 12 h

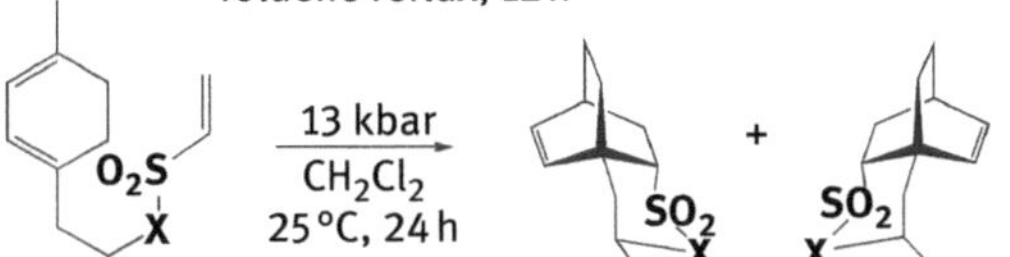

190 **191** **192**

	HP	Δ	HP	Δ
X = O	55.1	10.1	92%	58%
X = NBn	30:1	9:1	93%	77%

Scheme 2.36: Intramolecular Diels–Alder reactions of vinyl sulfonamides.

Synthetic studies toward the AB core of taxinine by IMDA reaction of substrate **193** catalyzed by boron trifluoride etherate afforded mixture of products **194** and **195** (Scheme 2.37) [101]. These compounds possess double bond at the $\Delta^{7,8}$ position,

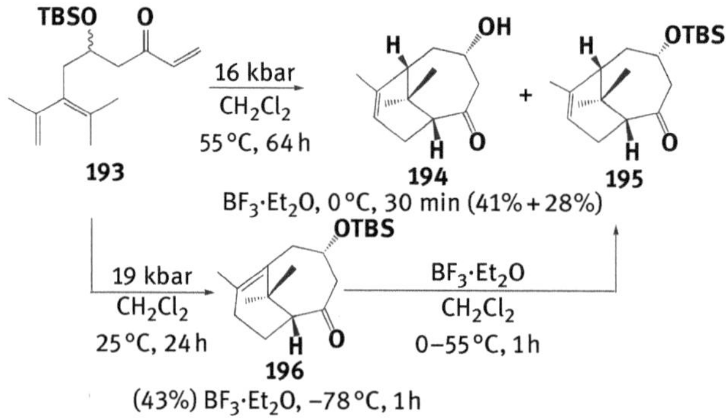

Scheme 2.37: Intramolecular Diels–Alder reaction of **193.**

arising from cationic double bond isomerization of the primary cycloadduct. Gratifyingly, compound **196** that requires bicyclo[4.3.1] ring system with the double bond in the $\Delta^{6,7}$ position could be obtained in HP conditions at room temperature and then isomerized by $BF_3.Et_2O$ to **195**. The use of higher temperature in conjunction with lower pressure resulted in the formation of mixture of products **194** and **195.**

Whereas in some instances, IMDAF cycloadducts are stable compounds, which could be further chemically transformed (as in the earlier example), IMDAF products are often unstable and revert to starting furan at ambient pressure ($t_{1/2}$ = 1,000 min/ 40 °C, 90 min/80 °C [102]. Similarly, IMDAF *endo,exo*-cycloadduct **198** was obtained under 8 kbar pressure, which quickly undergoes RDA reaction to bis-furan **197** ($t_{1/2}$ = 30 min/40°C) (Scheme 2.38) [103]. The formation of adduct **198** explains the failure of HP reaction of **197** with DMAD to produce 2:1 adduct **200**. Under the 8 kbar pressure, IMDAF reaction is faster than intermolecular addition of DMAD, forming **198**, which prevents the second DMAD addition. Upon depressurization, RDA process provides 1:1 adduct **201** in 25% yield. This is an example in which HP acts in two ways, to promote IMDA reaction and to prevent the intermolecular one. In contrast, when cycloaddition of **197** with DMAD was carried out in thermal conditions, mixture of 1:1 and 2:1 adducts **201** and **200** was obtained in 47% and 19% yields, respectively.

2.5 RDA reactions

RDA reactions (Alder–Rickert reaction, reverse DA) usually take place at elevated temperatures (above 200 °C), and there is no much information available on their kinetics [104]. The entropies of activation favor to be very small or slightly negative, whereas the enthalpy of reaction is large and positive. Due to large positive reaction, volume is not favored under HP conditions; however, some reports exist in the

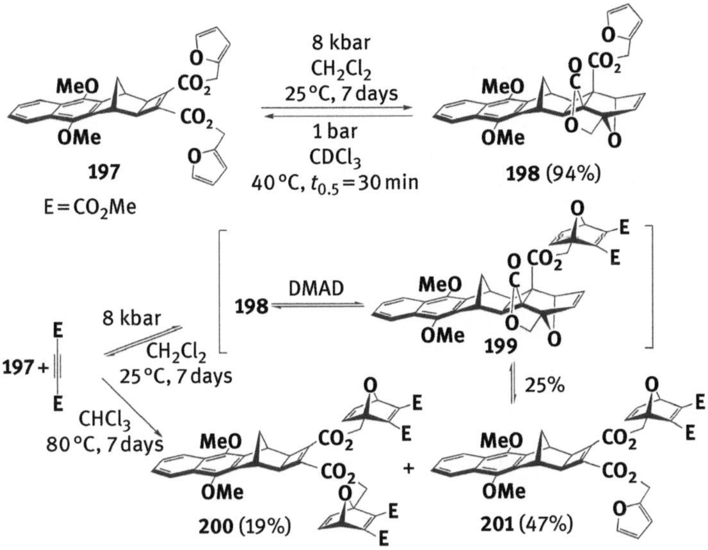

Scheme 2.38: Intramolecular Diels–Alder reactions of furyl diester **197**.

literature. Reaction of 1,4-cyclohexadiene with DMAD) at 10.5 kbar and 100 °C gave product **204**, which is accompanied with dimethyl phthalate **205** (Scheme 2.39) [105]. The formation of **204** is taking place in two steps, starting with ene reaction, which is followed by IMDA reaction. Product **204** subsequently rearranges to an intermediate, from which ethylene elimination by RDA route leads to **205**.

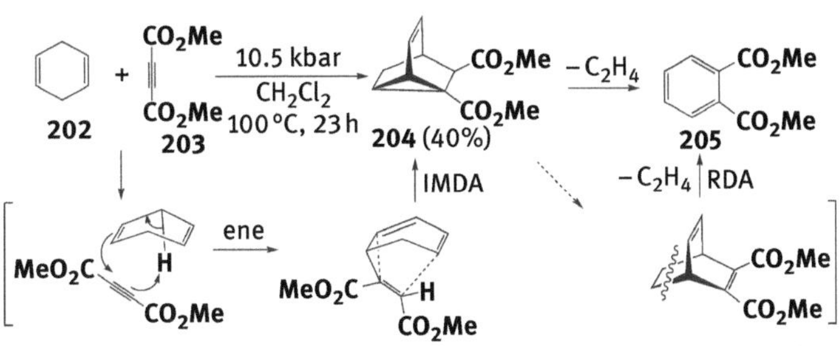

Scheme 2.39: Retro Diels–Alder reactions involving cyclohexadiene and DMAD.

Detailed study of effects of HP on several RDA reactions was carried out by Klärner (Scheme 2.40) [25]. The cleavage reaction of dihydrobarrelene **206** is decelerated by pressure and has a slightly positive activation volume ($\Delta V^{\neq}$ = + 3.1 cm^3 mol^{-1}), which is a similar value to the reaction of 2-cyanobarrelene **207**. On the other hand,

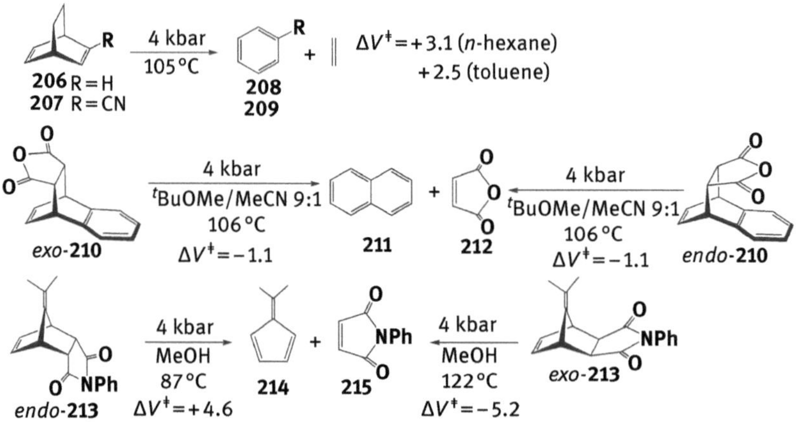

Scheme 2.40: Retro Diels–Alder reactions under high pressure.

the RDA reactions of *endo*-**210** and *exo*-**210** cycloadducts leading to naphthalene and maleic anhydride are slightly accelerated by pressure. The most interesting behavior was found for the RDA reactions of *endo*-**213** and *exo*-**213** adducts, which give dimethyl fulvene and *N*-phenyl maleimide. The reaction of *endo*-**213** is decelerated by pressure, showing $\Delta V^{\neq} > 0$, whereas the reaction of the *exo*-**213** is accelerated by pressure ($\Delta V^{\neq} < 0$). The packing coefficient η (of the pericyclic transition state calculated from the ratio of its van der Waals volume and its partial molar volume, $\eta = V_W/V$) calculated for *exo*-**213**$^{\neq}$ has to be larger than that of *endo*-**213**. Authors concluded that the packing of the reaction ensemble which consisted of solute and solvent molecules and its reorganization during the course of the reaction are the most important for the size of activation and reaction volumes. The changes of the intrinsic molecular volumes of the reactants to the products during the course of the reaction are less important. In the cleavage reactions with $\Delta V^{\neq} < 0$, the packing coefficients of TSs are calculated to be significantly larger than those of the corresponding cycloadducts. This is particularly evidenced in more polar systems possessing cyclic anhydride or imide functionalities. The RDA reaction of the *exo*-**213** has a peculiar pressure dependence behavior. The reaction is accelerated by the pressure rise up to 1.2 kbar, and then the rate of reaction is decreased at pressures higher than 1.5 kbar, while remained constant between 3 and 4 kbar.

Pyrrole DA adducts were prepared at pressures around 11 kbar at room temperature (Scheme 2.41) [106]. Rate measurements of RDA processes were carried at HPs up to 1.25 kbar. For compound **218**, pressure increase (from 1 bar to 1.25 kbar) effects the decomposition: $\Delta V^{\neq} = -8.3\ cm^3\ mol^{-1}$ and volume of reaction $V = +\ 22.4\ cm^3\ mol^{-1}$ were obtained at 38 °C in $CHCl_3$ [107].

Just slightly negative $\Delta V^{\neq}$ values (−2.0, −1.0 and −3.4 $cm^3\ mol^{-1}$ in cyclohexane, DCM and MeCN, respectively) were determined for the HP (2 kbar) decomposition of

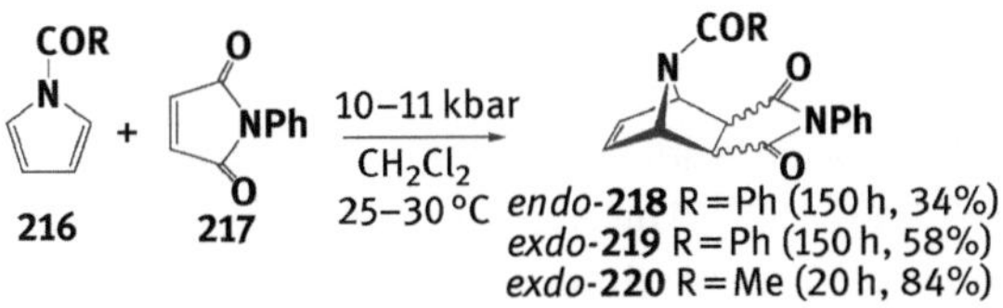

Scheme 2.41: Retro Diels–Alder reactions of pyrroles.

a mixture of *exo*- and *endo*-adducts of 2-methylfuran and acrylonitrile [108]. For tetracyanoethylene to 9-chloroanthracene addition, $\Delta V^{\neq}$ of forward and retro reactions are established to be −28.5 and −6.5 $cm^3\ mol^{-1}$ in DCE [109]. The activation volumes of this retro reaction are greatly affected by the solvent and vary from slightly positive to slightly negative values (+3.0, +1.5, −2.2, −6.5, −7.2 and −7.9 $cm^3\ mol^{-1}$, in BuOAc, EtOAc, PhCl, DCE, toluene and MeCN, respectively) [110].

The application of HP could give some additional information, leading to the postulation of new mechanism, as in DA addition of DMAD to azulene reported by Klärner (Scheme 2.42) [111]. The reaction carried out at 7 kbar allowed the isolation of an intermediate **225**, together with the final product, the heptalene–dicarboxylic acid dimethyl ester **223** and azulene diester side product **224**. Thermolysis of **225** under HP (7 kbar, toluene, 60 °C, 30 h) gives reactants (51%) via the RDA reaction and greater amount of **223** than thermolysis at atmospheric pressure. Under these conditions, addition of **221** and **222** takes place to obtain **225** (13%) and **223** (4%). Instead of dipolar mechanism deduced from thermal reactions, new reaction mechanism was postulated starting with DA cycloaddition leading to intermediate **225**, which subsequently rearranges to **223**, or reverts to reactants. Rearrangement includes the formation of biradical intermediate **226**, its transformation to norcaradiene **226** followed by valence bond isomerization. The influence of pressure on reaction was not informative in distinguishing between two mechanisms, since both dipolar and DA cycloadditions have large negative volumes of activation, and the isolation of intermediate **225** was decisive.

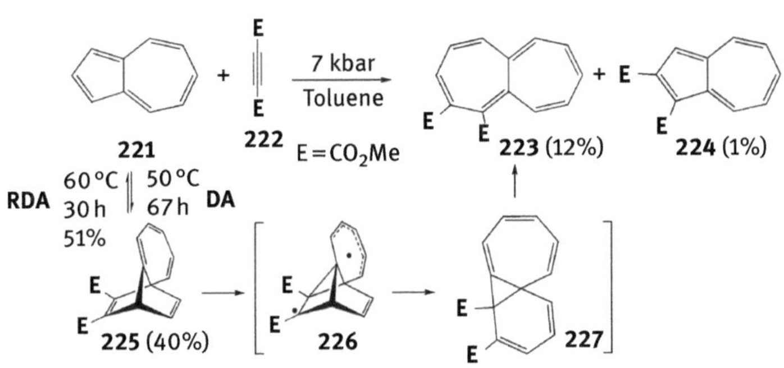

Scheme 2.42: Retro Diels–Alder reaction of azulene with DMAD.

2.6 Homo-Diels–Alder reactions

Mechanistic aspects of homo-DA reactions were studied in reaction of norbornadiene **228** with DMAD and TCNE (Scheme 2.43) [112]. The activation volumes $\Delta V^{\neq}$ are fairly negative, which indicate that homo-DA reactions are strongly pressure dependent and should be accelerated under HP. The variation of solvent polarity (diethyl ether, toluene, benzene, acetone, DCM and MeCN) has only minimal effect on the rate constants, which does not support the formation of dipolar intermediates in the transition state. Furthermore, an ionic stepwise mechanism was not supported by the independence $\Delta V^{\neq}$ on solvent polarity. These results indicate that the homo-DA reaction is likely a simultaneous bond forming nonpolar process like $[\pi_s{}^4 + \pi_s{}^2]$ cycloadditions.

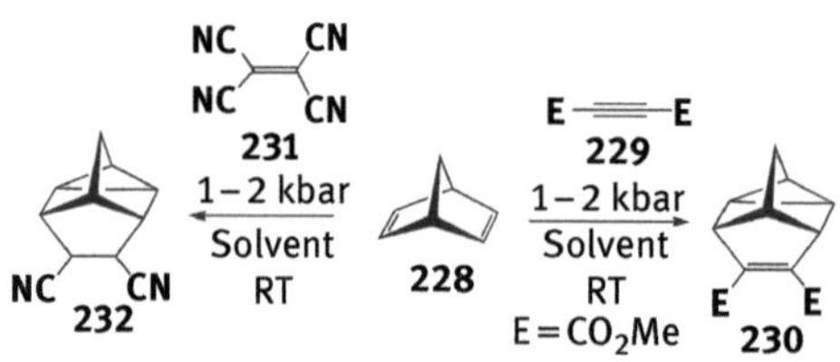

Scheme 2.43: Homo-Diels–Alder reactions of norbornadiene with DMAD and TCNE.

In the following years, other new systems were found to undergo this reaction under HP. Norbornadiene and very reactive dienophile 4-phenyl-1,2,4-triazolin-3,5-dione (PTAD) **233** undergo homo-DA [2π+2π+2π] reaction at elevated pressure to obtain cycloadduct **234** (Scheme 2.44) [113]. Kinetic studies under elevated pressure using the HP optical cell adjusted to a UV-spectrophotometer revealed very negative volume of activation ($\Delta V^{\neq}$ = –25.1 cm^3 mol^{-1}). The ratio of the volume of activation and the volume of reaction ($\Delta V_{r\text{-}n}$) of this reaction is considered as normal ($\Delta V^{\neq}/\Delta V_{r\text{-}n}$ = 0.81).

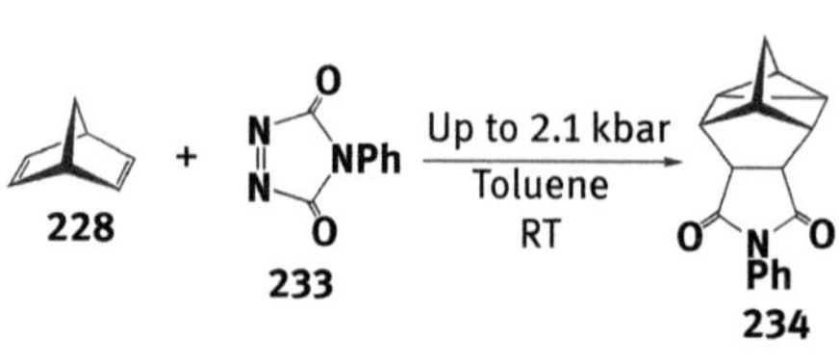

Scheme 2.44: Homo-Diels–Alder reaction of PTAD with norbornadiene.

At ambient temperature under 11 kbar pressure, the meta photocycloadduct **235** of anisole and *cis*-cyclooctene undergoes essentially quantitative addition to DMAD via [2π+2σ+2π] process (Scheme 2.45) [114]. The toluene *cis*-cyclooctene adduct **238** and the DMAD reacted to give the "ene" adduct **239** as the major product, which

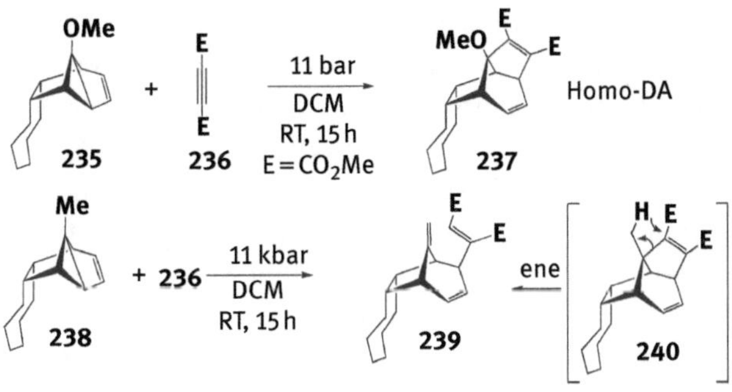

Scheme 2.45: Homo-Diels–Alder reactions of **235** and **238** with DMAD.

might have originated from homo-DA adduct **240** analogous to product **237**, which was subsequently rearranged.

Reaction of cycloheptatriene with DMAD initially gives an ene intermediate **242**, which is in cycloheptatriene–norcaradiene equilibrium (Scheme 2.46) [115, 116]. From intermediate **243**, product **246** was formed exclusively under 9 kbar, whereas product **244**, which accompanies reactions conducted at reflux was not found. Transformation of **243** into **246** involves IMDA reaction, followed by homo-[1,5]-sigmatropic shift. When methyl propiolate was applied (9 kbar, 80 °C, 68 h, 73%), mixture of four norcaradiene products was obtained. Products **247** and **248** are [4 + 2] cycloadducts on **241** and the initial ene adduct, whereas **249** and **250** are the result from a further homo-DA addition of the acetylenic bond to the adducts **247** and **248**. These cycloadditions occur only under HP conditions and exemplify an extension of the synthetic possibilities of DA reactions. Pressure has also

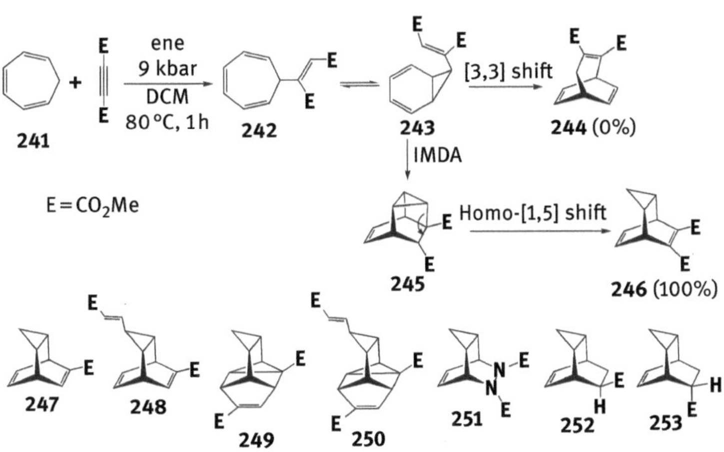

Scheme 2.46: Homo-Diels–Alder reactions of cycloheptatriene with DMAD.

beneficial effect on the reaction with diethyl diazodicarboxylate (9, kbar, RT, 3–5 h) affording the corresponding ene adduct on **241**, as well as norcaradiene product **251**, which is not formed at atmospheric pressure. Reaction of cyclohexatriene with methyl acrylate (and acrylonitrile, 9 kbar, 80 °C, 24 h) produced adducts **252** and **253** (1:1 and 1:5 ratio for CO_2Me and CN, in 80% and 62% yield, respectively), whereas the product resulting from a formal 1,6-addition which was reported for atmospheric condition was not detected. Results demonstrate that pressure does not affect the *exo–endo*-ratio which is obtained in classical conditions.

Complex polycyclic structures could be obtained by HP sequential (π^2+π^2+π^2) homo-DA and (π^2+σ^2+σ^2) bishomo-diene cycloaddition reactions of quadricyclane and methyl propiolate. Sequential (π^2+π^2+π^2) homo-DA process (neat) led to initial formation of product **256**, which in the second pressurization reaction gave **257**. Product **256** was previously obtained by $AlCl_3$ catalysis (in 5–12% yield) or at high temperature (38%), whereas **257** was found in traces (Scheme 2.47) [117]. The bishomo DA reaction of quadricyclane and **255** under HP afforded in quantitative yield mixture of mono- and bis-adducts **259** and **260**. By the extension of reaction time to 65 h, bis-adduct **260** is the exclusive product (**259:260** ratio 4:96). Cycloadditions at 1 bar afforded **156** only (100 °C, 50 h, 80%), showing the low reactivity of double bond in comparison to the acetylenic bond. Isomeric adducts **261** and **262** were obtained in high yield by the cross-addition reactions of monoadducts **256** and **259** with **254** and **258**, whereas these products could not be formed at ambient pressures. Similar homo-DA reactions of norbornadiene and quadricyclane with acetylenes are published by Jenner (Scheme 2.48) [118].

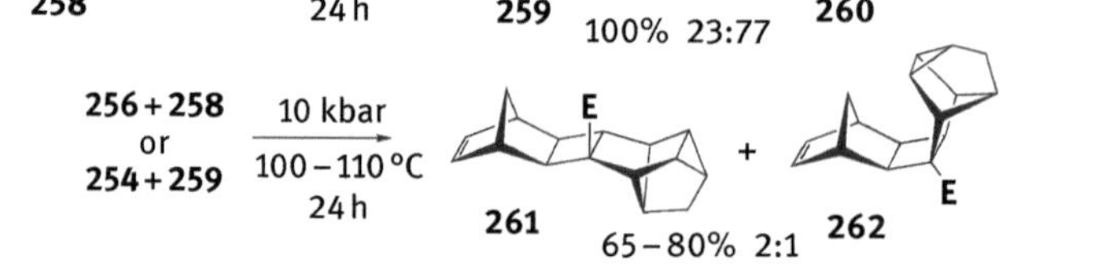

Scheme 2.47: Homo-Diels–Alder reactions of norbornadiene and quadricyclane.

Oxetanes could be effectively prepared by bishomo-HDA [2π+2σ+2σ] cycloadditions of quadricyclane with carbonyl compounds. For instance, phenylglyoxylate **269** gives 60% yield of two isomeric products **270/271**, whereas in atmospheric

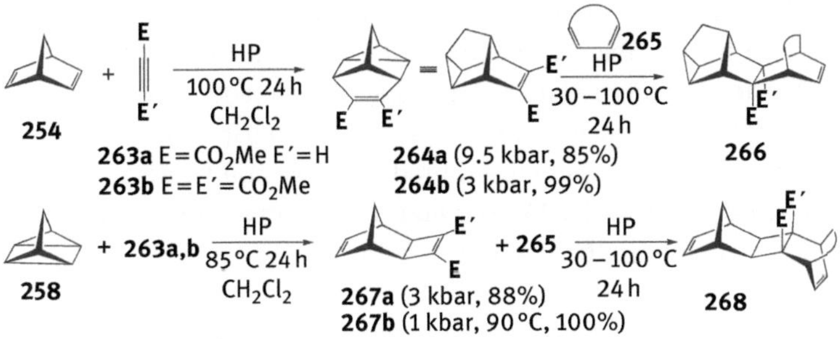

Scheme 2.48: Homo-Diels–Alder reactions of norbornadiene and quadricyclane.

conditions, product yield is low (13%), with a slight difference in isomeric ratio of two products (Scheme 2.49) [119, 120]. Reaction of quadricyclane with methylglyoxylate (R_2 = Me) proceeds in 90% yield and 50:50 ratio, whereas dibenzoyl (R_1 = COPh R_2 = Ph) afforded 59% of isomer **271** only, and dichloroacetyl chloride (R_1 = $CHCl_2$, R_2 = Cl) is less reactive affording at 8.5 kbar products in only 20%, with 40:60 ratio.

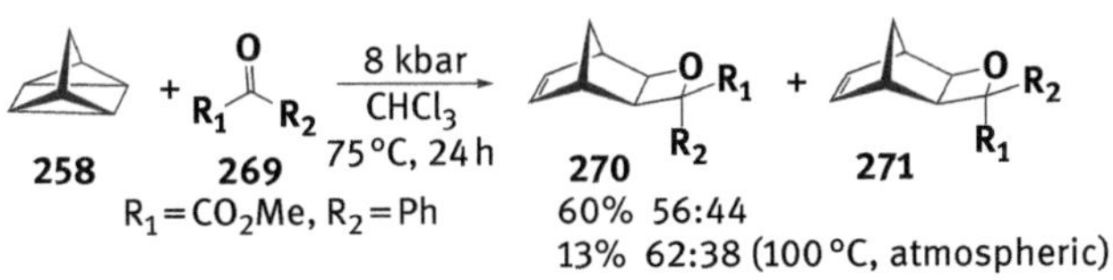

Scheme 2.49: Homo-Diels–Alder reaction of ketones with quadricyclane.

2.7 Tandem DA reactions

Tandem and domino cycloaddition reactions were carried out in different conditions (MW, HP, water) [121]. Similar to simple DA reactions, these are promoted by the use of extremely HPs. For instance, domino DA cycloaddition [122] reaction of [1,2]oxathiine-2,2-dioxide with dimethyl acetylenedicarboxylate advances via domino DA/RDA mechanism (with the subsequent SO_3 elimination at 13 kbar) [123].

Reactions of 2,4-bis(trifluoromethyl)-1,3,4-oxadiazole **272** with cycloalkenes proceed in tandem [4 + 2]/[3 + 2] cycloaddition process in a single reaction providing 7-oxabicyclo[2.2.1]heptane moiety (Scheme 2.50) [124]. In the first step, which is promoted by HP, DA reaction takes place. The intermediate DA adduct **273** undergoes an elimination of dinitrogen to form a very reactive 1,3-dipole intermediate **274** which is trapped by an cycloalkene by 1,3-dipolar cycloaddition process. It is not fully evidenced whether elimination of dinitrogen takes place under HP, or after the depressurization.

This reaction was used in synthesis of several polynorbornane derivatives, in which oxadiazole is incorporated into oxabicyclo[2.2.1]heptane fragment (Scheme 2.51) [125, 126]. Application of HP instead of thermal does not change the stereochemical outcome of this reaction, the C-bridged alkene reacted with *exo,exo*-stereospecificity, whereas the O-bridged alkene formed only *exo,endo*-fused products. Essentially, no 2-naphthol byproduct was produced from **278** under 14 kbar and room temperature, which is always present in significant amount in thermal OD couplings.

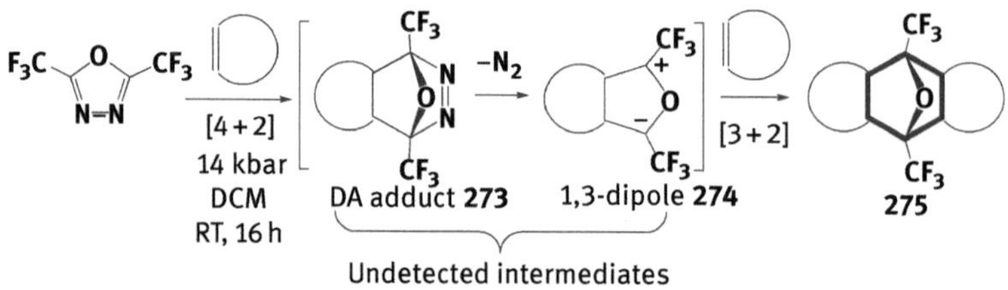

Scheme 2.50: Mechanism of Diels–Alder reactions of oxadiazoles.

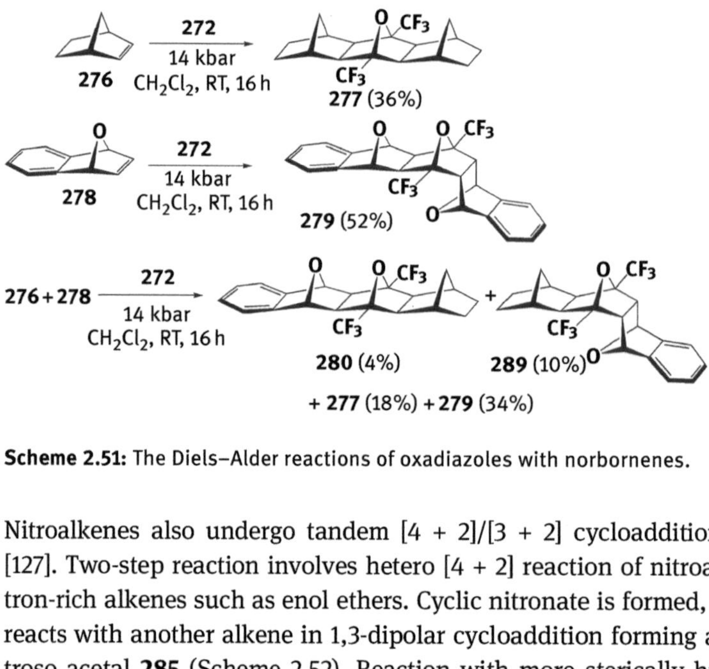

Scheme 2.51: The Diels–Alder reactions of oxadiazoles with norbornenes.

Nitroalkenes also undergo tandem [4 + 2]/[3 + 2] cycloadditions by HP activation [127]. Two-step reaction involves hetero [4 + 2] reaction of nitroalkene with the electron-rich alkenes such as enol ethers. Cyclic nitronate is formed, which subsequently reacts with another alkene in 1,3-dipolar cycloaddition forming a final product, a nitroso acetal **285** (Scheme 2.52). Reaction with more sterically hindered 2,3-dihydrofuran gave only monoadduct. Other substituents could be also employed [128]. This reaction was also carried as a three-component variant.

Similarly, in a tandem [4 + 2]/[3 + 2] sequence, 1-methoxycyclopent-1-ene and β-nitrostyrene have reacted under HP (15 kbar) to give the mixture of three nitroso acetal

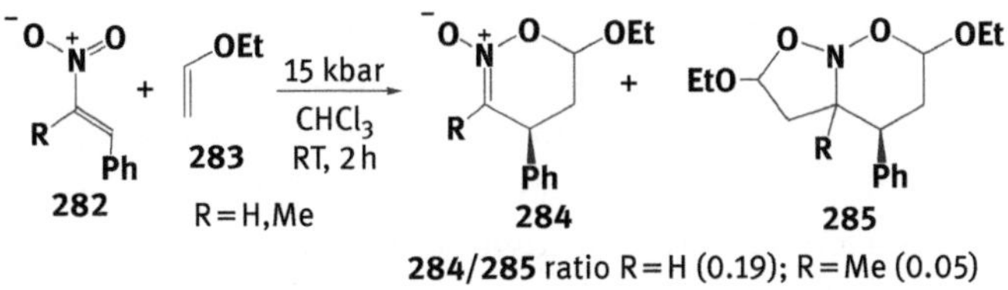

Scheme 2.52: Tandem Diels–Alder reaction of nitroalkene with ethyl vinyl ether.

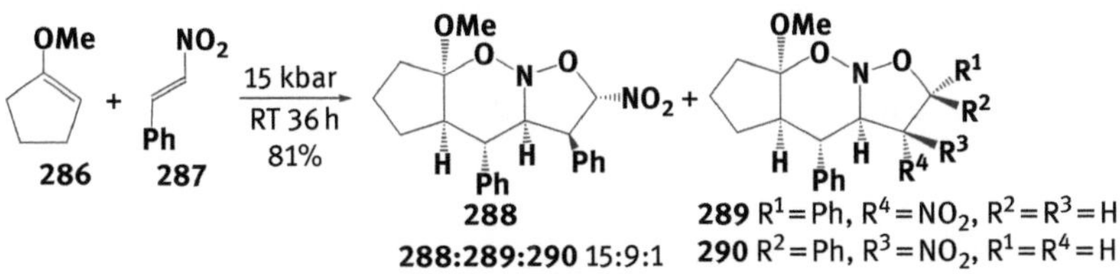

Scheme 2.53: Tandem Diels–Alder reactions of methoxycyclopentene with nitroalkene.

products **288–290** in a yield of 64% (Scheme 2.53) [129]. Tricyclic nitroso acetal **288** is the major component of the mixture, whereas only small amount of **290** was formed.

2.8 Selectivities of DA reactions

HP conditions may influence regioselectivities, *exo/endo* and *cis/trans* diastereoselectivity, enantioselectivity or the π-facial selectivity of DA reactions.

Tietze has found that regioselectivity of intramolecular HDA reaction of **291** is influenced by HP, as well as by the solvent (Scheme 2.54) [130]. This cycloaddition provides two adducts **292** (*ortho*) and **293** (*meta*) and acetonitrile solvent increases the *ortho/meta* ratio. The activation volumes for cycloaddition are found to be $\Delta V^{\neq}= -33.1$,

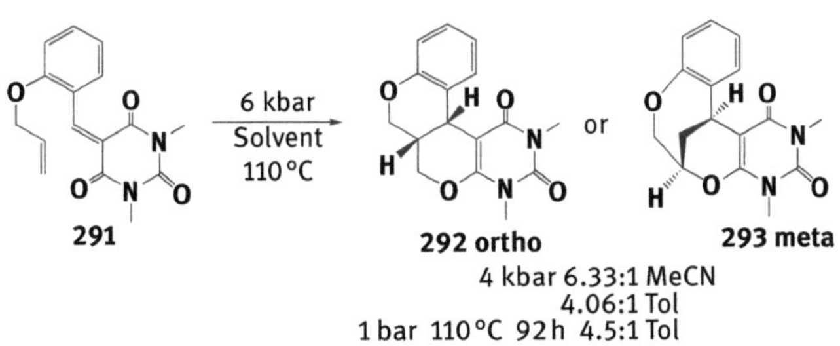

Scheme 2.54: Intramolecular hetero-Diels–Alder reaction of **291**.

−34.2 −13.4 and −17.0 cm^3 mol^{-1} in DCM, THF, toluene and acetonitrile, respectively. The significant solvent effects on $\Delta V^{\neq}$ were not proportionally reflected on the activation volume differences $\Delta\Delta V^{\neq}$ (−1.5, −2.0, −1.3, −2.1 cm^3 mol^{-1} in same solvents).

The change from atmospheric conditions to hyperbaric often has the influence on the regioselectivity and the *endo*-diastereoselectivity of cycloaddition reactions [131]. The regio- and stereoselectivity could be also improved by the simultaneous application of pressure and mild Lewis acid catalysts in comparison by the pressure alone. Pressure-promoted DA reaction of enol acetate **294** with methoxyquinone **295** is *endo*-stereoselective. However, it is not regioselective since mixtures of *endo*-adducts **296** and **297** are produced (Scheme 2.55) [132]. When Lewis acids were added, the rate of these HP cycloaddition reactions was substantially increased and the regioselectivity was also increased in favor of **297**.

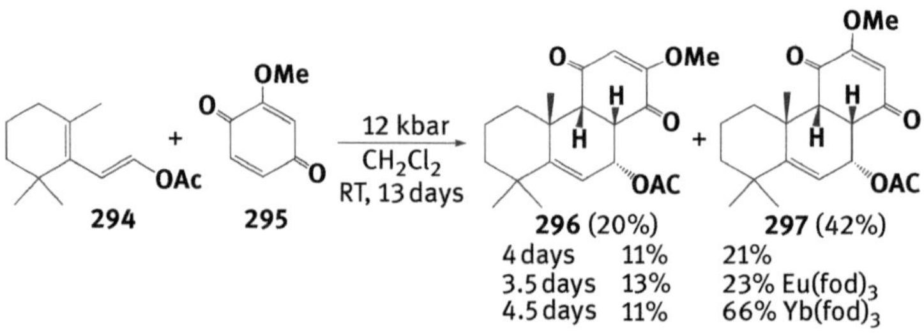

Scheme 2.55: Regioselectivity in Diels–Alder reaction of diene **294** with quinone **295**.

The significant effect of pressure on the *endo/exo* stereoselectivity of DA reactions could be illustrated by the reaction of 3,4-dimethoxy furan with benzoquinone. It provides mixtures of *endo/exo* cycloadducts **300** and **301**, whose ratio tends to increase in favor of the *exo*-product (Scheme 2.56) [133]. With the increase in temperature to 50 °C, at 19 kbar the *endo*-preference is completely altered and the *exo*-isomer is formed predominantly.

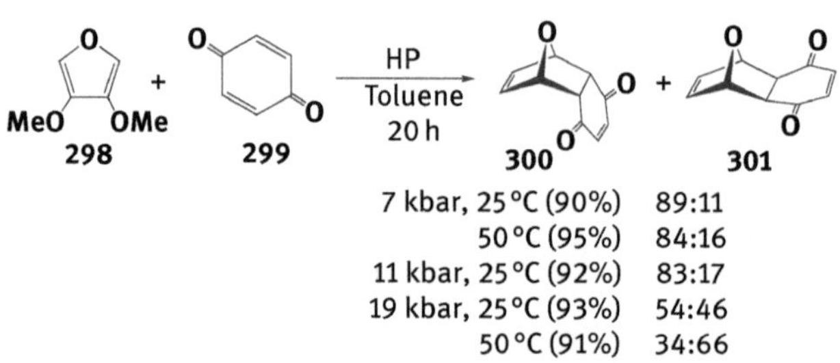

Scheme 2.56: *Endo/exo* stereoselectivity of Diels–Alder reaction of dimethoxy furan with benzoquinone.

The application of HP favors sterically demanded approach of cycloaddends in the transition state and the decrease or even an inversion of the diastereomeric excess under hyperbaric conditions was observed [134]. The diastereoselectivity of HDA reactions for instance can be increased significantly by applying HP [135–138]. An obvious pressure effect on the diastereoselectivity is observed for the intramolecular HDA reaction described by Tietze [139]. The ratio of *cis*-product is favored by increasing pressure: while at ambient pressure the *cis/trans*-ratio of cycloadducts is 5.2:1 (reflux in toluene), it increased to 6.75 at 5 kbar. From the values determined for each mode of addition ($\Delta V^{\neq}_{cis} = -19.4$ cm^3 mol^{-1}; $\Delta V^{\neq}_{trans} = -17.9$ cm^3 mol^{-1}), the difference in activation volumes could be deduced ($\Delta\Delta V^{\neq} = \Delta V^{\neq}_{cis} - \Delta V^{\neq}_{trans} = 1.5$ cm^3 mol^{-1}). Although this value is small, it became sufficient to favor *cis*-isomer. In the following example of HDA reaction, pressure has different effects on the *cis/trans*-diastereoselectivity: it either disfavors, does not influence or favors the *cis*-isomer ($\Delta\Delta V^{\neq}$ for three DA reactions are +4.1, −0.6 and −3.9 cm^3 mol^{-1}, CH_2Cl_2) with *cis/trans* ratios at 1 at 3 kbar are 6.58/5.37, 3.85/4.10 and 7.65/8.90, respectively [140]. By the addition of additives such as pyridine *cis*-diastereoselectivity can be induced [141].

2.9 Asymmetric reactions

Asymmetric natural product synthesis of such abietanoid *o*-quinones [142] and (+)-6-epi-mevinolin [143] was achieved using HP-promoted [4 + 2] cycloadditions. Chiral dienes [144, 145] or dienophiles [146] can be utilized in asymmetric cycloadditions, but sometimes just a small increase in percentage optical yield or moderate enantioselectivity is induced by pressure [147]. Profound effect of pressure on the diastereoselectivity of cycloaddition is exemplified in the reaction of 12-*N*-methyl (−)-cytisine **302** with *N*-phenylmaleimide (Scheme 2.57) [148]. At ambient pressure, the major product is **304** with a (3a*S*,4*R*,12a*R*,12b*S*) configuration, whereas cycloadduct **305** (3a*R*,4*S*,12a*S*,12b*R*) is a minor component of the reaction mixture

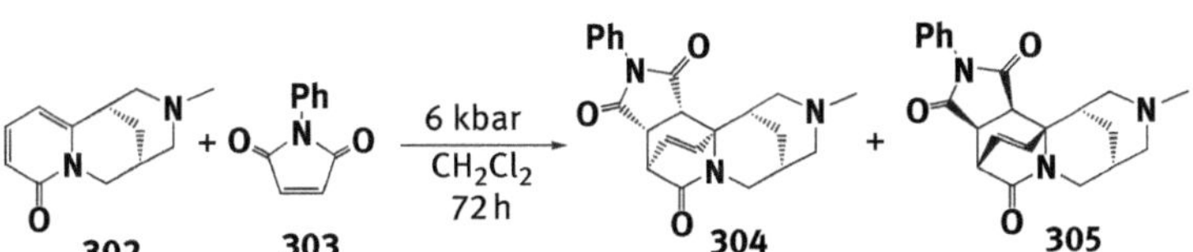

80 °C (80%)	1:5.5	
110 °C (63%)	1:5	Toluene
80 °C (72%)	1:5	Toluene
80 °C (79%)	1:5	MeCN
110 °C (87%)	2:1	Toluene, 1bar

Scheme 2.57: The Diels–Alder reaction of cytisine with *N*-phenylmaleimide.

(2:1 ratio). At 6 kbar, a complete inversion of diastereoselectivity occurs and diastereoisomer **305** is dominant (1:5 ratio), regardless of the solvent.

Furthermore, the asymmetric induction of hyperbaric DA reactions by the employment of metallosalen chiral complexes (salen)Cr(III) and (salen)Co(II) results in high diastereoselectivity and enantioselectivity [149]. Notable effects of pressure on enantioselectivity of the intramolecular HDA reaction of substrate **306** that is catalyzed by the chiral titanium catalyst were reported by Tietze (Scheme 2.58) [150]. However, the positive effect of pressure on the diastereoselectivity is not a general rule, and in some hetero-IMDA reactions decrease of enantioselectivity was observed due to the decomposition of titanium chiral catalyst [151]. Also, the effect on the π-facial selectivity of furan cycloaddition with chiral dienophiles is marginal in $ZnCl_2$ catalyzed reaction [152].

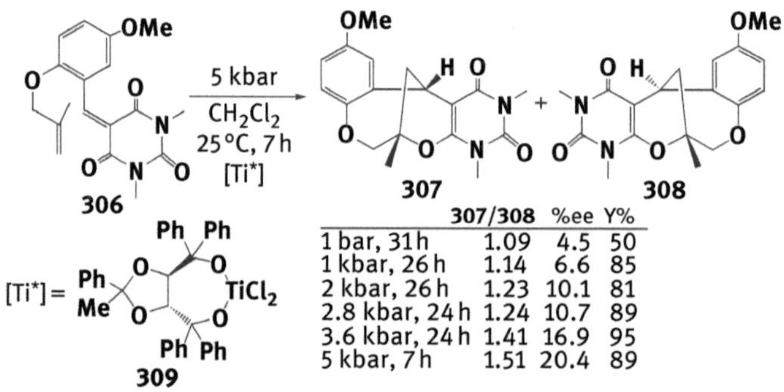

	307/308	%ee	Y%
1 bar, 31 h	1.09	4.5	50
1 kbar, 26 h	1.14	6.6	85
2 kbar, 26 h	1.23	10.1	81
2.8 kbar, 24 h	1.24	10.7	89
3.6 kbar, 24 h	1.41	16.9	95
5 kbar, 7 h	1.51	20.4	89

Scheme 2.58: Intramolecular hetero-Diels–Alder reaction of **306.**

2.10 Natural product synthesis

HP conditions are commonly used in the synthesis of natural products, as well as for chemical transformations of natural products such as pheophorbides [153]. Natural product synthesis often features furan cycloadditions as the key steps in the construction of polycyclic skeletons. When Smith applied 5 kbar of pressure to a neat 1:1 mixture of enone **312** and furan **310**, the *endo*- DA cycloadduct **313** was obtained in 80% yield, with the proper regiochemistry of the jatropholone skeleton (Scheme 2.59) [154, 155]. The results obtained under pressure are in stark contrast to thermal or Lewis acid conditions that proved unsuccessful, due to the instability of the diene component. The benzyloxy furan derivative **311** was distinctly less reactive, the adduct **314** was formed in low yield at the same pressure. The improved yield was obtained by increasing the pressure to 7 kbar and extending the reaction time. A combination of electronic effects and the steric hindrance inflicted by the benzyloxy substituent are proposed to diminish the reactivity of furan **311**. A single diastereomer was obtained

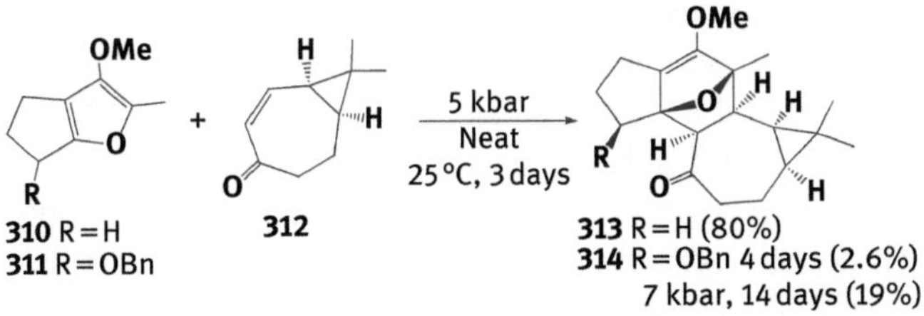

Scheme 2.59: Construction of jathropolone skeleton by the Diels–Alder reaction.

at both pressures applied, which suggests that a kinetic resolution takes place during the cycloaddition process. The discrimination of two possible *endo*-approaches of **312** is likely due to the shielding of one face of the diene by π-stacking of the furan and aromatic rings.

Synthetic approaches to phorbol diterpenes by stereoselective construction of a carbotricyclic core by IMDA reaction of furan in hyperbaric conditions are developed by Harwood. The *E*-configured *α,β*-unsaturated ester **318** was found to undergo IMDAF affording an inseparable mixture of cycloadducts and **318.** Just 5 min exposure of **318** to a pressure of 19 kbar was sufficient to nearly total conversion to two cycloadducts, which were isolated as hydrogenated products **319** and **320** (Scheme 2.60) [156]. In the

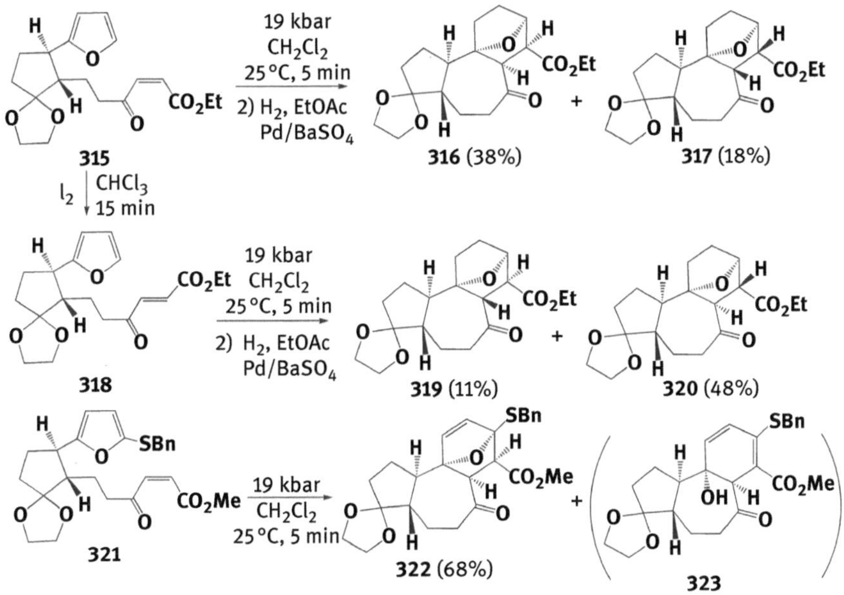

Scheme 2.60: IMDAF reactions toward phorbols.

identical fashion, the *Z*-ketoester **315** provided cycloadducts **316** and **317** within 5 min. When the IMDAF substrate **321** with and the *α'*-benzylthiofuran substituent was subjected to 19 kbar pressure (ca. 0.1 mol dm^{-3} solution) single *endo*-cycloadduct **322** was obtained [157]. The presence of benzylthio substituent greatly diminished its reactivity. The yield of **322** is much reduced in more dilute solutions (0.02 mol dm^{-3}) and another product **323** appeared in 1:2 ratio. The oxidative cleavage of the oxygen bridge may be due to the traces of acid present in dichloromethane.

The IMDAF strategy was also applied to access the tricyclic skeleton of the colletofragarones. Harwood has shown that the pressurization of *cis*-maleic diester **324** at 11 kbar led to intramolecular extra-annular addition to afford product **325** in 75% yield (Scheme 2.61) [158]. Reflux in benzene at atmospheric pressure was less satisfactory, and provided **325** in only 28% yield. Whereas the product has the ester substituents in a *trans*-relationship, the substrate possesses a *cis*-configuration on the dienophilic fragment, indicating that the epimerization step occurs subsequently to DA reaction.

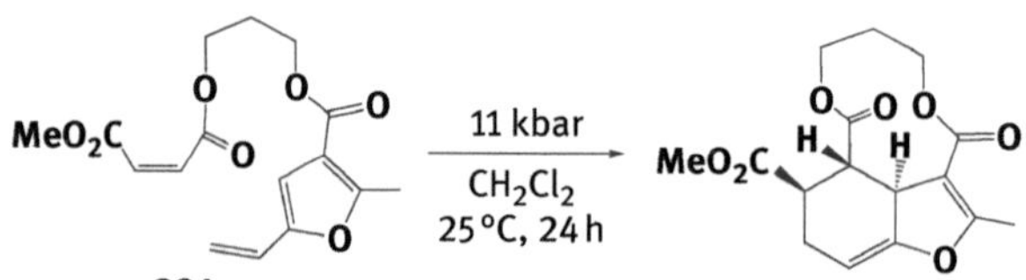

324

325 (75%)
1 bar, benzene, 80 °C, 24 h (28%)

Scheme 2.61: Application of IMDAF reaction to construct the colletofragarone skeleton.

Direct synthesis of cantharidin **330** by DA reaction of furan and dimethylmaleic anhydride failed, as anhydride does not add to furan even at pressures up to 60 kbar [159]. In view of the thermal instability of furan DA reaction products, this [4 + 2] cycloaddition could not be achieved by thermal activation, but instead at ambient temperature, with the aid of HP. In earlier HP synthetic studies, 3,4-dimethoxyfuran was employed to increase reactivity [160]. Dauben has found that 2,5-dihydrothiophene-3,4-dicarboxylic anhydride **327** is a suitable dienophile that overcomes electronic and steric problems present in the dimethyl maleic anhydride (Scheme 2.62) [161]. The reaction of **327** with furan proceeded well at 15 kbar providing the *exo*- and *endo*-adducts **328** and **329** in 85:15 ratio. Change of solvents for cycloadditions at 15 kbar in acetone and acetonitrile led to large amounts of unidentified products and the dicarboxylic acid **331** was formed at 8 kbar in acetonitrile. This hydrolysis which is accelerated by pressure comes from traces of water present in the solvent. The unknown products may be the result of the multiple addition of furan to the 1:1 adducts. The preparative-scale reaction was developed at lower pressure and

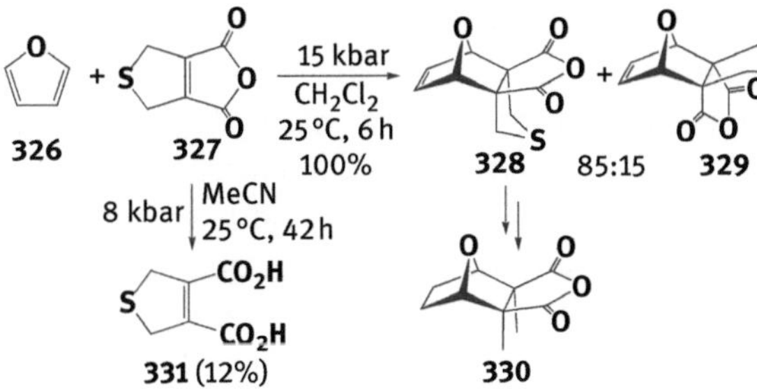

Scheme 2.62: Cantharidin synthesis by the Diels–Alder reaction.

with extended time (**327** (98 g), 8 kbar, 48 h) in larger volume HP reactor and furan as solvent [162].

HP technique was employed by Banwell for the preparation of key building blocks of several natural products. The preferential addition of the diol to the face of the diene *syn* to the oxygen substituents (*syn*-addition) takes place in the HP promoted DA reaction between the enantiomerically pure *cis*-1,2-dihydrocatechol **332** and cyclopentenone **333** (Scheme 2.63) [163]. This major *syn*-cycloadduct **334** was used in the synthesis of sesquiterpene (–)-hirsutene. Preferred *syn* stereochemistry is the same as obtained for related cycloaddition reactions of diol **332** at atmospheric pressure [164]. Another report from the same research group on the synthesis of phomopsidin gave different yields: applying pressure of 19 kbar to **332** and **333** at for 24 h afforded *syn*-DA adduct **334** in 56% as the major product and only 9% of the corresponding *anti*-adduct was isolated. The selective formation of *syn*-**334** (by dienophile approach to the sterically more hindered *syn* face of diene) is attributed to a stabilizing interaction between the C–O bonds within diene **332** and the π* orbital of dienophile **333** in the transition state [165].

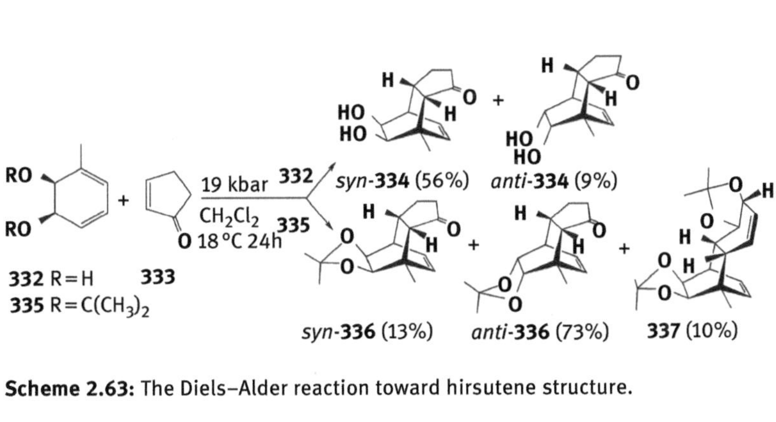

Scheme 2.63: The Diels–Alder reaction toward hirsutene structure.

Opposite *anti*-selectivity was observed in the reaction of cyclopentenone with the corresponding acetonide **335** leading to *anti*-adduct **336** (73%) [166]. This cycloadduct was used as a starting point in the synthesis of phellodonic acid. Reverse of the stereoselectivity could be attributed to steric bulkiness of the ketal protection group. Furthermore, synthesis of (+)-hirsutic acid and (−)-complicatic acid provided in the initial HP step a mixture of the desired DA *anti*-adduct **336** (73%), its *syn*-isomer **336** (13%) and a dimer **337** (10%) which is derived from the starting diene **335**. Products **336** are thought to be formed via an *endo*-transition state, where the dienophile is opposite to the face of the diene of the sterically demanding acetonide moiety [167]. Cycloadduct *anti*-**336** was also used as the precursor in the synthesis of deoxydihydrotsugicoline and radudiol [168]. For these *cis*-1,2-dihydrocatechol derivatives, promotion of the desired cycloaddition process by Lewis acid catalysts cannot be used as these dienes are sensitive to acid-catalyzed rearomatization. These conversions are affected at HP demonstrating the advantages of the technique.

Further examples of application of HP-DA reactions in natural product chemistry are synthesis of platencin featuring a chiral (*S*)-(−)-perillaldehyde and Danishefsky's diene cycloaddition in the first step (15 kbar, 16 h) [169], the preparation of ascididemine by aza-DA reaction of propenal *N,N*-dimethylhydrazone with quinones (10 kbar, 80 °C) [170] or wistarin synthesis by the regio- and enantioselective reaction of homochiral diene with spirocyclopentenone which provides a single enantiomer (14 kbar, 7 days) [171].

2.11 DA reactions of fullerenes

HP DA reactions also have found their application in material chemistry for fullerene functionalizations. Fullerene DA adducts are thermally unstable and undergo facile cycloreversion and the addition reaction is found to be more favorable under HP by retarding reverse reaction and opens a new route to fullerene functionalization. Komatsu had reported that under HP conditions (3–5 kbar) and simultaneous heating (90–120 °C), fullerene C_{60} readily undergoes [4 + 2] cycloaddition with 2*H*-pyran-2-one **339** to give a mixture of monoadduct **340** and bisadduct **341**, depending on the reaction conditions (Scheme 2.64) [172]. In all experiments, monoadduct is the dominant product and the reaction condition is adventitious over cycloaddition, which was carried out at atmospheric pressure (yielding product **340** in only 7.4%).

Addition of tropones **342** to fullerene under HP conditions led to the formation of 1:1 DA adducts **343** in high yields (Scheme 2.65) [173–175]. When reaction was carried out with the parent tropone in refluxing toluene, small amount of the product (<1%) was formed.

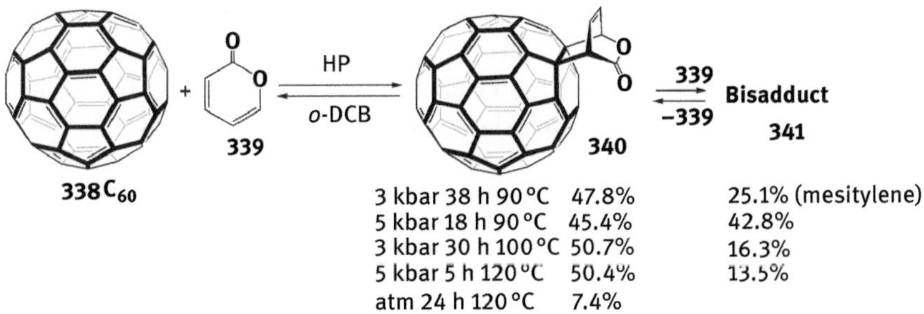

Scheme 2.64: The Diels–Alder reaction of fullerene with pyranone.

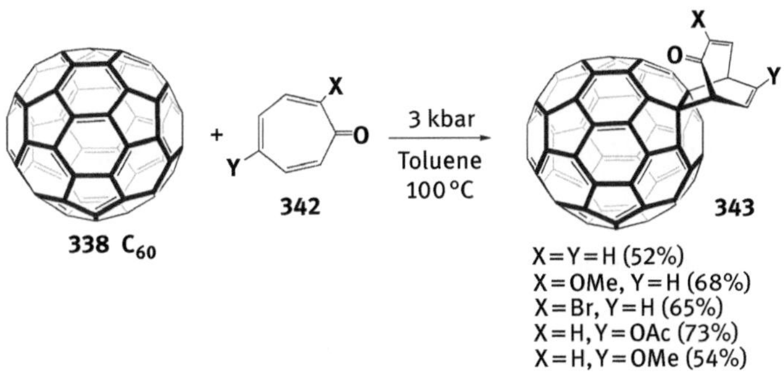

Scheme 2.65: The Diels–Alder reaction of fullerene with tropone.

When cycloheptatriene was used as a diene precursor in HP conditions, symmetrical fullerene [4 + 2] cycloadduct **345** was obtained as a major product (in 52% yield), whereas the valence isomer product **346** structurally related to the tropone cycloadduct **345** was a minor component (Scheme 2.66) [176].

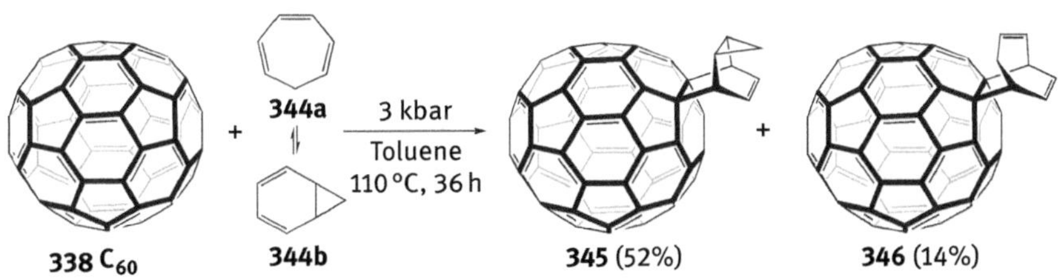

Scheme 2.66: The Diels–Alder reaction of fullerene with cycloheptatriene.

While 2,3-dioxy substituted-1,3-butadiene **350** is a reactive diene and readily undergoes exohedral functionalization of fullerene C_{60} by [4 + 2] cycloaddition, the thermal

DA reactions of the relatively less reactive 4,5-dimethylene-2,2-1,3-dioxolane **347** do not proceed under usual reaction conditions (1 atm, from 0 to 80 °C). Under HP conditions (3 kbar) it is possible to avoid the cycloreversion and to obtain the acyloin derivative **349** in 19% yield after hydrolysis of the primary adduct **348** during the purification (Scheme 2.67) [177].

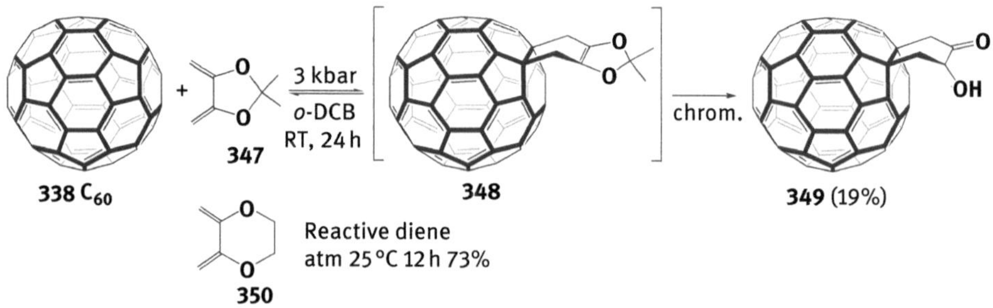

Scheme 2.67: The Diels–Alder reaction of fullerene with butadiene **347** and dioxolane **350.**

Takeshita reported that fullerene reaction with 2-cycloalkenones and 2-cycloalkenone acetals proceed smoothly at 3 kbar (Scheme 2.68) [178]. Reaction of C_{60} with **351** at atmospheric pressure in xylene (130 °C, 36 h) provided **355** in 82% yield, and HP reaction in 85%. Under applied HP conditions, reactivity of masked 4π-dienes 2,5-dioxaspiro[4.4]non-6-ene **351** and 2,5-dioxaspiro[4.5]dec-6-ene **352** was improved by the suppression of the cycloreversion reaction. This was achieved by the in situ transformation of the primary DA cycloadducts **353** and **354** bearing a 2-hydroxyethoxyethene group to the acetals under the reaction conditions.

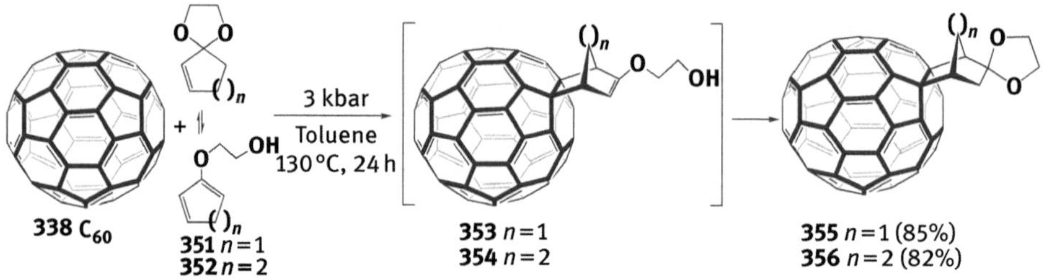

Scheme 2.68: The Diels–Alder reaction of fullerene with cycloalkenone acetals.

Besides in reactions with fullerene, HP was also applied to related carbon material, carbon nanotubes (Scheme 2.69) [179], which react with 2,3-dimethoxybutadiene **358**, 1,3-cyclohexadiene, 2,3-dimethylbutadiene and 9,10-dimethylanthracene in the presence of chromium catalyst.

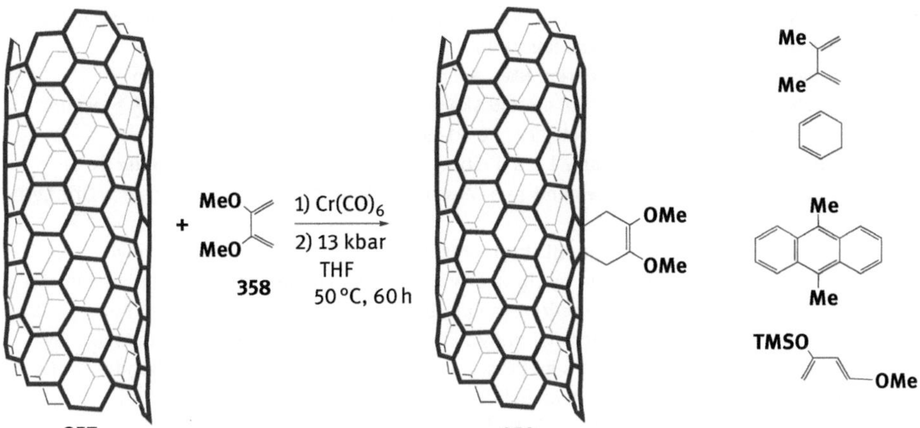

Scheme 2.69: The Diels–Alder reaction of carbon nanotube with 2,3-dimethoxybutadiene.

References

[1] Raistrick, B., Sapiro, RH., & Newitt, DM. Liquid-phase reactions at high pressures. Part V. The polymerisation of cyclopentadiene and α-dicyclopentadiene. J Chem Soc 1939, 1761–9.

[2] Margetić, D. High Pressure Cycloaddition Reactions in the Synthesis of Biologically Relevant Heterocycles. In G. Brahmachari, ed. Green Synthetic Approaches for Biologically Relevant Heterocycles, Elsevier, Amsterdam, 2015, 409–35.

[3] Plieninger, H., Wild, D., & Westphal, J. Substitutierte Naphthaline als Diene bei der Diels-Alder Reaktion Unter Hochdruck – II Mitteilung über Hochdruckversuche. Tetrahedron 1968, 25, 5561–8.

[4] Kiselev, VD., Shakirova, II., Kashaeva, EA., Potapova, LN., Kornilov, DA., Krivolapov, DB., & Konovalov, AI. Why can the Diels-Alder reaction of 9,10-diphenylanthracene with 4-phenyl-1,2,4-triazoline-3.5-dione pass by an abnormal way? Mendeleev Commun 2013, 23, 235–6.

[5] Jarre, W., Bieniek, D., & Korte, F. Benzene as dienophile in the Diels-Alder reaction. Angew Chem Int Ed 1975, 14, 181–2.

[6] Klärner, F-G., Ehrhardt, R., Bandmann, H., Boese, R., Bläser, D., Houk, KN., & Beno, BR. Pressure-induced cycloadditions of dicyanoacetylene to strained arenes: the formation of cyclooctatetraene, 9,10-dihydronaphthalene, and azulene derivatives; a degenerate [1,5] sigmatropic shift-comparison between theory and experiment. Chem Eur J 1999, 5, 2119–32.

[7] Matsumoto, K. Organic reactions at high pressure. Diels-Alder reactions of [2,2] paracyclophane. Chem Lett 1985, 1681–2.

[8] Brun, C., & Jenner, G. Etude des Réactions Pericycliques sour Pression-II. Contribution a l'Etude du Mechanisme de la Reaction de Diels-Alder. Tetrahedron 1972, 28, 3113–21.

[9] Gacem, B., & Jenner, G. Effect of pressure on sterically hindered reactions with late transition states. J Phys Org Chem 2004, 17, 221–5.

[10] Grieger, RA., & Eckert, CA. Mechanistic evidence for the Diels-Alder reaction from high-pressure kinetics. J Am Chem Soc 1970, 92, 7149-53.

[11] Isaacs, NS., Maksimović, Lj., & Laila, A. The effect of pressure on rates of some catalysed Diels-Alder reactions. High Press Res 1994, 13, 21–7.
[12] Gacs-Baitz, E., Minuti, L., Scheeren, HW., Selvaggi, R., & Taticchi, A. A one-step regioselective construction of the steroid skeleton by Lewis acid catalyzed high pressure Diels-Alder reaction. Nat Prod Lett 1993, 2, 91–5.
[13] Chataigner, I., Hess, E., Toupet, L., & Piettre, SR. Activation of the dienophilicity of indoles in normal electron demand [4 + 2] cycloadditions under high pressure. Org Lett 2001, 3, 515–8.
[14] Knol, J., Meetsma, A., & Feringa, BL. High pressure Diels Alder reactions of 1-vinyl-2,2,6-trimethylcyclohexene catalyzed by chiral Lewis acids; an enantioselective route to a drimane sesquiterpene precursor. Tetrahedron Asymm 1995, 6, 1069–72.
[15] Levina, IS., Kulikova, LE., Kamernitskii, AV., El'yanov, BS., & Gonikberg, EM. Catalysis by Lewis acids at high pressure as a method of involving hindered steroid dienophiles in the Diels-Alder reaction. Bull Russ Acad Sci Chem 1992, 41, 1253–7.
[16] Ballerini, E., Minuti, L., & Piermatti, O. High-pressure Diels-Alder cycloadditions between benzylideneacetones and 1,3-butadienes: application to the synthesis of (*R,R*)-(-)- and -(+)-Δ^8-tetrahydrocannabinol. J Org Chem 2010, 75, 4251–60.
[17] Vandenput, DAL., & Scheeren, HW. Synthesis of highly functionalized 3,4-dihydro-2H-pyrans by high pressure Lewis acid catalyzed cycloaddition of enol ethers and α,β-unsaturated aldehydes. Tetrahedron 1995, 51, 8383–8.
[18] Posner, GH., & Ishihara, Y. Lewis acid-catalyzed, high pressure, stereospecific, regiospecific, Diels-Alder cycloaddition of unsubstituted 2-pyrone: short synthesis of a racemic a-ring precursor to physiologically active 1-hydroxyvitamin D_3 steroids. Tetrahedron Lett 1994, 35, 7545–8.
[19] Mori, K., Maddaluno, J., Nakano, K., Ichikawa, Y., & Kotsuki, H. Efficient organocatalytic hetero-Diels-Alder reactions of activated ketones under high pressure for direct access to δ-lactones. Synlett 2009, 2346–50.
[20] Mielcarek, M., Barciszewska, MZ., Sałanski, P., Stobiecki, M., Jurczak, J., & Barciszewski, J. Native transfer RNA catalyzes Diels-Alder reaction. Biochem Biophys Res Commun 2002, 294, 145–8.
[21] Jenner, G. Multiactivation of organic reactions high pressure cycloaddition reactions catalyzed by lanthanides. High Press Res 1995, 13, 321–6.
[22] Vidiš, A., Laurenczy, G., Küsters, E., Sedelmeier, G., & Dyson, PJ. High-pressure effects on the Diels-Alder reaction in room temperature ionic liquids. J Phys Org Chem 2007, 20, 109–14.
[23] Kuster, GJ., van Berkom, LWA., Kalmoua, M., van Loevezijn, A., Sliedregt, LAJM., van Steen, BJ., Kruse, CG., Rutjes, FPJT., & Scheeren, HW. Synthesis of spirohydantoins and spiro-2,5-diketopiperazines via resin-bound cyclic α,α-disubstituted α-amino esters. J Comb Chem 2006, 8, 85–94.
[24] Brown, DS., & Paquette, LA. Synthetic studies on furanoheliangolides. stereocontrolled construction of the oxygen-bridged tricyclic framework. J Org Chem 1992, 57, 4512–21.
[25] Klärner, F-G., & Breitkopf, V. The effect of pressure on Retro Diels-Alder reactions. Eur J Org Chem 1999, 2757–62.
[26] Jones, WH., Mangold, D., & Plieninger, H. Diensynthesen unter hohem Druck. Tetrahedron 1962, 18, 267–72.
[27] Ashton, PR., Isaacs, NS., Kohnke, FH., Mathias, JP., & Fraser Stoddart, J. Stereoregular oligomerization by repetitive Diels-Alder reactions. Angew Chem Int Ed Engl 1989, 28, 1258–61.
[28] Ashton, PR., Isaacs, NS., Kohnke, FH., Stagno D'Alcontres, G., & Fraser Stoddart, F. Trinacrene – a product of structure-directed synthesis. Angew Chem Int Ed Engl 1989, 28, 1261–63.

[29] Benkhoff, J., Boese, R., & Klärner, F-G. Synthesis of sterically rigid macrocycles by the use of pressure-induced repetitive Diels-Alder reactions. Liebigs Ann 1997, 501–16.
[30] Benkhoff, J., Boese, R., F-G, Klärner., & Wigger, AE. Synthesis of sterically rigid macrocycles by the use of pressure-induced repetitive Diels-Alder reactions. Tetrahedron Lett 1994, 35, 73–6.
[31] F-G, Klärner., Polkowska, J., Panitzky, J., Seelbach, UP., Burkert, U., Kamieth, M., Baumann, M., Wigger, AE., Boese, R., & Bläser, D. Effect of substituents on the complexation of aromatic and quinoid substrates with molecular tweezers and clips. Eur J Org Chem 2004, 1405–23.
[32] Dauben WG, Kowalczyk BA, Lichtenthaler FW. Diels-Alder reactions of pyranoid enolone esters with cyclopentadiene under thermal, Lewis acid catalysis, and high-pressure conditions. J Org Chem 1990, 55, 2391–8.
[33] Srivastava, S., Marchand, AP., Vidyasagar, V., Flippen-Anderson, JL., Gilardi, R., George, C., Zachwieja, Z., & le Noble, WJ. Pressure effect on the product distribution in competing reactions: formation of a Bis Diels-Alder adduct via an aromatizable intermediate. J Org Chem 1989, 54, 247–9.
[34] Tian, GR., Sugiyama, S., Mori, A., & Takeshita, H. Novel preparation of 1- and 3-substituted bicyclo[3.2.2]nona-3,6,8-trien-2-ones from tropones and 2,3-bis(methoxycarbonyl)-7-oxabicyclo[2.2.1]heptadiene by high-pressure cycloaddition-thermal cycloreversion procedure. Bull Chem Soc Jpn 1988, 61, 2393–9.
[35] Dauben, WG., & Kozikowski, AP. Organic reactions at high pressure. cycloadditions of enamines and dienamines. J Am Chem Soc 1974, 96, 3664–6.
[36] Breitkopf, V., Bubenitschek, P., Hopf, H., Jones, PG., F-G, Klärner., Schomburg, D., Witulski, B., & Zimny, B. The effect of pressure on the cycloadditions of cyanoacetylene to furan derivatives – [2.2](2,5)furanoparacyclophane, [8](2,5)furanophane, and furan. Liebigs Ann 1997, 127–37.
[37] Paquette, LA., Branan, BM., & Rogers, RD. High-pressure Diels-Alder reactions of l-Oxa[4.4.4]propella-5,7-diene proceed with framework isomerization. J Org Chem 1995, 60, 1852–5.
[38] Kotsuki, H., Kataoka, M., Matsuo, K., Suetomo, S., Shiro, M., & Nishizawa, H. High pressure organic chemistry. Part 17. Diels-Alder reaction of methyl palustrate with maleic anhydride and *N*-phenylmaleimide. J Chem Soc Perkin Trans 1 1993, 2773–6.
[39] Jurczak, J., Belniak, S., Koźluk, T., & Sałański, P. High-pressure synthesis of the bis adduct of dimethyl acetylene-dicarboxylate and 2,5-dimethylfuran. Bull Pol Acad Sci 1984, 32, 135–41.
[40] Sera, A., Ohara, M., Kubo, T., Itoh, K., Yamada, H., & Mikata, Y. High pressure mediated Diels-Alder reaction of furan with dialkyl (Acetoxymethylene) malonate. J Org Chem 1988, 53, 5460–4.
[41] Okamoto, Y., Giandinoto, S., & Bochnik, MC. High pressure [4+2] cycloaddition of dichloromaleic anhydride and furan. J Org Chem 1983, 48, 3830–1.
[42] Margetić, D., & Warrener, RN. *Ab initio* and semiempirical modelling of stereoselectivities of Diels-Alder cycloadditions of furan and cyclopentadiene with norbornenes. Croat Chem Acta 2003, 76, 357–63.
[43] Margetić, D., Warrener, RN., & DN, Butler. High-pressure facilitated cycloaddition of furan to 1,4-epoxynaphthalene. The Sixth International Electronic Conference on Synthetic Organic Chemistry (ECSOC-6), http://www.mdpi.org/ecsoc-6.htm, September 1–30, 2002.
[44] Jurczak, J., Kawczyński, AL., & Koźuk, T. Stability of cycloadducts obtained by high-pressure Diels-Alder reaction between 3,4-dimethoxyfuran and 1,4-benzoquinones: kinetic studies of Retro-Diels-Alder reaction. J Org Chem 1985, 50, 1106–7.

[45] Kotsuki, H., Nishizawa, H., Ochi, M., & Matsuoka, K. High pressure organic chemistry V. Diels-Alder reaction of furan with acrylic and maleic esters. Bull Chem Soc Jpn 1982, 55, 496–9.
[46] Kotsuki, H., Kondo, A., Nishizawa, H., Ochi, M., & Matsuoka, K. High-pressure Diels-Alder reactions of vinyl furans. J Org Chem 1981, 46, 5454–5.
[47] Sugiyama, S., Tsuda, T., Mori, A., Takeshita, H., & Kodama, M. High pressure cycloaddition reaction of tropone with furans. Bull Chem Soc Jpn 1987, 60, 3633–8.
[48] Beusker, PH., Aben, RWM., Seerden, J-PG., Smits, JMM., & Scheeren, HW. Exploration of high-pressure cycloadducts of furans and citraconic anhydride as precursors for CD-ring fragments of paclitaxel and its analogues. Eur J Org Chem 1998, 2483–92.
[49] Dauben, WG., Lam, JYL., & Guo, ZR. Total synthesis of (-)-palasonin and (+)-palasonin and related chemistry. J Org Chem 1996, 61, 4816–9.
[50] Kawamura, N., Li, YM., Engel, JL., Dauben, WG., & Casida, JE. Endothall thioanhydride: structural aspects of unusually high mouse toxicity and specific binding site in liver. Chem Res Toxicol 1990, 3, 318–24.
[51] Ando, K., Kankake, M., Suzuki, T., & Takayama, H. New building blocks, 3,5-dihydro-1*H*-thieno[3,4-*c*]pyrrole 2,2-dioxides; preparation and their Diels-Alder reaction with dimethyl acetylenedicarboxylate. Chem Commun 1992, 100–2.
[52] Suzuki, T., & Takayama, H. Intramolecular Diels-Alder reaction of new building blocks, *N*-substituted 3,5-dihydro-1*H*-thieno[3,4-*c*]pyrrole *S,S*-dioxides; a general route to the tricyclic azanorbornane framework. Chem Commun 1995, 807–8.
[53] Chrétien, A., Chataigner, I., & Piettre, SR. 1-Triflylpyrroles as efficient dienophiles in normal electron demand [4+2] cycloaddition reactions under pressure. Chem Commun 2005, 1351–3.
[54] Seerden, J-PG., Tulp, MTM., Scheeren, HW., & Kruse, CG. Synthesis and structure-activity data of some new epibatidine analogues. Bioorg Med Chem 1998, 6, 2103–10.
[55] Warrener, RN. New adventures in the synthesis of hetero-bridged *syn*-facially fused norbornadienes ("[*n*]polynorbornadienes") and their topological diversity. Eur J Org Chem 2000, 3363–80.
[56] Margetić, D. Warrener, RN. & Dibble, PW. Diels-Alder reactivity of benzannulated isobenzofurans as assessed by density functional theory. J Mol Model 2004, 10, 87–93.
[57] Warrener, RN., Margetić, D., Sun, G., & Russell, RA. Position-addressable nano-scaffolds. I. The preparation of *N,O*-, *N,C*- and *N,N*-bridged sesquinorbornadiene succinimides as compact, highly functionalized addressable building blocks. Aust J Chem 2003, 56, 263–7.
[58] Warrener, RN., Malpass, JR., Butler, DN., & Sun, G. Isoindole cycloadditions. Part III: the synthesis of "Windscreen Wiper" and other N-bridged cavity systems. Struct Chem 2001, 12, 291–304.
[59] Malpass, JR., Sun, G., Fawcett, J., & Warrener, RN. Novel 'Windscreen Wiper' cavity structures formed by the cycloaddition of N-substituted isoindoles onto molrac bis-Alkenes. Tetrahedron Lett 1998, 39, 3083–6.
[60] Kotsuki, H., Kitagawa, S., Nishizawa, H., & Tokoroyama, T. Diels-Alder Reaction of Thiophene with Maleic Anhydride at Very High Pressure. J Org Chem 1978, 43, 1471–2.
[61] Kotsuki, H., Nishizawa, H., Kitagawa, S., Ochi, M., Masaki, N., Matsuoka, K., & Tokoroyama, T. High pressure organic chemistry III. Diels-Alder reaction of thiophene with maleic anhydride. Bull Chem Soc Jpn 1979, 52, 544–8.
[62] Kumamoto, K., Fukada, I., & Kotsuki, H. Diels-Alder reaction of thiophene: dramatic effects of high-pressure/solvent-free conditions. Angew Chem Int Ed Engl 2004, 43, 2015–7.
[63] Margetić, D., Murata, Y., & Komatsu, K. X-ray and DFT Study of 7-thianorbornenes: 4-aza-4-phenyl-10-thiatricyclo[5.2.1.0.^{2,6}]deca-8-ene-3,5-dione and 4-aza-1,4,7-trimethyl-10-thiatricyclo[5.2.1.0.^{2,6}]deca-8-ene-3,5-dione. Struct Chem 2007, 18, 279–86.

[64] McCluskey, A., Keane, MA., Walkom, CC., Bowyer, MC., Sim, ATR., Young, DJ., & Sakoff, JA. The first two cantharidin analogues displaying PP1 selectivity. Bioorg Med Chem Lett 2002, 12, 391–3.

[65] Thiemann, T., Ohira, D., Li, Y., Sawada, T., Mataka, S., Rauch, K., Noltemeyer, M., & de Meijere, A. [4+2] Cycloaddition of thiophene *S*-monoxides to activated methylenecyclopropanes. J Chem Soc Perkin Trans 1 2000, 2968–76.

[66] Margetić, D., Butler, DN., & Warrener, RN. Sulfur-bridged molecular racks: *O, S*-sesquinorbornadienes, *CNS*-[3] and *CNOS*-[4]polynorbornanes. Arkivoc 2002, vi, 234–6.

[67] Eckert-Maksić, M., & Margetić, D.. High pressure synthesis of novel organometallic electron donor-acceptor polycyclic dyads. High Press Res 2007, 27, 109–14.

[68] Kirin, SI., Vikić-Topić, D., Meštrović, E., Kaitner, B., & Eckert-Maksić, M. Diels-Alder reactions of C-phenylated siloles with 1,4-epoxy-1,4-dihydronaphthalene. J Organomet Chem 1998, 566, 85–91.

[69] Kirin, SI., F-G, Klärner., & Eckert-Maksić, M. High pressure synthesis of new silicon containing heteroatom analogues of fused norbornenes. Synlett 1999, 351–3.

[70] Margetić, D., Prugovečki, B., Đilović, I., & Eckert-Maksić, M. Structure of Diels-Alder adduct of phenylated germole and 1,4-epoxy-1,4-dihydro naphthalene revisited. Struct Chem 2006, 17, 301–6.

[71] Eckert-Maksić, M., Margetić, D., Kirin, S., Milić, D., & Matković-Čalogović, D. High pressure assisted synthesis of fused norbornenes containing two 7-metallonorbornene units. Eur J Org Chem 2005, 21, 4612–20.

[72] Margetić, D., Murata, Y., Komatsu, K., & Eckert-Maksić, M. Synthesis, X-ray and DFT study of the double bond pyramidalization in 1,7,8,9-tetraphenyl-4,10,10-trimethyl-4-aza-10-silatricyclo[5.2.1.0.2,6]deca-8-ene-3,5-dione and its germanium analogue. Organometallics 2006, 25, 111–7.

[73] Sekiguch, A., Maruki, I., Ebata, K., Kabuto, C., & Sakurai, H. High-pressure synthesis, structure and novel photochemical reactions of 7,7,8,8-tetramethyl-7,8-disila bicyclo[2.2.2] octa-2,5-diene. Chem Commun 1991, 341–3.

[74] Ibata, T., Nakawa, H., Isogami, Y., & Matsumoto, K. The high pressure Diels-Alder reaction of 5-methoxy-2-methyl-4-(*p*-nitrophenyl)oxazole with ethylenic dienophiles. Bull Chem Soc Jpn 1986, 59, 3197–200.

[75] Pfaff, E., & Plieninger, H. Synthese einer Reihe von 3-Oxo-2-oxabicyclo[2.2.2]oct-7-en-Derivaten und Versuche zu deren Umwaldung in 3-Oxo-2-oxabicyclo[2.2.2]octa-5,6-dien. Chem Ber 1982, 115, 1967–81.

[76] Hatsui, T., Hashiguchi, T., & Takeshita, H. Total a synthesis of (±)-sativene *via* high pressure Diels-Alder route. Chem Lett 1994, 1415–6.

[77] Boger, DL., & Brotherton, CE. Thermal reactions of cyclopropenone ketals. application of the cycloaddition reactions of delocalized singlet vinylcarbenes: three-carbon 1,l-/1,3-dipoles. an alternative synthesis of deacetamidocolchiceine: formal total synthesis of colchicine. J Am Chem Soc 1986, 108, 6713–9.

[78] Larsson, R., Scheeren, HW., Aben, RWM., Johansson, M., & Sterner, O. Total synthesis of transtaganolide E and F: insight in the biosynthesis of the transtaganolides. Eur J Org Chem 2013, 6955–60.

[79] Matsumoto, K., Hamada, K., Uchida, T., & Yoshida, H. Formation of novel 1:3 adduct in the high pressure reaction of 2(1H)-pyridones with dimethylacetylenedicarboxylate: an X-ray crystal structure determination. Heterocycles 1989, 29, 21–4.

[80] Matsumoto, K., & Ikemi-Kono, Y. High-pressure cycloaddition of 2(1*h*)-pyridones with dimethyl acetylenedicarboxylate. Chem Commun 1979, 1091–2.

[81] Matsumoto, K., Ikemi, Y., Nakamura, S., Uchida, T., & Acheson, RM. Formation of novel 1:3 molar adducts in the high pressure reaction of 2(1*H*)-pyridones with dimethyl acetylenedicarboxylate. Heterocycles 1982, 19, 499–502.
[82] Warrener, RN., Margetić, D., & Russell, RA. The preparation of rigid alicyclic molecules bearing effector groups from alkene BLOCKs using *s*-tetrazines and 1,3,4-triazines as stereoselective coupling agents. Synlett 1998, 585–7.
[83] Warrener, RN., Margetić, D., Amarasekara, AS., Butler, DN., Mahadevan, I., & Russell, RA. Building BLOCK strategies for the synthesis of molecular clefts with inside functionality. Org Lett 1999, 1, 199–202.
[84] Golić, M., Johnston, MR., Margetić, D., Schultz, AC., & Warrener, RN. Use of a 9,10-dihydrofulvalene pincer cycloadduct as a cornerstone for molecular architecture. Aust J Chem 2006, 59, 899–914.
[85] Johnston, MR., Latter, MJ., & Warrener, RN. Porphyrin-containing molecular capsules: metal mediated dimerization of a bis-porphyrin cavity. Org Lett 2002, 4, 2165–8.
[86] Boger, DL., & Panek, JS. Inverse electron demand Diels-Alder reactions of heterocyclic azadienes: formal total synthesis of streptonigrin. J Am Chem Soc 1985, 107, 5745–54.
[87] Martin, JC., & Muchowski, JM. One-step synthesis of 3,8-methano-l-aza[l0]annulene derivatives via Diels-Alder reaction of benzocyclopropene with 1,2,4-triazines. J Org Chem 1984, 49, 1040–3.
[88] Margetić, D., Murata, Y., Komatsu, K., & Marinić, Ž. Rigid alicyclic molecules from bicyclo [2.2.1]hept-2-enes (=8,9,10-trinorbornenes) and 1,4-dipyridin-2-ylphthalazines as stereoselective coupling agents. Helv Chim Acta 2009, 92, 298–312.
[89] Boger, DL., & Robarge, KD. Accelerated inverse electron demand Diels-Alder reactions of 1-oxa-1,3-butadienes: [4+2] cycloaddition reactions of *β*,γ-unsaturated *α*-keto esters, J Org Chem 1988, 53, 3373–7.
[90] Al-Badri, H., Maddaluno, J., Masson, S., & Collignon, N. Hetero-Diels-Alder reactions of *α*-carbonylated styrylphosphonates with enol ethers. High-pressure influence on reactivity and diastereoselectivity. J Chem Soc Perkin Trans 1 1999, 2255–66.
[91] Boger, DL., Corbett, WL., Curran, TT., & Kasper, AM. Inverse electron demand diels-alder reactions of *N*-sulfonyl *α*,*β*-unsaturated imines: a general approach to implementation of the 4π participation of 1-aza-1,3-butadienes in Diels-Alder reactions. J Am Chem Soc 1991, 113, 1713–29.
[92] Bouaziz, Z., Nebois, P., & Fillion, H. Additions of crotonaldehyde *N,N*-dimethylhydrazone to *p*-quinones under ultrasonic and thermal conditions. Tetrahedron 1995, 51, 4057–64.
[93] Junge, H., & Oehme, G. Hetero Diels-Alder reaction between 2,3-dimethyl-l,3-butadiene and perfluorooctanonitrile under high pressure. Tetrahedron 1998, 54, 11027–32.
[94] Al-Badri, H., Collignon, N., Maddaluno, J., & Masson, S. Efficient synthesis of new phosphono-substituted dihydrothiopyrans via hetero diels-alder reaction, under thermal and high pressure conditions. Tetrahedron 2000, 56, 3909–19.
[95] Bell, SI., & Weinreb, SM. Lewis acid and high pressure promoted Diels-Alder cycloadditions of *N*-alkyl-*N*-sulfinyl dienophiles. Tetrahedron Lett 1988, 29, 4233–6.
[96] Ito, S., Okujima, T., Kabuto, C., & Morita, N. Preparation, characterization, and cycloaddition reaction of the heterocumulenes attached directly to azulenes. An efficient strategy for the preparation of azulene-substituted heterocycles. Tetrahedron 2003, 53, 4651–9.
[97] Isaacs, NS., & Van der Reeke, P. Reactions at high pressure, Part 10: the volume profile of an intramolecular Diels-Alder reaction. Tetrahedron Lett 1982, 23, 2147–8.
[98] Diedrich, MK., & Klärner, F-G. The effect of pressure on cyclizations. the ring-size dependent reaction volumes of the cyclization of 1-alkenes to cycloalkanes. Experimental measurement of activation and reaction volumes of the intramolecular Diels-Alder reaction of 1,3,8-

nonatriene and 1,3,9-decatriene. Temperature dependence of activation and reaction volumes. J Am Chem Soc 1998, 120, 6212–8.

[99] Treus, M., Harwood, LM., Estévez, JC., Salas, C., Drew, MGB., & Estévez, RJ. Intramolecular Diels-Alder Furan-mediated synthesis of 8-aryl-3,4-dihydroisoquinolin-1(2*H*)-ones, convenient precursors of indeno[1,2,3-*ij*]isoquinolines. Synlett 2013, 24, 2221–4.

[100] Plietker, B., Seng, D., Fröhlich, R., & Metz, P. High pressure intramolecular diels-alder reactions of vinylsulfonates and vinylsulfonamides. Tetrahedron 2000, 56, 873–9.

[101] Phillips, AJ., Morris, JC., & Abell, AD. Progress towards an intramolecular Diels-Alder ring-expansion approach to taxinine: the interplay of Lewis acids and high pressure. Tetrahedron Lett 2000, 41, 2723–7.

[102] Burrell, SJ., Derome, AE., Edenborough, MS., Harwood, LM., & Leeming, SA. The high pressure mediated intramolecular diels-alder reaction of furan dienes: a synthetic approach towards daphnanes. Tetrahedron Lett 1985, 26, 2229–32.

[103] Margetić, D., & Warrener, RN. Container systems II: An experimental study of the high-pressure intramolecular cycloaddition of tethered furans and anthracenes onto norbornane cyclobutene-1,2-diesters. J Heterocycl Chem 2014, 51, 1369–79.

[104] Csende, F., Stájer, G., & Fülöp, F. Retro Diels-Alder Reactions. In Comprehensive Organic Synthesis: Second Edition, Vol. 5, Elsevier Ltd. 2014, pp. 518–94.

[105] Jenner, G., & Papadopoulos, M. High-pressure extension of the scope of the homo-Diels-Alder reaction. High Pressure Res 1988, 1, 67–75.

[106] Drew, MGB., George, AV., Isaacs, NS., & Rzepa, HS. High-pressure synthesis, structures, and conformational properties of some derivatives of 7-azabicyclo[2.2.1] heptane. X-ray determination of *endo*-10-benzoyl-4-phenyl-4,10-diazatricyclo[5.2.1.0^{2,6}]dec-8-ene- 3,5-dione and *exo*-10-acetyl-4-phenyl-4,10-diazatricyclo[5.2.1.0^{2,6}]decane-3,5-dione. J Chem Soc Perkin Trans 1 1985, 1277–84.

[107] George, AV., & Isaacs, NS. Reactions at high pressure. Part 15. Rates, activation parameters, and a volume profile for Retro-Diels-Alder reactions in the pyrrole series. J Chem Soc Perkin Trans 2 1985, 1845–7.

[108] Jenner, G., Papadopoulos, M., & Rimmelin, J. Activation parameters and location of the transition state in the retro-Diels-Alder reaction of a 7-oxabicyclo[2.2.1]hept-5-ene derivative. J Org Chem 1983, 48, 748–9.

[109] Kiselev, VD., Kashaeva, EA., & Konovalov, AI. Pressure effect on the rate and equilibrium constants of the Diels-Alder reaction 9-chloroanthracene with tetracyanoethylene. Tetrahedron 1999, 55, 1153–62.

[110] Kiselev, VD., Kashaeva, EA., Bolotov, AV., Shakirova, II., & Konovalov, AI. Energy and volume activation parameters of the retro-Diels-Alder reaction in different solvents. Russ Chem Bull 2009, 58, 21–4.

[111] Klärner, F-G., Dogan, B., Roth, WR., & Hafner, K. Cycloadditions of azulene at high pressure. Angew Chem Int Ed Engl 1982, 21, 708–9.

[112] Jenner, G., & Papadopoulos, M. The pressure effect in the homo-Diels-Alder reaction. Tetrahedron Lett 1982, 23, 4333–6.

[113] Kiselev, VD., Shakirova, II., Kornilov, DA., Kashaeva, HA., Potapova, LN., & Konovalov, AI. Homo-Diels-Alder reaction of a very inactive diene, bicyclo[2.2.1]hepta-2,5-diene, with the most active dienophile, 4-phenyl-1,2,4-triazolin-3,5-dione. Solvent, temperature, and high pressure influence on the reaction rate. J Phys Org Chem 2012, 26, 47–53.

[114] Fenton, GA., Isaacs, NS., & Gilbert, A. [2π+2σ+2π] Cycloaddition under high pressure. Tetrahedron Lett 1985, 26, 1597–8.

[115] Jenner, G., & Papadopoulos, M. High-pressure addition reactions involving cycloheptatriene-norcaradiene equilibrium. J Org Chem 1986, 51, 585–9.

[116] Papadopoulos, M., & Jenner, G. High pressure [4+2] and [2+2+2] reactions between cycloheptatriene and methyl propiolate, Tetrahedron Lett 1985, 26, 3335–8.
[117] Jenner, G. High pressure sequential ($\pi^2+\pi^2+\pi^2$) and ($\pi^2+\sigma^2+\sigma^2$) cycloaddition reactions. Synthesis of polycyclic bisadducts from C_7H_8 hydrocarbons. Tetrahedron Lett 1987, 28, 3927–30.
[118] Jenner, G. Synthesis of polycyclic structures from C_7H_8 hydrocarbons via high pressure repetitive cycloaddition reactions. High Press Res 1996, 15, 51–7.
[119] Papadopoulos, M., Jost, R., & Jenner, G. Synthesis of annelated oxetans via cycloaddition reactions of quadricyclane under thermal and high pressure conditions. J Chem Soc Chem Commun 1983, 221–2.
[120] Jenner, G., & Papadopoulos, M. Pressure effect on reactions of C_7H_8 hydrocarbons. Physica 1986, 129&140B, 729–31.
[121] Jiang, B., S-J, Tu., & Li, G. Domino Processes under Microwave Irradiation, High Pressure, and in Water in Domino Reactions: Concepts for Efficient Organic Synthesis. Tietze LF, Wiley VCH, Weinheim 2014, Chapter 12, pp. 463–94.
[122] Tietze, LF., Brasche, G., & Gericke, KM. Domino Reactions in Organic Synthesis. Wiley, Weinheim 2006.
[123] Gaitzsch, J., Rogachev, VO., Metz, P., Yusubov, MS., Filimonov, VD., & Kataeva, O. The first example of a domino Diels-Alder/retro-Diels-Alder reaction of 1,3-dienic δ-sultones with alkynes: a simple synthesis of *m*-terphenyl dicarboxy derivatives from 4,6-diphenyl-[1,2] oxathiine 2,2-dioxide. J Sulfur Chem 2009, 30, 4–9.
[124] Margetić, D., Trošelj, P., & Johnston, MR. Tandem [4+2]/[3+2] cycloadditions of 1,3,4-oxadiazoles with alkenes. Mini Rev Org Chem 2011, 8, 49–65.
[125] Warrener, RN., Margetić, D., Tiekink, ERT., & RA, Russell. The 1,3,4-oxadiazole and 1,3,4-thiadiazole coupling of norbornenes and 7-oxanorbornenes under high pressure: new structures, mechanistic detail and synthetic applications. Synlett 1997, 196–8.
[126] Margetić, D., Eckert-Maksić, M., Trošelj, P., & Marinić, Ž. Reaction of 2,5-bis-trifluoromethyl-1,3,4-oxadiazole with 7-oxanorbornenes revisited: experimental and quantum-chemical study of reaction stereospecificity. J Fluor Chem 2010, 131, 408–16.
[127] Uittenbogaard, RM., Seerden, J-PG., & Scheeren, HW. High-pressure promoted stereoselective tandem [4+2]/[3+2] cycloadditions of nitroalkenes and enol ethers. Tetrahedron 1997, 53, 11929–36.
[128] Kuster, GJ., Kalmoua, F., de Gelder, R., & Scheeren, HW. A simple entry towards novel bi- and tricyclic *N*-oxy-*β*-lactams by high pressure promoted tandem [4+2]/[3+2] cycloadditions of enol ethers and *β*-nitrostyrene. Chem Commun 1999, 855–6.
[129] LWA, van Berkom., Kuster, GJT., de Gelder, R., & Scheeren, HW. Synthesis and rearrangement of *N*-organyloxy *β*-lactams derived from a (4+2)/(3+2) sequential cycloaddition reaction involving enol ethers and nitro alkenes. Eur J Org Chem 2004, 4397–404.
[130] Buback, M., Gerke, K., Ott, C., & Tietze, LF. Kinetics and selectivity of the intramolecular Hetero Diels-Alder reaction of 1 -oxa-l,3-butadienes using a benzylidenebarbituric acid derivative effect of pressure, temperature, and solvent. Chem Ber 1994, 127, 2241–8.
[131] Minuti, L., Taticchi, A., Gats-Baitz, E., & Wenkert, E. High-pressure Diels-Alder reactions of bicyclic dienones. Tetrahedron 1995, 51, 10033–40.
[132] Engler, TA., Sampath, U., Naganathan, S., Vander Velde, D., & Takusagawa, F. A new general synthetic approach to diterpenes: application to syntheses of (±)-taxodione and (±)-royleanone. J Org Chem 1989, 54, 5712–27.
[133] Jurczak, J., Koźluk, T., Tkacz, M., & Eugster, CH. Cycloaddition of 3,4-dimethoxyfuran with 1,4-benzoquinones under high pressure. Helv Chim Acta 1983, 66, 218–221.

[134] Rahm, A., Ferkous, F., Degueil-Castaing, M., Jurczak, J., & Golebiowski, A. High pressure organometallic chemistry: diastereoselectivity in Diels Alder reactions of 2,3-dimethyl-1,3-butadiene with (*z*)-3-tributyl-stannylacrylic acid (-) menthyl ester. Synth React Inorg Met-Org Chem 1987, 17, 937–45.
[135] Tietze, LF., Hübsch, T., Voβ, E., Buback, M., & Tost, W. Intermolecular hetero-diels-alder reactions of enamino ketones at high pressure. the first significant pressure-induced diastereoselectivity in organic transformations. J Am Chem Soc 1988, 110, 4065–6.
[136] Buback, M., Tost, W., Hübsch, T., Voβ, E., & Tietze, LF. Diastereoselectivity and kinetics of intermolecular hetero diels-alder reactions under high pressure. a significant pressure-induced increase in stereoselectivity. Chem Ber 1989, 122, 1179–86.
[137] Tietze, LF., Hübsch, T., Oelze, J., Ott, C., Tost, W., Wörner, G., & Buback, M. Effect of high pressure on the stereoselectivity Diels-Alder reactions. Chem Ber 1992, 125, 2249–58.
[138] Tietze, LF., Hübsch, T., Buback, M., & Tost, W. Diastereoselectivity and kinetics of hetero Diels-Alder reactions under high pressure. High Press Res 1990, 2, 638–40.
[139] Buback, M., Abeln, J., Hübsch, T., Ott, C., & Tietze, LF. Effect of high pressure on the kinetics and selectivity of the intramolecular Hetero Diels-Alder reaction of an l-oxa-l,3-butadiene. Liebigs Ann 1995, 9–11.
[140] Buback, M., Kuchta, G., Niklaus, A., Henrich, M., Rother, I., & Tietze, LF. Intermolecular hetero diels-alder reactions of enamino ketones with highly substituted vinyl ethers effect of high pressure on the kinetics and diastereoselectivity. Liebigs Ann 1996, 1151–8.
[141] Al-Badri, H., Collignon, N., Maddaluno, J., & Masson, S. Switching stereocontrol in the high pressure-induced hetero-Diels-Alder reaction. Chem Commun 2000, 1191–2.
[142] Haiza, M., Lee, J., & Snyder, JK. Asymmetric syntheses of *Salvia miltiorrhiza* abietanoid *o*-quinones: methyl tanshinonate, tanshinone IIB, tanshindiol B, and 3-hydroxytanshinone. J Org Chem 1990, 55, 5008–13.
[143] Araki, Y., & Konoike, T. Enantioselective total synthesis of (+)-6-*epi*-mevinolin and its analogs. Efficient construction of the hexahydronaphthalene moiety by high pressure-promoted intramolecular Diels-Alder reaction of (*R*,2*Z*,8*E*,10*E*)-1-[(*tert*-butyldimethylsilyl)oxy]-6-methyl-2,8,10-dodecatrien-4-one. J Org Chem 1997, 62, 5299–309.
[144] Butz, T., & Sauer, J. High-pressure and thermally induced intramolecular Diels-Alder reactions of furfuryl fumarates. Influence of tether substituents on diastereoselectivity. Tetrahedron Asymm 1997, 8, 703–14.
[145] Jurczak, J. Stereochemistry of Diels-Alder reactions at high pressure II. Influence of high pressure on asymmetric induction in condensation of (-)-di-(*R*)-menthyl fumarate with butadiene and isoprene. Bull Chem Soc Jpn 1979, 52, 3438–42.
[146] Araújo, N., Gil, MV., Román, E., & Serrano, JA. New molecular scaffolds: synthesis of (2*R*,3*R*)- and (2*S*,3*S*)-3-nitro-7-oxabicyclo[2.2.1]hept-5-ene-2-carbaldehyde from furan. Synthesis 2006, 15, 2503–6.
[147] Mori, K., Yamauchi, T., Maddaluno, J., Nakano, K., Ichikawa, Y., & Kotsuki, H. Organocatalytic asymmetric Hetero-Diels-Alder reaction of oxindoles under high pressure. Synlett 2011, 2080–4.
[148] Tsypysheva, IP., Borisevich, SS., Lobov, AN., Kovalskaya, AV., Shamukaev, VV., Safiullin, RL., & Khursan, SL. Inversion of diastereoselectivity under high pressure conditions: Diels-Alder reactions of 12-*N*-substituted derivatives of (-)-cytisine with *N*-phenylmaleimide. Tetrahedron Asymm 2015, 26, 73–7.
[149] Kwiatkowski, P., Chaładaj, W., Malinowska, M., Asztemborska, M., & Jurczak, J. The high-pressure [4+2]cycloaddition of 1-methoxybuta-1,3-diene to the glycolaldehyde-derived heterodienophiles, catalyzed by chiral metallosalen complexes. Tetrahedron: Asymm 2005, 16, 2959–64.

[150] Tietze, LF., Ott, C., Gerke, K., & Buback, M. The first example of an increase in the enantioselectivity of a chemical reaction in the presence of a chiral Lewis acid under high pressure, Angew Chem Int Ed Engl 1993, 32, 1485–6.
[151] Tietze, LF., Ott, C., & Frey, U. Enantioselective Diels-Alder reactions: NMR spectroscopy of a chiral titanium Lewis acid under high pressure. Liebigs Ann 1996, 63–7.
[152] Dols, PPMA., Klunder, AJH., & Zwanenburg, B. 4-Hydroxycyclopent-2-en-1-one and derivatives as chiral synthetic equivalents of cyclopentadienone in asymmetric Diels-Alder reactions. Tetrahedron 1994, 50, 8515–38.
[153] Briš, A., Marinić, Ž., Chen, Z-L., & Margetić, D. Synthesis of chlorins by diels-alder cycloadditions of pheophorbide a and its derivatives. Synlett 2015, 26, 991–4.
[154] Smith III, AB., Liverton, NJ., Hrib, NJ., Sivaramakrishnan, H., & Winzenberg, K. Total synthesis of (+)-jatropholones A and B: exploitation of the high-pressure technique. J Am Chem Soc 1986, 108, 3040–8.
[155] Smith III, AB., Liverton, NJ., Hrib, NJ., Sivaramakrishnan, H., & Winzenberg, K. Total Synthesis of (+)-Jatropholone A and B. J Org Chem 1985, 50, 3239–41.
[156] Harwood, LM., Jackson, B., Jones, G., Prout, K., Thomas, RM., & Witt, FJ. Synthetic approaches to phorbols via the intramolecular diels-alder reaction of furans: influence of the dienophile upon the stereochemical control. Chem Commun 1990, 608–10.
[157] Harwood, LM., Ishikawa, T., Phillips, H., & Watkin, D. Synthetic approaches to phorbols via the intramolecular Diels-Alder reactions of furans (IMDAF): oxygen bridge cleavage and high stereocontrol during cycloaddition using α'-benzylthio substituted furan substrates. Chem Commun 1991, 527–30.
[158] Drew, MGB., Jahans, A., Harwood, LM., & Apoux, SABH. A Diels-Alder study of the diene reactivity of 2-methyl-5-vinyl-3-furoate esters. Eur J Org Chem 2002, 3589–94.
[159] Matsumoto, K., Hashimoto, S., Ikemi, Y., & Otani, S. The high pressure mediated reaction of π-excessive heterocycles with maleic anhydride derivatives: A synthetic approach to cantharidin analogues. Heterocycles 1986, 24, 1835–9.
[160] Jurczak, J., Koźluk, T., Filipek, S., & Eugster, CH. High-pressure [2+4]-cycloaddition of dimethylmaleic anhydride to 3,4-dimethoxyfuran; synthesis of dimethoxycantharidin. Helv Chim Acta 1982, 65, 1021–24.
[161] Dauben, WG., Kessel, CR., & Takemura, KH. Simple, efficient total synthesis of cantharidin via a high-pressure Diels-Alder reaction. J Am Chem Soc 1980, 102, 6894–6.
[162] Dauben, WG., Gerdes, JM., & Smith, DB. Organic reactions at high pressure. The preparative scale synthesis of cantharidin. J Org Chem 1985, 50, 2576–8.
[163] Banwell, MG., Edwards, AJ., Harfoot, GJ., & Jolliffe, KA. A chemoenzymatic synthesis of (-)-hirsutene from toluene. J Chem Soc Perkin Trans 1 2002, 2439–41.
[164] Gillard, JR., & Burnell, DJ. Facial selectivity in the Diels-Alder reactions of *cis*-cyclohexa-3,5-diene-1,2-diol and derivatives with *N*-phenylmaleimide. J Chem Soc Chem Commun 1989, 1439–40.
[165] Banwell, MG., Edwards, AJ., McLeod, MD., & Stewart, SG. A chemoenzymatic synthesis of the *cis*-decalin core associated with the novel anti-mitotic agent phomopsidin: some observations concerning a high-pressure-promoted Diels-Alder cycloaddition reaction of (1*S*,2*R*)-3-methyl-*cis*-1,2-dihydrocatechol and the anionic oxy-cope rearrangement of compounds Derived from the Adduct. Aust J Chem 2004, 57, 641–4.
[166] Reekie, TA., Austin, KAB., Banwell, MG., & Willis, AC. The chemoenzymatic total synthesis of phellodonic acid, a biologically active and highly functionalized hirsutane derivative isolated from the tasmanian fungus *Phellodon melaleucus*. Austral J Chem 2008, 61, 94–106.

[167] Banwell, MG., Austin, KAB., & Willis, AC. Chemoenzymatic total syntheses of the linear triquinane-type natural products (+)-hirsutic acid and (-)-complicatic acid from toluene. Tetrahedron 2007, 63, 6388–403.
[168] Chang, EL., Bolte, B., Willis, AC., & Banwell, MG. Chemoenzymatic total syntheses of the enantiomers of the protoilludanes 8-deoxydihydrotsugicoline and radudiol. J Org Chem 2016, 81, 2078–86.
[169] Waalboer, DCJ., Schaapman, MC., van Delft, FL., & Rutjes, FPJT. High-pressure entry into platencin. Angew Chem Int Ed Engl 2008, 47, 6576–8.
[170] Álvarez, M., Feliu, L., Ajana, W., Joule, JA., & Fernández-Puentes, JL. Synthesis of ascididemine and an isomer. Eur J Org Chem 2000, 849–55.
[171] Knappwost-Gieseke, C., Nerenz, F., Wartchow, R., & Winterfeldt, E. High-pressure selectivity studies-a simple route to a homochiral wistarin precursor. Chem Eur J 2003, 9, 3849–58.
[172] Mori, S., Karita, T., Komatsu, K., Sugita, N., & Wan, TSM. High-pressure synthesis of cycloadduct of fullerene C_{60} with 2H-pyran-2-one. Synth Commun 1997, 27, 1475–82.
[173] Takeshita, H., Liu, J-F., Kato, N., Mori, A., & Isobe, R. High-pressure Diels-Alder reaction of [60] fullerene with several tropones. Characterization of the 1:1 cycloadducts. J Chem Soc Perkin Trans 1 1994 1433–37.
[174] Takeshita, H., Liu, J., Kato, N., & Mori, A. High-pressure Diels-Alder reaction between buckminsterfullerene and tropone. Chem Lett 1993, 22, 1697–8.
[175] Liu, J-F., Mori, A., Kato, N., & Takeshita, H. High-pressure Diels-Alder reaction of buckminsterfullerene (C_{60}) with several tropone derivatives. Fullerene Sci Technol 1995, 3, 45–58.
[176] Liu, JF., Kato, N., Mori, A., Takeshita, H., & Isobe, R. The high-pressure Diels-alder reaction of [5,6]-fullerene-C_{60} with cycloheptatriene. Formation of Diels-Alder adducts with improved thermal-stability. Bull Chem Soc Jpn 1994, 67, 1507–9.
[177] Torres-García, G., & Mattay, J. Exohedral functionalization of [60]fullerene by [4+2] cycloadditions. Diels-Alder reactions of [60]fullerene with electron rich 2,3-dioxysubstituted-l,3-butadienes. Tetrahedron 1996, 15, 5421–6.
[178] Takeshita, H., Liu, J-F., Kato, N., Mori, A., & Isobe, R. Highly thermally-stable diels-alder adducts of [60]fullerene with 2-cycloalkenones and their acetals. Chem Lett 1995, 24, 377–8.
[179] Ménard-Moyon, C., Dumas, F., Doris, E., & Mioskowski, C. Functionalization of single-wall carbon nanotubes by tandem high-pressure/$Cr(CO)_6$ activation of Diels-Alder cycloaddition. J Am Chem Soc 2006, 128, 14764–5.

3 1,3-Dipolar cycloaddition reactions

These reactions are also partially covered in the chapters dealing with tandem and multicomponent reactions. 1,3-Dipolar cycloaddition reactions carried under high pressure are important type of reactions, in which various heterocycles are produced. Cycloadditions of azides, nitrones, diazo compounds and nitrile oxides are covered, providing triazolines, triazoles, pyrazolines, oxadiazolines, isoxazole, isoxazolidines, isoxazoline and nitrosoacetals. Successful application of high pressure to this type of reactions is related to their very large negative $\Delta V^{\neq}$ [1].

3.1 Azides

In classical conditions at room temperature, these cycloadditions reactions take weeks, or even months, to achieve only partial completion. Heating is usually precluded due to triazoline instability.

The addition of 1,3-dipolar azide to the electron-rich double bond of *tert*-butyldimethylsilyl enol ethers **1** gives a Δ^2-1,2,3-triazoline cycloadducts **2** in high yields (Scheme 3.1) [2]. These adducts spontaneously fragment via a diazonium betaine to imidate esters, which are moisture sensitive and undergo the ring contraction to stable cyclopentane ring of **3**. Dauben has reported high regioselectivity of the addition of 1,3-dipolar azides to the polarized enol π-bond of a variety of silyl enol ethers possessing varying degrees of substitution and the reaction rate acceleration in comparison to ambient pressure additions.

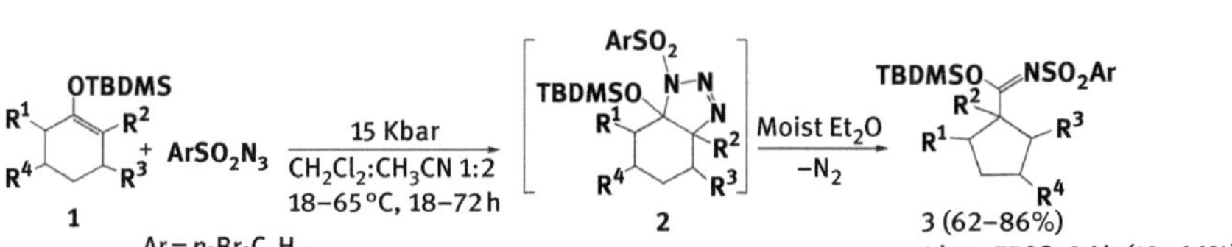

Scheme 3.1: The addition of azide to TBDMS enol ethers.

The electron-deficient olefin dipolar cycloadditions with azides studied by Weinreb showed a significant rate acceleration by the implementation of the pressure (Scheme 3.2) [3]. Azides with methacrylate derivatives **4** at 12 kbar at room

https://doi.org/10.1515/9783110556841-003

temperature cycloadd rapidly to afford predominantly regioisomeric triazolines **5** and smaller amounts of isomers **6.** In general, good yields of cycloadducts obtained in several different solvents can be used (dichloromethane, diethyl ether, methanol, ethyl acetate). At ambient pressure, the same cycloadditions are much slower, usually with lower yields, which might be the result of triazoline decomposition over the long reaction periods at 1 atm. The reaction of more functionalized methacrylate substrate **7** afforded at 12 kbar 5:95 regioisomeric mixture of triazolines **8** and **9** in 83% yield. No selectivity and lower yield (30%) were found at 1 bar reaction.

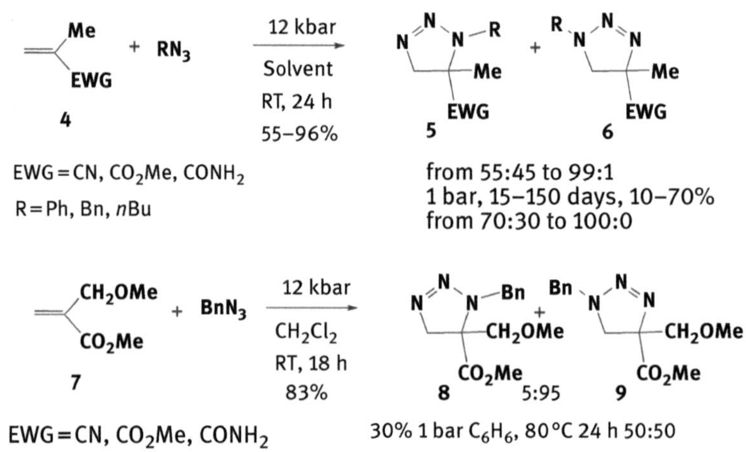

Scheme 3.2: Azide reaction with alkenes.

Pressure increased the rates of the reactions of 2-morpholinobut-1-en-3-yne **10** with azides. At pressures of 7–8 kbar, the yield of triazoline **11** could be improved from 37% to 54%, which was formed as the only product. Furthermore, triazoline **12,** which was not obtained in the cycloaddition reactions at the atmospheric pressure, was formed under high-pressure conditions in high yield (Scheme 3.3) [4]. Klärner has also found that the reaction of phenyl azide with **10** is strikingly dependent on the applied reaction conditions. At atmospheric pressure and room temperature, the reaction proceeded sluggishly, and only by heating in refluxing toluene triazoline **15** was obtained as the sole product in very low yield. On the other hand, under high-pressure conditions at 10 kbar, a single product, triazole **14,** was obtained (54%). Triazoline **13,** the initial cycloadduct, was obtained as a single product by pressurizing the reaction mixture at 6.7 kbar, when the reaction was stopped before the complete conversion of the reactants.

Benzonorbornadiene cyclobutene-1,2-diester **16** was converted to triazoline **17** by 1,3-dipolar addition of benzyl azide under high pressure. This 1,3-dipolar cycloaddition is π-face selective, with the formation of the adduct in which azide approached π-

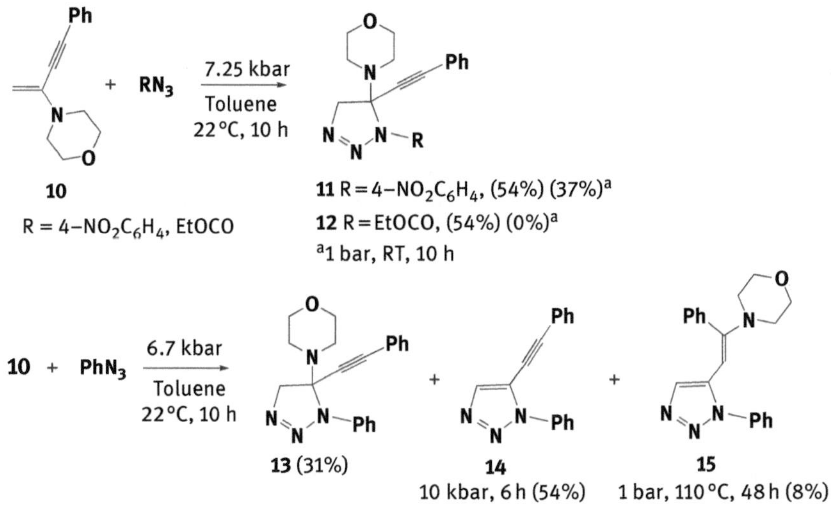

Scheme 3.3: The addition of azide to 2-morpholinobut-1-en-3-yne.

bond of **16** exclusively from the *endo*-face with respect to norbornane. Photochemical dinitrogen elimination from triazolines and subsequent thermal dipolar cycloaddition of azomethine ylide were shown to be synthetically useful route to 7-azanorbornane skeleton. Although triazoline **17** could be produced either by thermal [5] or high-pressure methods, the cyclobutene-1,2-diester groups in **18** are reluctant to add benzyl azide under thermal conditions, but do so satisfactorily when compressed at 14 kbar. The resultant mixture consists of almost equal amounts of C_2-isomer **19a** and σ-isomer **19b** (Scheme 3.4) [6]. When the pressurization was carried out in neat benzyl azide, the **19a:19b** ratio has changed to 96:4, with almost exclusive formation of **19a**. An explanation for this observation is that benzyl azide solidified under high pressure and the reaction occurred in the solid phase. Similar mixture of C_2- and σ-isomers was formed by the reaction with anhydride **20**, and even more complex mixture of products **23** was obtained at room temperature reaction with 7-azanorbornane substrate **22**, due to the invertomerization of the *N*-benzyl substituents and the carbamate-type isomerization of the NCbz groups [7].

High-pressure 1,3-dipolar cycloaddition of benzyl azide onto strained cyclobutene norbornene double bonds possessing electron withdrawing substituents is very general, and a number of cyclobutene-1,2-diesters **24** are reactive, possessing carbon or heteroatom at the position 7- of norbornane moiety. Furthermore, cyclobutene derivatives with only one carbmethoxy group are still reactive under high-pressure conditions [8], and the esters could be also replaced by CF_3 substituents. Besides benzyl azide [9], other substituted benzyl azides, pyridyl, carbmethoxymethyl [7], allyl, benzhydryl, di-(*p*-methoxybenzhydryl), 9-fluorenyl [10] and methoxymethyl [11] azides were employed (Scheme 3.5) [12].

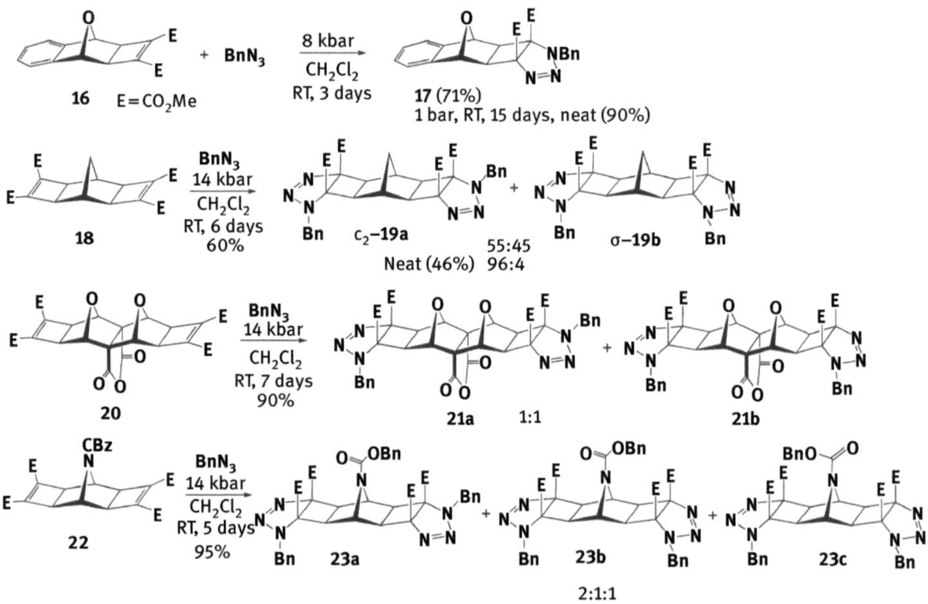

Scheme 3.4: The addition of azide to norbornenes.

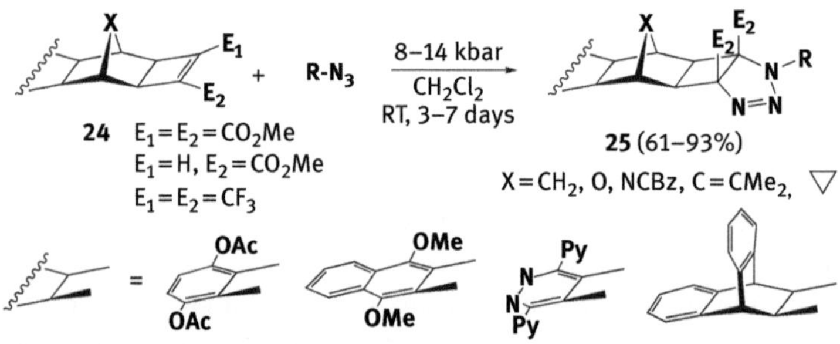

R=Bn, CH_2-4-OMeC_6H_4, CH_2-4- OMeC_6H_4, CH_2-4-$NO_2C_6H_4$, 4-Py, 3-Py, MOM, allyl, CH_2CO_2Me, $CH_2CH=CH_2$

Scheme 3.5: The addition of azide to norbornane cyclobutene-1,2-diesters.

Norbornene π-bond is also reactive in 1,3-dipolar cycloadditions of azides, with the *exo*- π-facial selectivity and regioselectivity due to employing the unsymmetrical 6-azabicyclo[3.1.0]hexane substrates **26a–d** (Scheme 3.6) [13]. Katagiri reported that under high-pressure conditions, triazolines **27a–d** were formed as the major regioisomeric product.

High regioselectivity of 1,3-dipolar cycloadditions of azides was found in reactions with 16-dehydropregnenolone acetate at 10 kbar (Scheme 3.7) [14]. Zavarzin

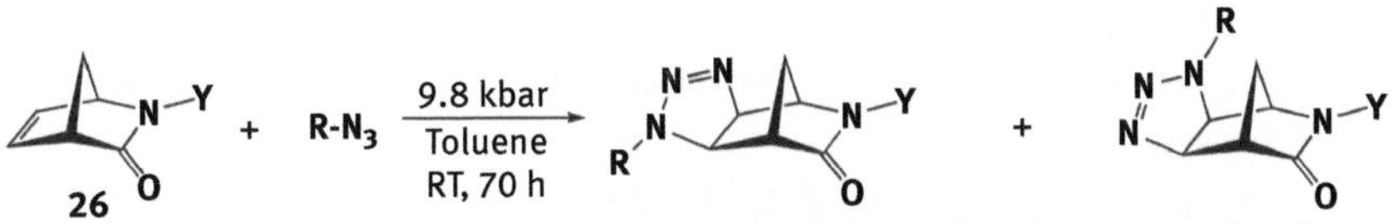

26

Y=Boc, Cbz

27a R=CO_2ET, Y=Boc (52%) **28a** (21%)
27b R=$P(O)(OPh)_2$, Y=Boc (56%) **28b** (14%)
27c R=Ph, Y=Boc (60%) **28c** (16%)
27d R=$P(O)(OPh)_2$, Y=CBz **28d**

Scheme 3.6: The addition of azide to norbornene derivative.

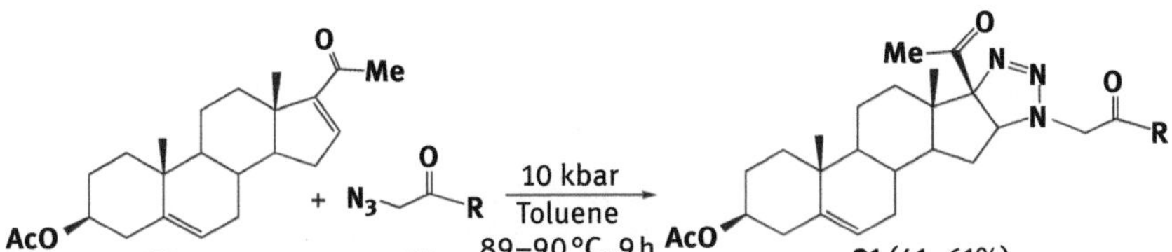

29 30 31 (41–61%)

R=OEt, Me, PhNH, 4-MeC_6H_4NH, 4-$MeOC_6H_4NH$

Scheme 3.7: The addition of azide to 16-dehydropregnenolone acetate.

has prepared condensed pregnano[17,16-*d*]triazolines **31** in high yields, where cycloaddition reactions occurred only at the 16-π-bond activated by the acetyl group, while the nonactivated π-bond at the position 5- remained intact.

Several research groups have put significant efforts toward developing copper-free variant of azide-terminal alkyne cycloaddition (CAAC) reaction, the most important reaction in [3 + 2] Huisgen cycloadditions, often called "click" chemistry. High-pressure conditions were applied in an effort to avoid the toxicity of the copper ions. Zanirato has shown that under high pressure, the 1,2,3-triazole products **33a–d** are formed in a few hours at room temperature by reacting azides with trimethylsilyl acetylene (Scheme 3.8) [15]. These reaction rates are in orders of magnitude faster than the corresponding transformations at ambient conditions. Kinetic studies revealed that 10 kbar is a pressure above which aryl azides immediately

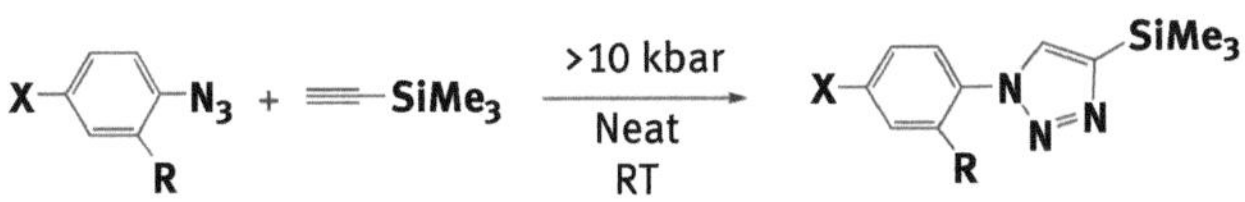

32a X=NO_2, R=H
32b X=Cl, R=H
32c X=H, R=Ph
32d X=OMe, R=H

33a 1 kbar, 2 h (1bar, 20 days)
33b 1 kbar, 4 h (1bar, 25 days)
33c 2 kbar, 15 h (1bar, 35 days)
33a 3 kbar, 1.5 h (1bar, 35 days)

Scheme 3.8: Cycloaddition of aryl azides with trimethylsilyl acetylene.

transform to the corresponding 1,2,3-triazoles. All reactions were carried out in the absence of solvent, using an excess of trimethylsilyl acetylene both as the reactant and hydrostatic pressure medium.

Hessel has carried out optimization of the 1,3-dipolar cycloaddition of 2,6-difluorobenzyl azide **34** and methyl propiolate, leading to the 4-substituted 1,2,3-triazole **36**, a precursor of Rufinamide, an antiepileptic drug (Scheme 3.9) [16]. Triazole **36** and related 1,2,3-triazoles were synthesized under catalyst-free conditions in microcapillary flow reactor in *N*-methyl-2-pyrrolidone in high yield with regioselectivity toward 1,4-cycloadduct.

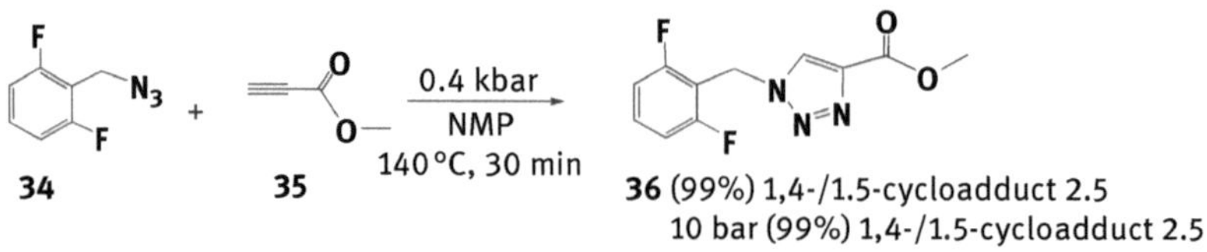

Scheme 3.9: The addition of azide to methyl propiolate.

When azide and alkyne were preorganized in the solid state, copper-free 1,3-dipolar cycloaddition was accelerated, likely by the azide and alkyne groups in an arrangement that is close to the transition state of the 1,4-disubstituted cycloaddition. Thus, the reaction of supramolecular system of 4-ethynylaniline **37** and 4-azido-2,3,5,6-tetrafluorobenzoic acid **38** in crystalline state when subjected to 10 kbar was completed overnight with an 80–90% yield in a regioselective manner (Scheme 3.10) [17]. Crystal engineering combined with high pressure was also applied to the solid-state 1,3-dipolar cycloaddition of substrate **40** to obtain triazoline polymer **41** in very short

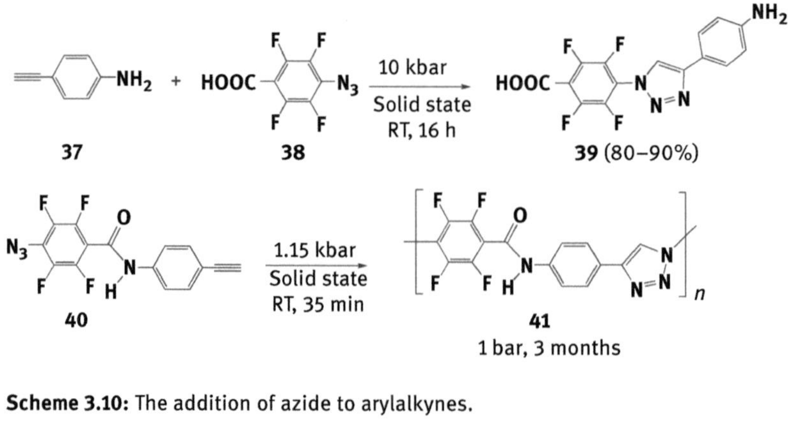

Scheme 3.10: The addition of azide to arylalkynes.

time [18]. This is a striking reaction rate acceleration; for the complete conversion of **40** to polymer **41** at ambient pressure, three months were required.

When dipolarophile is a nitrile bond, tetrazole rings are obtained. The α-ketonitriles react with hydrazoic acid and organic azides under high pressure to afford the preparation of acyl- and benzoyl-*N*-tetrazoles in high yields (Scheme 3.11) [19]. The employment of high-pressure conditions significantly shortened reaction times, lowered temperature and increased the product yield, also allowing the use of thermally unstable ketonitriles and azides. A significant difference in the product yields was noticed, for instance, **43c** was prepared in 35% at 5 kbar after 12 h, whereas by the pressurization at 10 kbar, the yield was increased to 90%. Similarly, methyldiazoacetate reacts in hyperbaric conditions (10 kbar, 110 °C, 8 h, MeCN or in the presence of Bu_2SnO at 5 kbar, 70–80 °C) with dithiazoleimines to afford tetrazole derivatives in high yields [20].

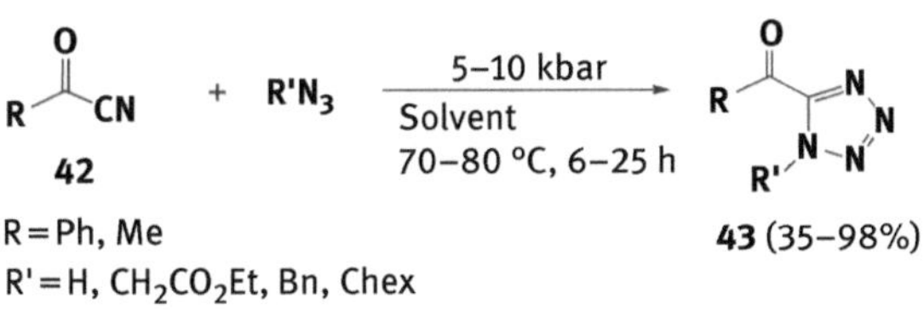

Scheme 3.11: The addition of azide to nitrile bond.

3.2 Diazo compounds

Kinetic studies of 1,3-dipolar cycloaddition reactions carried out by Huisgen [21] and Isaacs [22] revealed very negative volumes of activation and volumes of reaction, and hence the acceleration by the increased pressure, which was employed by synthetic chemists. For instance, cycloaddition reactions of diphenyldiazomethane with dipolarophiles DMAD, diethyl fumarate and diethyl maleate in chlorobenzene have shown values of $\Delta V^{\neq}$ = −17.8,20.9 and −23.6 $cm^3\ mol^{-1}$, respectively, and the corresponding volumes of reaction ΔV_R = −26.4,−27.0 and −25.1 $cm^3\ mol^{-1}$. The influence of solvent on the kinetics is illustrated by cycloadditions with DMAD in *n*-hexane and toluene, where different $\Delta V^{\neq}$ were determined (−24 and −23.2 $cm^3\ mol^{-1}$). Reactions of DMAD and maleic anhydride that were carried out in acetonitrile have $\Delta V^{\neq}$ = −15.3 and −23.6 $cm^3\ mol^{-1}$, respectively, and ΔV_R = −27.8 and −16.7 $cm^3\ mol^{-1}$. Even larger values are measures for the reaction of diethyl azodicarboxylate with *n*-butyl vinyl ether in toluene ($\Delta V^{\neq}$ = −45.9 $cm^3\ mol^{-1}$) and diphenyldiazomethane with 2,3-diethyldiazanorbornene (−30.0 $cm^3\ mol^{-1}$).

Weiler has reported that diazomethane under high pressure undergoes 1,3-dipolar cycloaddition reaction (Scheme 3.12) [23]. Reactions with alkenes and Schiff bases at 5 kbar afforded in high yields of corresponding pyrazolines **45** and 1,2,3-

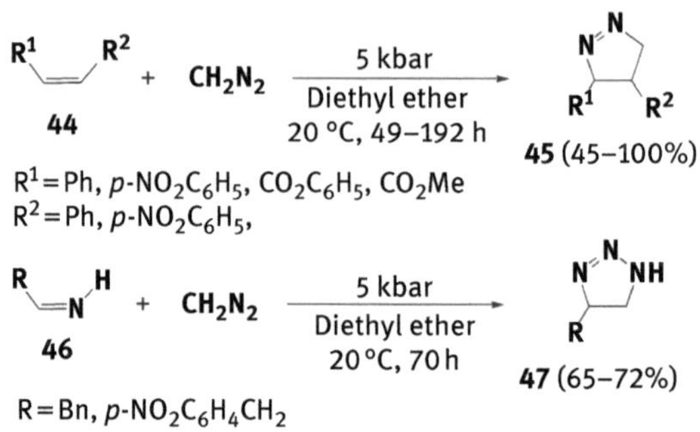

Scheme 3.12: Cycloadditions of diazomethane with alkenes and Schiff bases.

triazolines **47**, respectively. At atmospheric pressure, these cycloadditions do not take place, or reactions are low yielding.

The reaction of 16-dehydropregnenolone acetate **48** with ethyl diazoacetate at 14 kbar proceeded regioselectively at the 16-π-bond to afford two pyrazoline products: Δ^1-pyrazoline **49** and Δ^2-pyrazoline **50** (Scheme 3.13) [24]. The mixture of two isomeric steroido[16α,17α]pyrazolines was not reported for thermally conducted reaction at atmospheric pressure at 120 °C, but instead, only pyrazoline **50** as a single product. This is due to the thermal lability of **49** and its facile transformation to more stable **50** by 1,3-prototropic shift. The isolation of pyrazoline product **49** was explained by the change of thermal to acceleration by pressure and lowering of the reaction temperature, which inhibits this process. Identical π-site selectivity and regioselectivity was observed in the 1,3-dipolar cycloaddition reactions of **48** with nitronates **51** at 14 kbar affording 1,2-oxazolidines **52** [25].

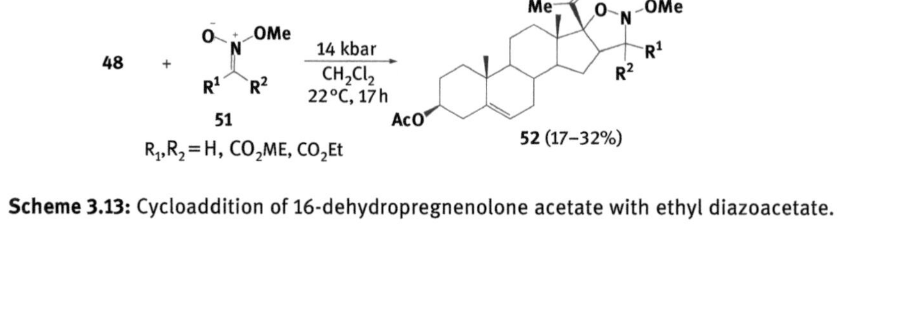

Scheme 3.13: Cycloaddition of 16-dehydropregnenolone acetate with ethyl diazoacetate.

The 1,3-dipolar reaction of 9-diazo-4,5-diazafluorene **54** with Smith's diene **53** in benzene produces the pyrazoline **55** (Scheme 3.14) [26]. Cycloaddition could be effectively carried out at 14 kbar to afford 1:1 adduct **55** with a cyclobutene π-bond preference. The exclusive reaction having occurred at the cyclobutene π-bond is accompanied by the *endo*-π facial selectivity and production of single isomer. In the reactions of **54** with norbornene π-bond, products such as **57** were obtained by the *exo*-face approach of the dipole. Analogous cycloadducts were obtained under high-pressure reactions of norbornenes **58–61.** Polycyclic norbornene dienes **63** and **64** were used to prepare the corresponding bis-adducts [27].

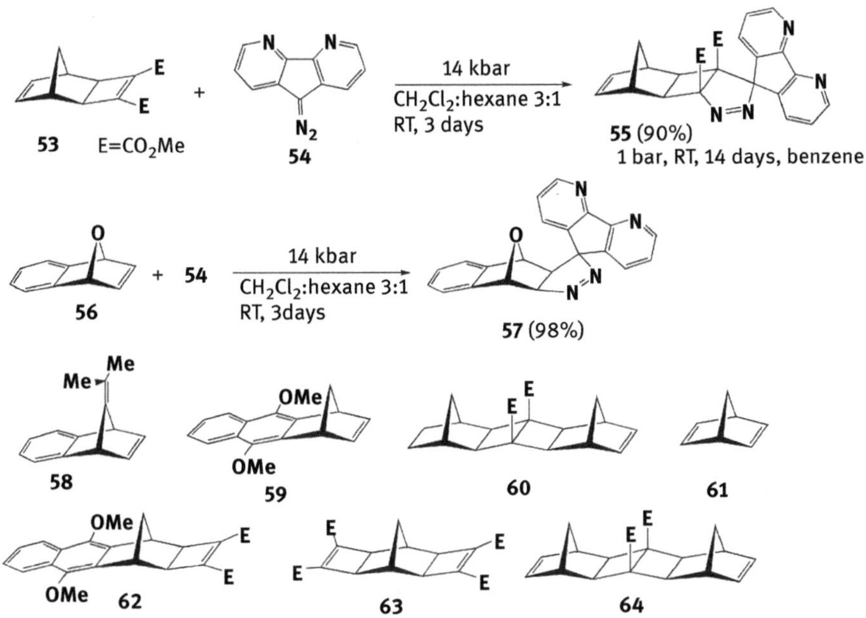

Scheme 3.14: Cycloadditions of -diazo-4,5-diazafluorene with norbornenes.

3.3 Nitrones

DeShong has demonstrated that the problems of the application of thermally unstable nitrones in 1,3-dipolar cycloaddition reactions with electron-rich olefins could be circumvented when used under high-pressure conditions [28]. Pressurization of nitrone **65** and neat ethyl vinyl ether provided phenylisoxazolidine **68** in 83% yield, as a mixture of two stereoisomers (Scheme 3.15). A similar yield (78%) was obtained by heating at room pressure (80 °C, 72 h); however, high-pressure conditions are advantageous, since all unreacted nitrone could be recovered. Analogously, it was necessary to use high-pressure conditions to synthesize product **72** in high yield,

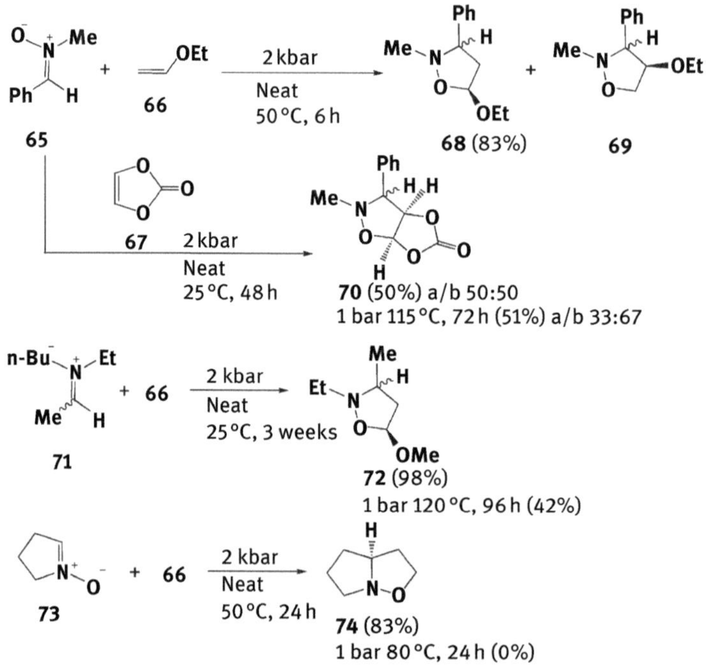

Scheme 3.15: Nitrone cycloadditions to phenyl vinyl ether.

since thermally conducted cycloaddition was low yielding. Synthetic usefulness of the high-pressure promoted cycloaddition reaction is well demonstrated in the reaction involving nitrone **73** and ethyl vinyl ether. Product **74** was formed in high yield at 2 kbar, whereas the thermal reaction at ambient pressure produced only a polymeric material, due to thermal decomposition of cycloadduct **74**. To attain **74** would not be possible without employing high pressures. An obvious influence of pressure on the stereochemical outcome of the reaction was noted in the reaction of **65** with vinylene carbonate. At ambient temperature, stereoisomers **70a/b** were obtained in equal amounts, whereas at 2 kbar **70a/b** the ratio has changed to 33:67.

In addition, improved stereoselectivity of the cycloaddition of nitrone **75** with vinylene carbonate was obtained by employing hyperbaric conditions (Scheme 3.16) [29]. Although in thermal conditions the ratio of isoxazolidines **76:77** is 1:2, when carrying out the reaction at 12 kbar, the ratio improves in favor of the isomer **77**.

In contrast, high-pressure conditions did not help to improve stereoselectivity of the asymmetric 1,3-dipolar cycloaddition of nitrones **78** with ethyl vinyl ether catalyzed by chiral oxazaborolidines (Scheme 3.17) [30]. Complete regioselectivity with formation of the isoxazolidine products **79** was obtained, as mixtures of *cis*- and *trans*-isomers **79a/79b**. The *cis-/trans*-ratios do not change with the application of

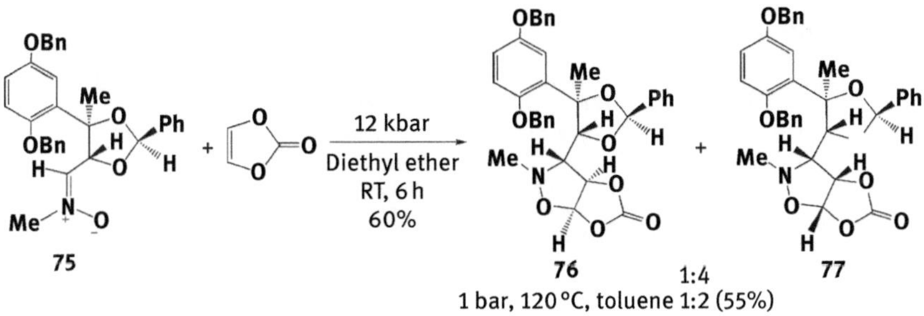

Scheme 3.16: Cycloaddition of nitrone with vinylene carbonate.

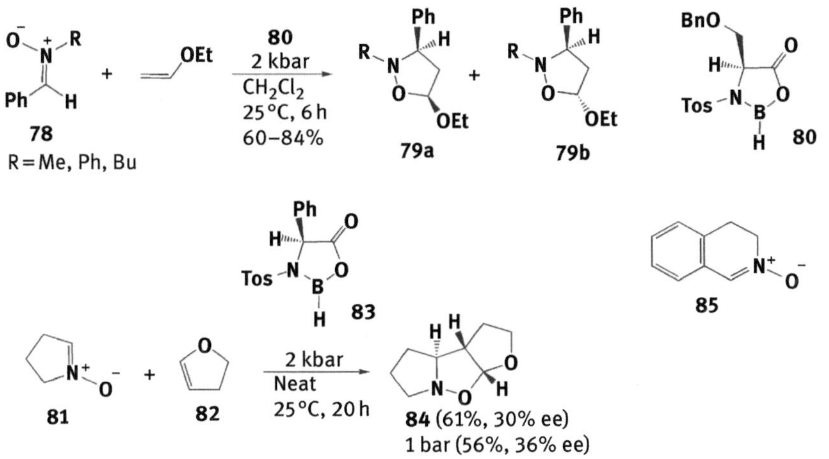

Scheme 3.17: Cycloaddition of nitrones with ethyl vinyl ether and dihydrofuran.

pressure of chiral Lewis catalyst **80**, and the enantiomeric excesses are low (0% ee). Application of various chiral Lewis acid catalysts did not induce any significant chirality in cycloaddition of nitrone **85** with ethyl vinyl ether or 2,3-dihydrofuran, or reaction of nitrone **81** to 2,3-dihydrofuran, which was catalyzed by catalyst **83**.

The other method to increase stereoselectivity is the use of chiral nitrones for asymmetric induction. The 1,3-dipolar cycloaddition of chiral nitrone such as **86** that was prepared from L-valine with methyl acrylate provided the corresponding isoxazolidines **87a–c** as a mixture of diastereoisomers in good yield (Scheme 3.18) [31]. The ratio of diastereoisomers has only slightly changed when high pressure was applied.

High-pressure 1,3-dipolar cycloaddition of nitrones was also carried out on a maleimide-modified monolayer-protected gold nanoparticle **88**, which is soluble in dichloromethane (Scheme 3.19) [29]. Quantitative formation of isoxazolidines **90**

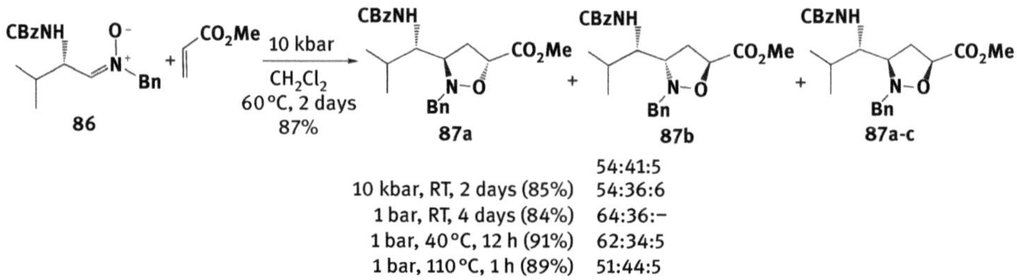

Scheme 3.18: 1,3-Dipolar cycloaddition of chiral nitrone with methyl acrylate.

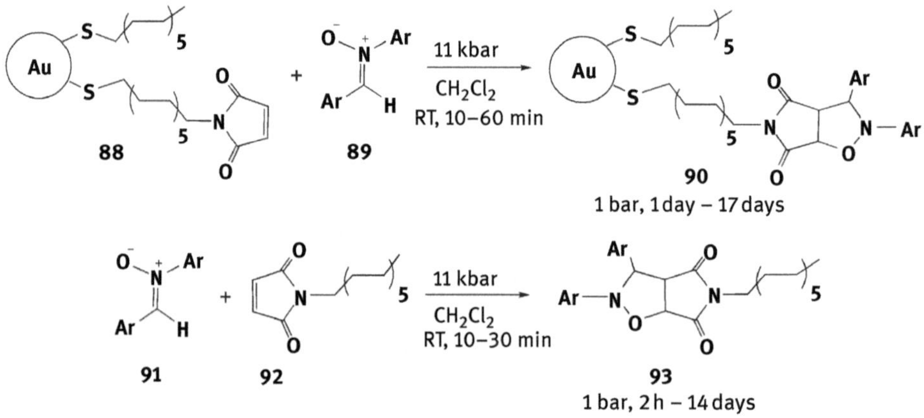

Scheme 3.19: Nitrone cycloaddition with maleimide-modified gold nanoparticle.

was achieved within 1 h, which was much faster than reactions at atmospheric pressure (as shown by times required for total conversion of nitrones). Similar acceleration of reaction rates was obtained for the model maleimide **92**.

When nitriles were used as dipolarophiles, 1,3-dipolar cycloadditions with nitrones afforded Δ^4-1,2,4-oxadiazolines **96** (Scheme 3.20) [32]. Reaction time was

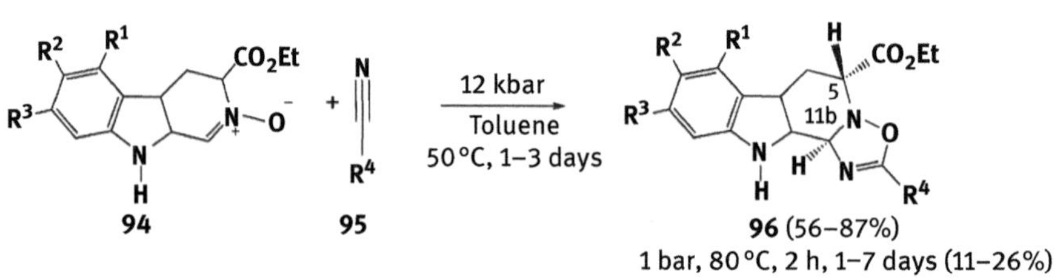

Scheme 3.20: Cycloaddition of nitrones with nitriles.

shortened, temperatures were lowered and yields increased when hyperbaric conditions were employed. Cycloaddition reactions were completely stereoselective with *trans-* orientation of C_5 and C_{11b} substituents.

The method to overcome difficulties in low reactivity of cyclic 3-substituted nitronates in classical conditions was developed by Tabolin and includes combined activation by Lewis catalyst and high pressure. Formal [3 + 3] cycloaddition of cyclic 3-substituted nitronates **97** with cyclobutanes **98** (possessing at the same time electron-donating and electron-withdrawing groups) at 8 kbar in the presence of Yb $(OTf)_3$ catalyst afforded nitrosoacetal cycloadducts **99** in moderate-to-high yields with good stereoselectivity (Scheme 3.21) [33]. The reaction proceeds via Lewis acid activation of cyclopropane to formal 1,3-dipole **100**, which subsequently reacts with nitronates.

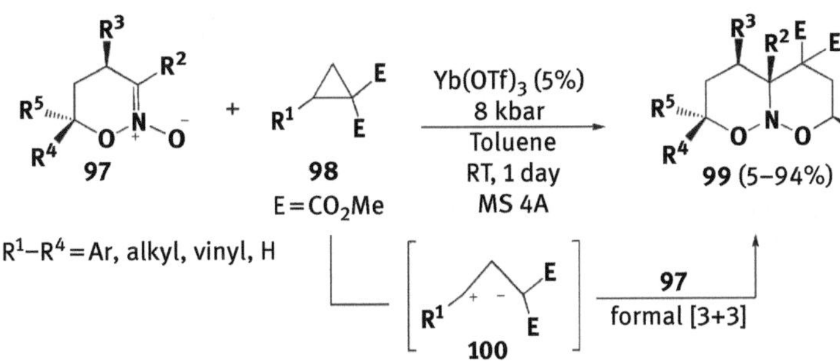

Scheme 3.21: Lewis acid-catalyzed reactions of nitronates with cyclobutanes.

3.4 Nitrile oxides

Exemplary enhancement in reaction yield caused by high-pressure activation is the dipolar cycloaddition reactions of nitrile oxides published by Krompiec (Scheme 3.22) [34].

Cycloadditions of terephthalobis(nitrile *N*-oxide) **102** with (*E/Z*) 5-(1-propenyl)-2,2′-bithiophene **101** provided at 10 kbar bis-isoxazoline product **103** regioselective and in quantitative yield. When acetylenic dipolarophile (5-ethynyl-2,2′-bithiophene) **104** was used for the reaction with of terephthaloyldinitrile, the corresponding bis-isoxazole cycloadduct **105** was obtained. The best conditions for the reaction with **101** at the ordinary pressure afforded 60% of bis-adduct, contaminated with 10% of the monoadduct, whereas the reaction of **104** did not proceed even at 100 °C.

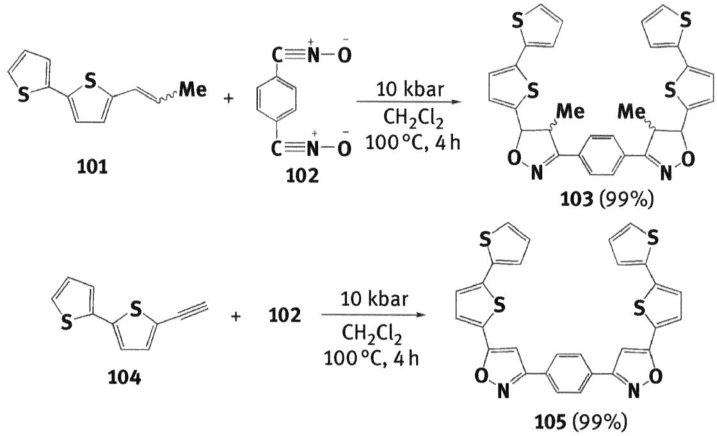

Scheme 3.22: Cycloadditions of terephthalobis(nitrile *N*-oxide) with bithiophene.

References

[1] Isaacs, NS. Liquid Phase High Pressure Chemistry, Wiley, Chichester, 1981.

[2] Dauben, WG., & Bunce, RA. Organic reactions at high pressure. Dipolar cycloaddition-ring contraction reactions of hindered silyl enol ethers and arylsulfonyl azides. J Org Chem 1982, 47, 5042–5044.

[3] Anderson, FT., Henry, JR., & Weinreb, SM. High-pressure induced 1,3-dipolar cycloadditions of azides with electron-deficient olefins. J Org Chem 1991, 56, 6946–6948.

[4] Brunner, M., Maas, G., & Klärner, F-G. 1,3-Dipolar cycloaddition reactions of organic azides with morpholinobuta-1,3-dienes and with an *α*-ethynyl-enamine. Helv Chim Acta 2005, 88, 1813–1825.

[5] Butler, DN., Hammond, MLA., Johnston, MR., Sun, G., Malpass, JR., Fawcett, J., & Warrener, RN. In-line proximity effects in extended 7-azanorbornanes. 1. A new concept for modifying effector group separation based on the control of *N*-invertomer geometry. Org Lett 2000, 2, 721–724.

[6] Butler, DN., Malpass, JR., Margetic, D., Russell, RA., Sun, G., & Warrener, RN. A 1,3-dipolar cycloaddition route to 7-azanorbornanes: application to the synthesis of *syn*-facial *N*-bridged polynornorbornanes. Synlett 1998, 588–589.

[7] Margetić, D., Warrener, RN., Butler, DN., & Holland, CM. Dipolar cycloaddition reactions of azomethine ylides generated by endocyclic-exocyclic 1,3-dipole rearrangement. J Het Chem 2015, 52, 1447–1456.

[8] Margetić, D., Butler, DN., & Warrener, RN. Photolysis of fused cyclobutatriazolines: an adventitious route to 3,6-Di(2-pyridyl)pyridazonorbornadiene. Aust J Chem 2000, 53, 959–963.

[9] Warrener, RN. New adventures in the synthesis of hetero-bridged *syn*-facially fused norbornadienes ("[*n*]Polynorbornadienes") and their topological diversity. Eur J Org Chem 2000, 3363–3380.

[10] Margetić, D., Butler, DN., & Warrener, RN. Current status of synthesis of deprotected Tridents by substituent variations of aza-ACE reaction, The Sixth International Electronic

Conference on Synthetic Organic Chemistry (ECSOC-6), September 1–30, 2002, MDPI, Basel, Switzerland.
[11] Warrener, RN., Margetić, D., Butler, DN., & Sun, G. Neighbouring group participation in *N*-methoxymethyl 7-azanorbornanes 1: the synthesis of *N,N'*-methano-bridged diazasesquinorbornanes, N^3-[3]polynorbornanes and CN^3-[4]polynorbornanes. Synlett 2001, 202–205.
[12] Margetić, D., Warrener, RN., Butler, DN., & Jin, C-M. A cascade thermal isomerisation of cyclobutane di-(carbomethoxy) Δ^2-1,2,3-triazolines with intramolecular 1,3-dipolar cycloreversion as the key step. Tetrahedron 2012, 68, 3306–3318.
[13] Ishikura, M., Kudo, S., Hino, A., Ohnuki, N., & Katagiri, N. A concise access to 6-azabicyclo [3.1.0]hexanes *via* high-pressure promoted cycloaddition reaction of azide to ABH. Heterocycles 2000, 53, 1449–1504.
[14] Sedishev, IP., Zharov, AA., Levina, IS., Tyurin, AYu., Volkova, YuA., Aksenov, AN., Kachala, VV., Tikhonova, TA., & Zavarzin, IV. Synthesis of condensed pregnano[17,16-*d*]triazolines under high pressure. Russ Chem Bull 2018, 67, 308–312.
[15] Melai, V., Brillante, A., & Zanirato, P. High pressure assisted 1,3-dipolar cycloadditions: formation of 1,2,3-triazoles from aryl azides and (trimethylsilyl)acetylene. J Chem Soc Perkin Trans 1998, 2, 2447–2449.
[16] Borukhova, S., Seeger, AD., Noël, T., Wang, Q., Busch, M., & Hessel, V. Pressure-accelerated azide-alkyne cycloaddition: micro capillary versus autoclave reactor performance. ChemSusChem 2015, 8, 504–512.
[17] Chen, H., Ni, B-B., Gao, F., & Ma, Y. Pressure-accelerated copper-free cycloaddition of azide and alkyne groups pre-organized in the crystalline state at room temperature. Green Chem 2012, 14, 2703–2705.
[18] Ni, B-B., Wang, K., Yan, Q., Chen, H., Ma, Y., & Zou, B. Pressure accelerated 1,3-dipolar cycloaddition of azide and alkyne groups in crystals. Chem Commun 2013, 49, 10130–10132.
[19] Zavarzin, IV., Zhulin, VM., Yarovenko, VN., & Krayushkin, MM. Synthesis of 5-ketotetrazoles under high pressure. Russ Chem Bull 1988, 1027–1029.
[20] Yarovenko, VN., Es'kov, AA., Konrashev, PA., Ignatenko, AV., Zavarzin, IV., Vorontsova, LG., Sedishev, IP., Krayushkin, MM., & Starikova, ZA. Synthesis and reactivity of *N*-vinyl-1,2,3-dithiazolimines. Heterocycles 2005, 65, 1601–1608.
[21] Zhu, J., Lines, BM., Ganton, MD., Kerr, MA., & Workentin, MS. Efficient synthesis of isoxazolidine-tethered monolayer-protected gold nanoparticles (MPGNs) via 1,3-dipolar cycloadditions under high-pressure conditions. J Org Chem 2008, 73, 1099–1105.
[22] Isaacs, NS., & Rannala, E. Reaction studies at high pressure. Part I. activation volumes of some [2+2] and dipolar cycloadditions. J Chem Soc Perkin Trans 1975, 2, 1555–1560.
[23] de Suray, H., Leroy, G., & Weiler, J. Effet de la haute pression sur les additions dipolaires 1,3-du diazomethane– Tetrahedron Lett 1974, 25, 2209–2210.
[24] Kamernitskii, AV., Galakhova, TN., Levina, IS., & El'yanov, BS. Cycloaddition of diazoacetic ester to 16-dehydropregnenolone acetate. Russ Chem Bull 1977, 2207–2209.
[25] Kamernitzky, AV., Levina, IS., Mortikova, EI., Shitkin, VM., & El'yanov, BS. Transformed steroids – part 95. High pressure induced 1,3-dipolar cycloadditions of nitronic esters to 16-dehydro-20-oxo steroids. Tetrahedron 1977, 33, 2135–2139.
[26] Warrener, RN., Ferreira, ABB., & Tiekink, ERT. The preparation of rigidly-linked 4,5-diazafluorenes: new molrac bidentate ligand systems. Tetrahedron Lett 1996, 37, 2161–2164.
[27] Warrener, RN., Ferreira, ABB., Schultz, AC., Butler, DN., Keene, RF., & Kelso, LS. Space-Separated 1,l0-phenanthroline, 4,5-diazafluorene, or 3,6-Di(2-pyridyl)pyridazine units as

ligands in diruthenium complexes: preliminary studies of metal-metal interactions. Angew Chem Int Ed Engl 1996, 35, 2485–2487.

[28] Dicken, CM., & DeShong, P. Reactions at high pressures. [3+2] dipolar cycloaddition of nitrones with vinyl ethers. J Org Chem 1982, 47, 2047–2051.

[29] DeShong, P., Li, W., Kennington, Jr J., Ammon, HL., & Leginus, JM. A nitrone-based cycloaddition approach to the synthesis of the glycosyl system of nogalomycin, menogaril, and their congeners. J Org Chem 1991, 56, 1364–1373.

[30] Seerden, J-PG., Boeren, MMM., & Scheeren, HW. 1,3-Dipolar cycloaddition reactions of nitrones with alkyl vinyl ethers catalyzed by chiral oxazaborolidines. Tetrahedron 1997, 53, 11843–11852.

[31] Blanáriková-Hlobilová, I., Kubánová, Z., Fišera, L., Cyranski, MK., Salanski, P., Jurczak, J., & Prónayová, N. Stereoselectivity of 1,3-dipolar cycloadditions of L-valine-derived nitrones with methyl acrylate. Tetrahedron 2003, 59, 3333–3339.

[32] Hermkens, PHH., V Maarseveen, JH., Kruse, CG., & Scheeren, HW. 1,3-Dipolar cycloaddition of nitrones with nitriles. Scope and mechanistic study. Tetrahedron 1988, 44, 6491–6504.

[33] Tabolin, AA., Novikov, RA., Khomutova, YuA., Zharov, AA., Galina, A., Stashina, GA., Nelyubina, YuV., Tomilova, YuV., & Ioffe, SL. Formal [3+3]-cycloaddition of 3-methyl-5,6-dihydro-*4H*-1,2-oxazine-*N*-oxides with cyclopropane dicarboxylates under hyperbaric conditions. Tetrahedron Lett 2015, 56, 2102–2105.

[34] Krompiec, S., Filapek, M., Grudzka-Flak, I., Slodek, A., Kula, S., Malecki, JG., Malarz, J., Szafraniec-Gorol, G., Penkala, M., Schab-Balcerzak, E., Paluch, M., Mierzwa, M., Matussek, M., Szlapa, A., Pajak, M., Blach, D., Marcol, B., Danikiewicz, W., Boharewicz, B., & Iwan, A. Multifaceted strategy for the synthesis of diverse 2,2′-bithiophene derivatives. Molecules 2015, 20, 4565–4593.

4 C–C bonds

The importance and the scope of the C–C bond-forming reactions are reflected in large number of publications featuring hyperbaric conditions. The aldol and Knoevenagel condensations as well as Michael and Morita–Baylis–Hillman reactions could be emphasized. A number of the C–C bond-forming reactions were performed on the aromatic compounds as substrates involving Friedel–Crafts reaction, aromatic substitutions and palladium-catalyzed coupling reactions (Suzuki and Heck).

4.1 Aldol condensation

Robinson annulation method, which combines Michael and aldol reactions, was also accelerated under high-pressure conditions. Dauben has demonstrated that the outcome of the reaction could be controlled by selecting the temperature (Scheme 4.1) [1]. When 2-carbomethoxycyclohexanone **1**, a six-membered ring ketoester, was pressurized at 15 kbar at 20 °C with methyl vinyl ketone **2**, a simple Michael adduct, 2-carbomethoxy-2-(3-oxobutyl)cyclohexanone **3**, was obtained as the sole product in the excellent yield. With the temperature increased to 35–40 °C, the bicyclic ketol **4** was produced. Finally, a fused enone (octalone) **5** was formed at 60–70 °C. By increasing the temperature of the reaction mixture of the Michael adduct **3**, it could be interconverted to the bicyclic ketol **4**. Analogously, the fused enone **5** could be produced from the bicyclic ketol **4**.

Employing these mild reaction conditions (35–40 °C), various enone Michael acceptors were converted to bicyclic ketols **4**. All reactions are carried out in

https://doi.org/10.1515/9783110556841-004

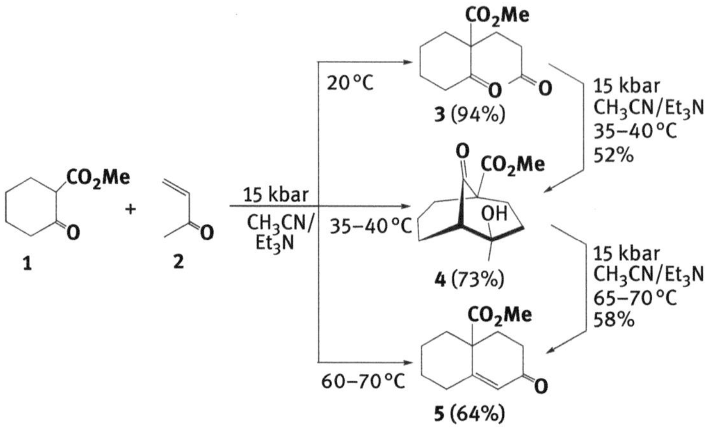

Scheme 4.1: Aldol reaction of 2-carbomethoxycyclohexanone with methyl vinyl ketone.

acetonitrile containing an excess of nonnucleophilic base as a cosolvent. Triethyl amine was strong enough, whereas for the less reactive substrates, 1,5-diazabicyclo [4.3.0]non-5-ene proved to be a superior base. At ambient pressure, condensations of 2-carbomethoxy cycloalkanones stopped after the initial Michael addition and subsequent aldol condensation were not observed.

The reaction of the ketolactam **6** with a stabilized Wittig reagent methyl (triphenylphosphoranylidene)acetate **7** was facilitated by high pressure, providing an unexpected product, tetracycline aminolactame **8** (Scheme 4.2) [2]. In the process, the pyrido-benzazonine nucleus of the substrate **6** was rearranged to quinolinone structure, and both nitrogen were methylated. The cyclic product **8** exhibits the backbone of the [ABCD] ring system of *Melodinus* alkaloids such as meloscine. This transformation is thought to originate from the transannular aldol reaction of the

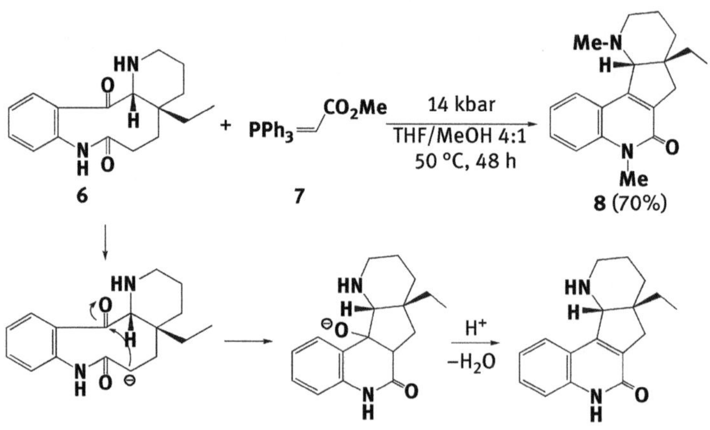

Scheme 4.2: Aldol reaction of ketolactam with methyl (triphenylphosphoranylidene) acetate.

nine-membered ketolactam (Camps reaction). Methylation of nitrogen was proposed to occur from **7** generating active alkoxytriphenylphosphonium salts species.

L-Proline was used by Kotsuki as a chiral catalyst for the asymmetric aldol reaction of ketones with aldehydes. The optimal high-pressure conditions encompass 2 kbar and ambient temperature (Scheme 4.3) [3]. In model system, the reaction of acetone with benzaldehyde at ambient pressure afforded aldol product in 62% yield (and 60% ee). Pressure elevation to 2 kbar increased the reaction yield to 90% and ee to 72%; however, reactions at 5 and 8 kbar showed deterioration of ee values (to 64% and 65%, respectively). This behavior shows that discrimination of the delicate interactions in transition states where benzaldehyde approaches by the *re-* or *syn-* face is overruled when pressure exceeds 2 kbar.

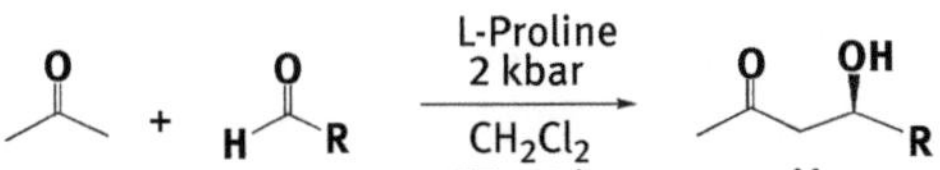

9 10 11

R = Ar, alkyl 62–95%, 57–92% ee

Scheme 4.3: Aldol reaction of ketones with aldehydes.

Mukaiyama aldol reaction is usually carried out that is catalyzed by Lewis acids, but Yamamoto has shown that reaction of silyl enol ether **12** with aldehydes could be carried out in neutral conditions (Scheme 4.4) [4]. Very clean conversion to products **14** was obtained, and the stereoselectivity was reversed in comparison with $TiCl_4$-catalyzed Mukaiyama reaction. At 10 kbar, for instance, the *erythro/threo* ratio for the reaction of cyclohexenyl silyl enol ether with benzaldehyde was 75/25, whereas $TiCl_4$-catalyzed reaction provided 25/75 *erythro/threo* ratio. The *erythro* selectivity was rationalized by a boat transition state being favored at high pressure and differences of $\Delta V^{\neq}$ for boat and chair TSs.

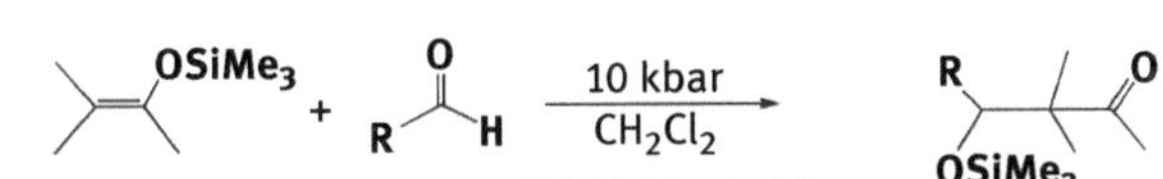

12 13 14 (20–90%)

R^1, R^2 = H, Me

R^3 = Et, Me

erythro/threo 11/89–75/25

Scheme 4.4: Aldol reaction of silyl enol ether with aldehydes.

By using a high-pressure technique, the neutral aldol condensation of aldehydes and ketones takes place with O-silylated ketene acetals **114** (Scheme 4.5) [5]. High yields of products were obtained, but with low stereoselectivity (*threo/erythro* ratio).

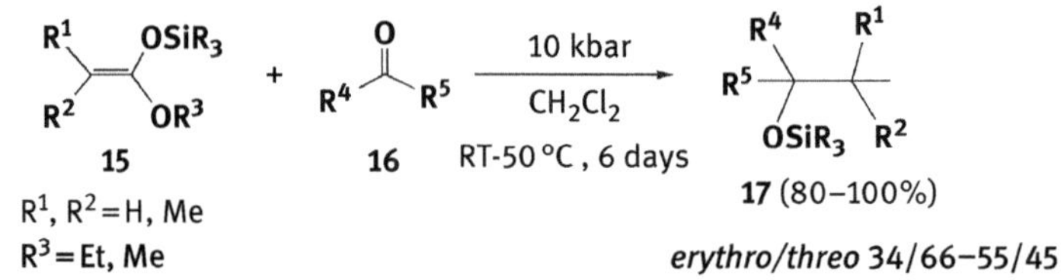

Scheme 4.5: Aldol reaction of O-silylated ketene acetals with aldehydes and ketones.

Mukaiyama reaction of bis-silyl ketene acetals with aromatic aldehydes was studied by Dumas. The reaction of 1,1-bis(trimethylsiloxy)-l-propene **18** with benzaldehyde to yield bis-silyl aldols **20** showed moderate diastereoselectivity, regardless of solvent, pressure and temperature (Scheme 4.6) [6]. Diastereoselectivity was found to be sensitive to the steric bulk of the substituent in bis-silyl ketene acetals **20**: the replacement of methyl group by isopropyl led to the increase of the *syn/anti* ratio to 84:16 (85% yield, 8 kbar, 64 °C, 3 days). The preference for the formation of the *syn*-aldols **20** was attributed to transition states structures with minimal steric interactions between the reactants. Aldol condensations of unsaturated bis-silyl ketene acetal **21** showed the correlation of the regioselectivity on pressure. At lower pressure, the *γ*-adduct **22** is dominant product, whereas with the increase in temperature, *α*-adduct **23** is major one. These results suggest that the linear adduct **22** has a less compact transition state in comparison to the one leading to the formation of branched product.

Henry or nitroaldol reaction was carried out under high-pressure conditions by Matsumoto. It was found that the reaction of cyclohexanones with nitroalkanes is

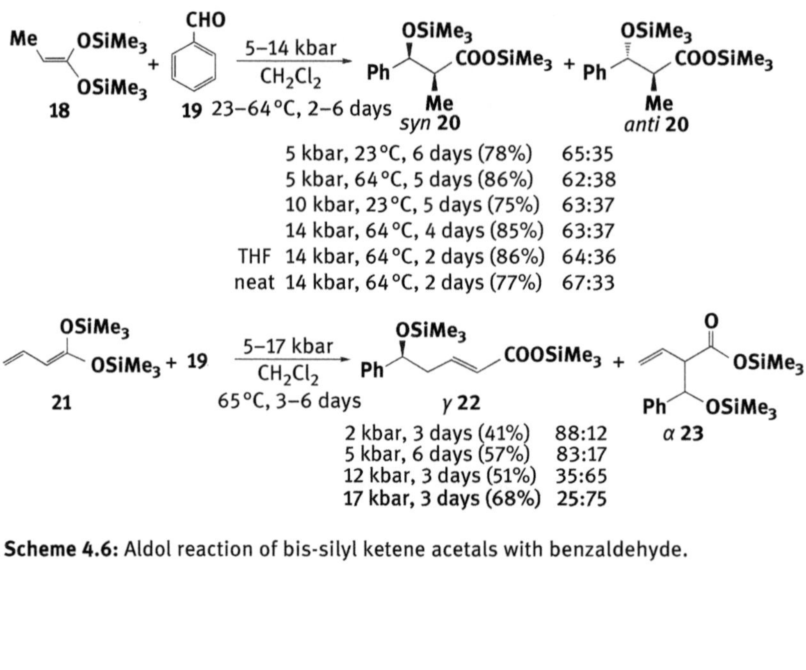

Scheme 4.6: Aldol reaction of bis-silyl ketene acetals with benzaldehyde.

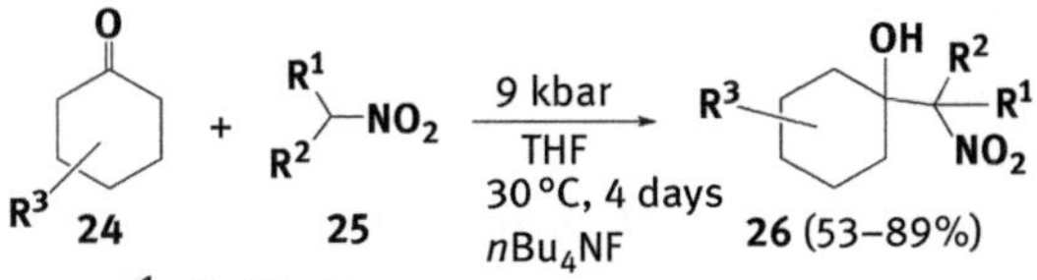

24 25 26 (53–89%)

R^1 = H, Me, Et
R^2 = H, Me
R^3 = H, Me

Scheme 4.7: Henry reaction of cyclohexanones with nitroalkanes.

promoted by pressure (Scheme 4.7) [7]. High yields of nitroalcohol adducts **26** were obtained, and reactions yields are much higher than that at room pressure. Notably, the reaction of 2-methylcyclohexanone with nitropropane proceeds in 40% yield, whereas at ambient temperature the reaction does not proceed at all.

Significant kinetic effect of pressure on the Henry reaction was noted by Jenner in nitroaldol condensations of acetone, cyclopentanone and 3-pentanone with nitromethane or nitroethane (Scheme 4.8) [8]. In neat ketones used as a solvent, a large negative activation volumes ranging from −20.5 to −25.2 cm^3 mol^{-1} were determined.

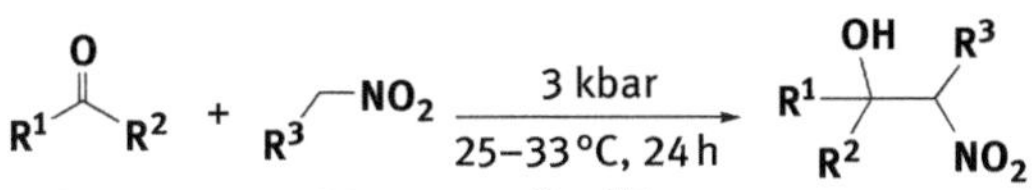

27 28 29 (14–100%)

R^3 = H, Me

Scheme 4.8: Henry reaction of ketones with nitromethane and nitrorethane.

Diastereoselectivity of nitro-aldol reactions could be improved by applying pressure. Matsumoto has shown that the addition of nitromethane to benzaldehyde in the presence of quinidine (3 mol%) proceeds in much better yields and ee's (Scheme 4.9) [9]. The optimal reaction conditions are established at 2 kbar, whereas further pressure increase causes a sharp drop in ee's, with accompanying improvement of the reaction yield. This behavior is associated with the sterically more congested transition state leading to *R*-enantiomer, which is more affected by pressure.

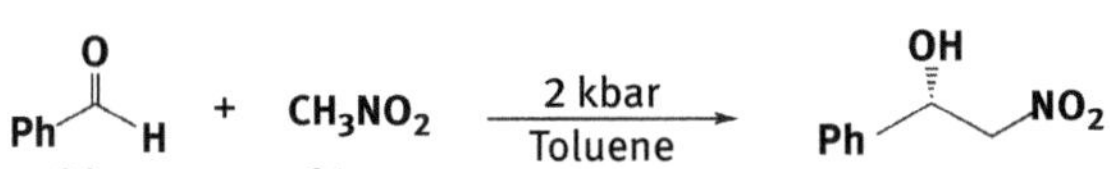

30 31 *S*-**32** (9%, 35% ee)

1 bar (4%, 18% ee)
3 kbar (38%, 6% ee)
5 kbar (66%, 4% ee)
7 kbar (80%, 3% ee)
2 h 7 kbar (37%, 11% ee)

Scheme 4.9: Diastereoselectivity of nitroaldol reaction of benzaldehyde with nitromethane.

The asymmetric nitroaldol reactions of nitromethane with acetophenone or 2,2,2-trifluoroacetophenone provided ee's below 1%, showing that there is no positive pressure effect on the diastereoselectivity.

By employing *N,N*-dibenzyl *α*-amino aldehydes **33** as substrates in nitro-aldol reaction with nitroalkanes, 3-amino-2-hydroxy acids **35** were obtained in a diastereoselective manner without using a catalyst (Scheme 4.10) [5]. Matsumoto noted that solvents influence the outcome of this asymmetric Henry reaction, and acetonitrile being the most suitable one, whereas reactions performed in dichloromethane and toluene do not give any product. High yields, ee's and the (2*R*,3*S*)-**35**/(2*S*,3*S*)-*epi*-**35** ratio varying from 71:29 to 89:11 were obtained, whereas at atmospheric pressure **33** (R=Ph) did not react with nitromethane.

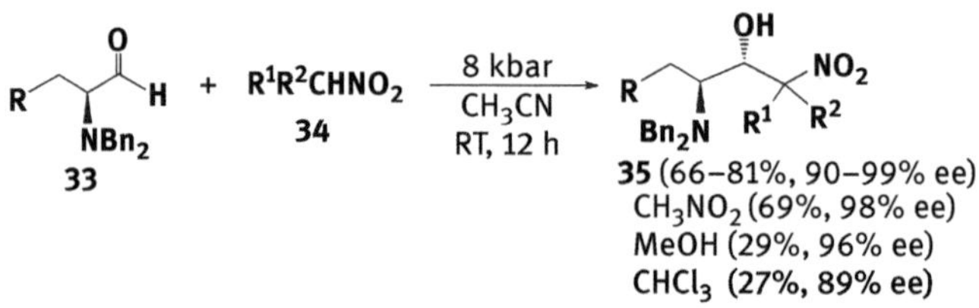

Scheme 4.10: Diastereoselective nitro-aldol reaction of aldehydes with nitroalkanes.

For Henry reaction between aromatic aldehydes and nitromethane, various effects of pressure were observed, depending on the electronic nature of substituent on the aromatic ring [10]. At 1 bar, yields are generally higher for aldehydes with strong acceptor substituents, while at 10 kbar yields increase for donor aldehydes. On the other hand, the yields for acceptor aldehydes are lower at 10 kbar.

4.2 Knoevenagel reaction

Pressure affects the Knoevenagel reaction in several ways: by assisting the formation of anion in the initial step, by favoring the formation of anionic intermediate adduct in the following step and by retardation of the dehydration process in the last step (as E1cB type eliminations are decelerated by pressure).

Newitt et al. reported that yields of Knoevenagel condensation of cyclohexanone with ethyl cyanoacetate catalyzed by piperidine were significantly improved by pressure (up to 72% yield) [11]. Without piperidine catalyst, condensation does not take place and pressures as high as 5 kbar were used in conjunction to heating at 60 °C to obtain ethylcyclohexylcyanoacetate in 11–26% yield.

For the quantification of rate acceleration by pressure of Knoevenagel condensation between ketones and ethyl cyanoacetate, Jenner used the β value (ratio of yields obtained at 1bar and at 3 kbar), which was calculated to vary from 1 to >11 (Scheme 4.11) [12]. The highest increase in yield was obtained for the sterically demanding substrates, which is useful for synthetic purpose to provide products that could not be synthesized at atmospheric conditions. At the same time, the isomer ratio remained identical as at atmospheric pressure. The yield of reactions with hindered ketones could be further improved by the pressure elevation up to 9 kbar at 50–65 °C. In the follow-up publication, Jenner studied some sterically problematic ketones that do not undergo Knoevenagel condensation at atmospheric pressure, even 3 kbar was not enough to initiate reaction, and reacted only at 8.5 kbar (R,R′=Et, *i*Pr, Pr, 33–52%) [13].

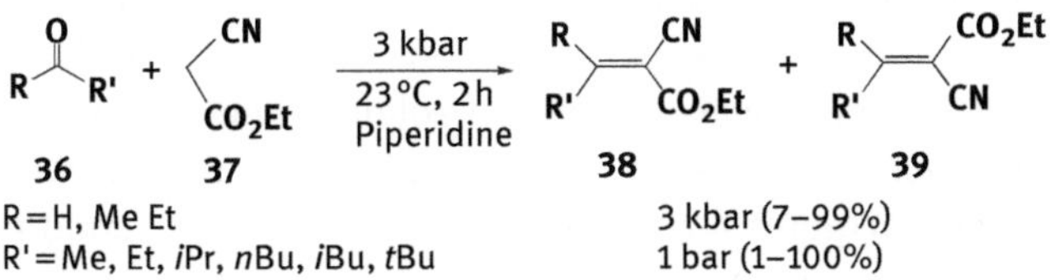

Scheme 4.11: The Knoevenagel reaction of ketones with ethyl cyanoacetate.

4.3 Cope rearrangement

Klärner's kinetic study of the Cope rearrangement of 1,3,5-hexatriene **40** to 1,3-cyclohexadiene **41** established the negative activation volume $\Delta V^{\neq} = -9.8\ cm^3\ mol^{-1}$, (101 °C, toluene) and the acceleration of the process under pressure (Scheme 4.12) [14]. The negative activation volumes obtained for the Cope rearrangement were rationalized with the larger packing coefficients of the pericyclic transition structure due to their cyclic geometry than that of the acyclic ground-state structures of reactants. The activation volumes of the Cope rearrangement of *meso*-**42** to *E,Z*-**43** and *E,E*-**43** were found to be $\Delta V^{\neq} = -13.3$ and $-8.8\ cm^3\ mol^{-1}$, respectively (127 °C, toluene). At ambient pressure, the ratio of *E,Z*-**43** and *E,E*-**43** was 60.5:39.5, while the pressure

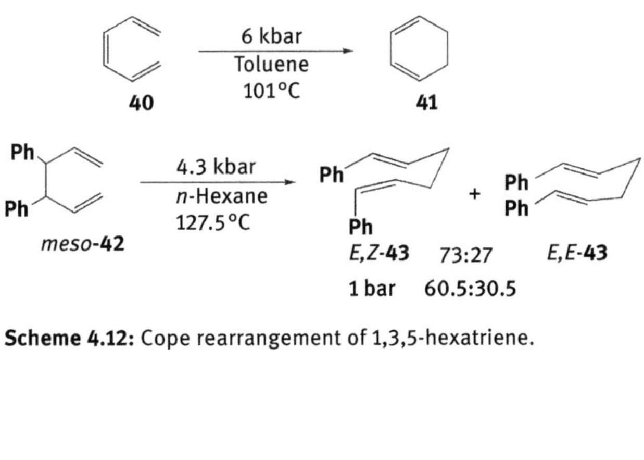

Scheme 4.12: Cope rearrangement of 1,3,5-hexatriene.

increase accelerated both reactions as well as the selectivity in favor of *E,Z*-**6** [15]. Higher steric hindrance in the chair-like transition state having one phenyl group in the axial position could be associated with the observed $\Delta\Delta V^{\neq}$ difference of two competing processes.

The elucidation of the reaction mechanism of the Cope rearrangement (concerted pericyclic vs diradical process) could be supported by the determination of the volumes of activation of the two processes. The $\Delta V^{\neq}$ of cyclic transition states are smaller than those of corresponding acyclic transition states. Having a negative $\Delta V^{\neq}$ of cyclic TSs, these reactions are accelerated by the increase in pressure. A positive $\Delta V^{\neq}$ and a reaction rate deceleration upon an increase in pressure are characteristic of a homolytic mechanism. Accordingly, model test system *cis*-1,2-divinylcyclobutane **44** rearranges to cycloocta-1,5-diene via the concerted pericyclic Cope mechanism, with the rate acceleration by a factor of 4.5 and negative activation volume (–13.4 cm^3 mol^{-1}) (Scheme 4.13) [16]. On the other hand, because of the geometrical constraints, *trans*-1,2-divinylcyclobutane **44** initially requires homolytic cleavage of the C_1–C_2 bond to rearrange to **45**. This process is retarded by pressure and possesses a positive $\Delta V^{\neq}$ (+ 4.2 cm^3 mol^{-1}), indicating consistency with a diradical mechanism.

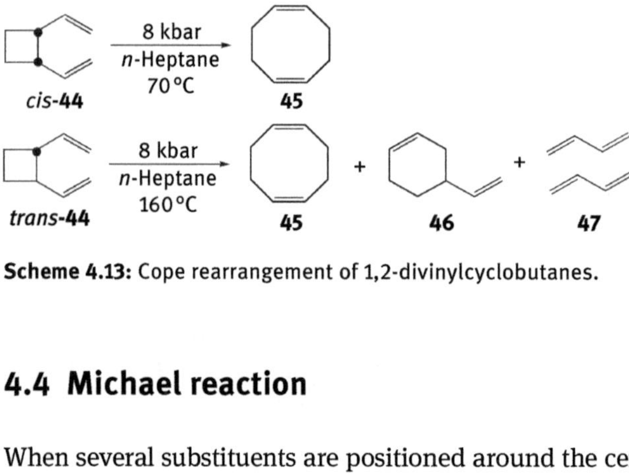

Scheme 4.13: Cope rearrangement of 1,2-divinylcyclobutanes.

4.4 Michael reaction

When several substituents are positioned around the center of the Michael donor, deprotonation and addition reaction steps are hindered. Sterically congested Michael condensations were successfully achieved by employing high pressure and aprotic reaction conditions.

In the pioneering study of Matsumoto, several reactions that were not successful at room pressure were promoted at 10–15 kbar within 42–95 h in the presence of triethyl amine. Michael products were obtained in high yields (42–95%). Notable examples shown are the reactions of Michael acceptors methyl-2-butenoate **48** and 1,3-diphenylprop-2-en-1-one (chalcone) **51** with 1-nitropropane **49** and diethyl ethyl malonate **52** acting as Michael donors (Scheme 4.14) [17]. Difficulties in the deprotonation step could be circumvented by applying nBu_4NF catalyst in THF at 10 kbar

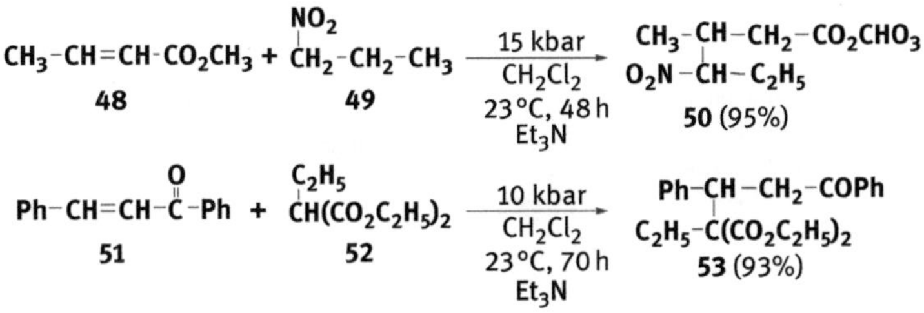

Scheme 4.14: Michael reactions of methyl-2-butenoate and 1,3-diphenylprop-2-en-1-one with 1-nitropropane and diethyl ethyl malonate.

[18]. These results that show that the pressure promotes the Michael reaction are supported by experimental determination of activation volumes by Jenner [8]. Determined $\Delta V^{\neq}$ for Michael reactions are highly negative and depend on the solvent and catalyst. For reaction involving nitromethane and methyl vinyl ketone $\Delta V^{\neq}$ is $-31.5\ cm^3\ mol^{-1}$ (H_2O, no catalyst), $-46.5\ cm^3 mol^{-1}$ (chloroform, $Eu(fod)_3$) and $-19.7\ cm^3\ mol^{-1}$ (THF, Bu_4NF).

Sterically demanding Michael enone acceptors **54** at 15 kbar reacted efficiently with activated Michael donors diethylmalonate, diethyl methylmalonate and ethyl 2-methylacetoacetate **55** using 1,5-diazabicyclo[4.3.0]non-5-ene as the base (Scheme 4.15) [19]. At ambient pressure, in refluxing acetonitrile reaction provided little or no formation of Michael adducts, demonstrating the requirement of mild conditions under high pressure in aprotic solvent.

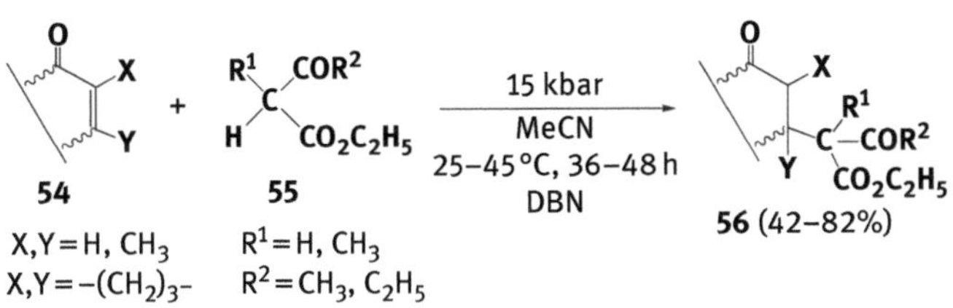

Scheme 4.15: Michael reactions of enones with ethyl 2-methylacetoacetate.

Michael addition of nitromethane on the 4-en-3-one system of testosterone **58** is difficult to achieve at atmospheric pressure due to steric hindrance and occurs with poor yields. However, the addition is achievable at high pressures using the TBAF or DBU (Scheme 4.16) [20]. Depending on the nitromethane reagent ratio, mixtures contains variable ratios of the 1:1 **59** and 2:1 adducts **60** with 5α-nitromethyl and 3β,5α-bis(nitromethyl) configurations. Adducts **60** arose from the initial Michael addition, followed by nitro-aldol reaction.

Kotsuki has shown that $Yb(OTf)_3$ could also be used to catalyze high-pressure Michael addition of β-ketoesters with a variety of α,β-unsaturated carbonyl compounds.

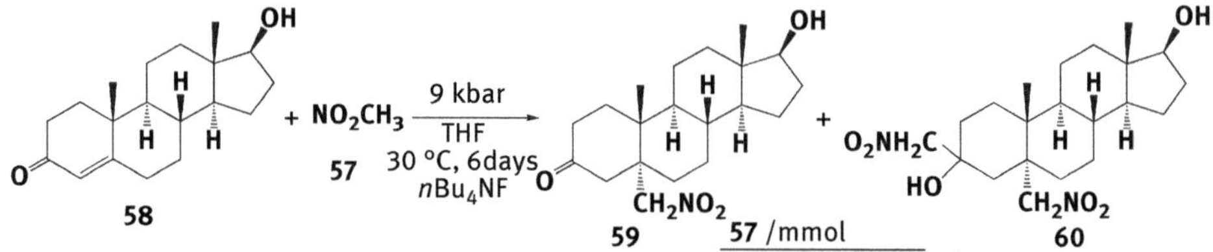

57 /mmol	59 : 60	
50	8 : 5	- 1 bar
50	5 : 79	
10	22 : 36	
5	25 : 15	
30	4 : 57	- DBU/MeCN
7	38 : 7	DBU/MeCN

Scheme 4.16: Michael addition of nitromethane with testosterone.

Optimal high-pressure reaction conditions were established using the reaction of β-ketoester **61** with 2-cyclopentenone **63** (Scheme 4.17) [21]. The addition of small amount of water was beneficial to higher yields, while atmospheric reaction did not afford product **63**. Ytterbium trifluoromethanesulfonate reactions (39–89% yield) were compared to room pressure reactions on silica gel under solvent-free conditions. In the majority of cases presented in the study, atmospheric pressure reactions on silica gel provided Michael adducts in higher yields. However, there are examples where high-pressure conditions were better yielding, such as the reaction of **61** with ethyl acrylate providing 81% of Michael adduct at 8 kbar, whereas in atmospheric conditions employing silica gel adduct was not found.

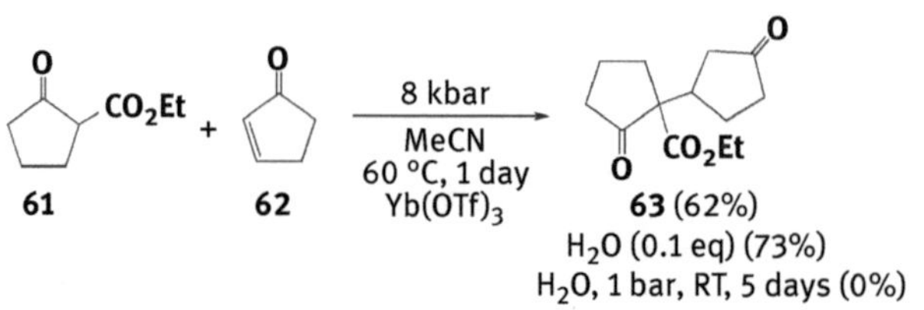

Scheme 4.17: Michael reaction of β-ketoesters with an α,β-unsaturated carbonyl compounds.

Significant efforts were made to develop the asymmetric Michael reactions. The addition of methyl oxoindancarboxylate **64** to methyl vinyl ketone, catalyzed with (–)-quinine (QN) yielded (S)-**66** in high yields; however, applying pressure decreased the ee's in all catalyzed addition conditions applied (Scheme 4.18) [22]. This may be a result of the positive influence of pressure on the less favorable transition state (leading to (R)-isomer) where the pro-R center of the anion facing to the protonated QN suffers severe steric interactions. Matsumoto also observed the decrease of ee's in the Michael reaction of chalcone **68** with nitromethane catalyzed by

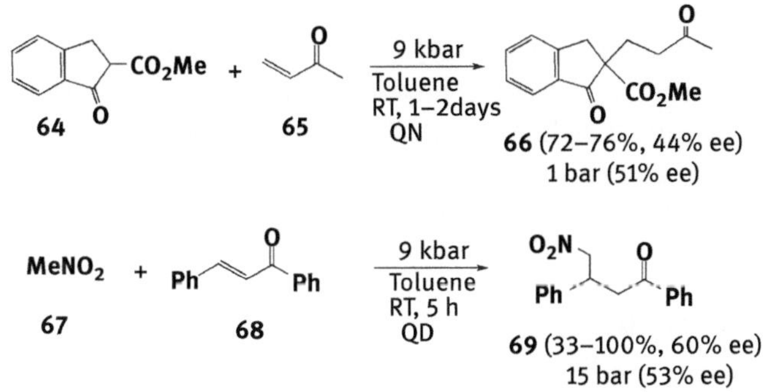

Scheme 4.18: Asymmetric Michael reactions catalyzed by (–)-quinine and (+)-quinidine.

(+)-quinidine (QD), affording 1,3-diphenyl-4-nitrobutan-1-one **69**, when going from 9 to 15 kbar. The partial solidification of toluene or partial crystallization of the catalyst was proposed to cause this decrease in ee's. This reaction does not proceed at atmospheric pressure at all.

In variance, applying high pressure has a significant effect on the reaction rate acceleration, with enantioselectivity kept on a very high level in the Michael addition of nitromethane to 3-methylcyclohexenone **70** (Scheme 4.19) [23]. Several chiral catalyst amine derivatives of cinchona alkaloids were successfully applied, and model work with 9-amino-9-deoxy-*epi*-cinchonine **72** catalyst produced *R*-**71** with excellent stereoselectivity. These optimized conditions were applied to a number of cyclic and acyclic enones with high yields and usually >95% ee. Kwiatkowski also successfully employed the chiral thiourea derivatives of cinchona alkaloids as catalysts to obtain highly stereoselective high-pressure (8–10 kbar) addition of nitromethane to sterically congested *β*,*β*-disubstituted enones [24].

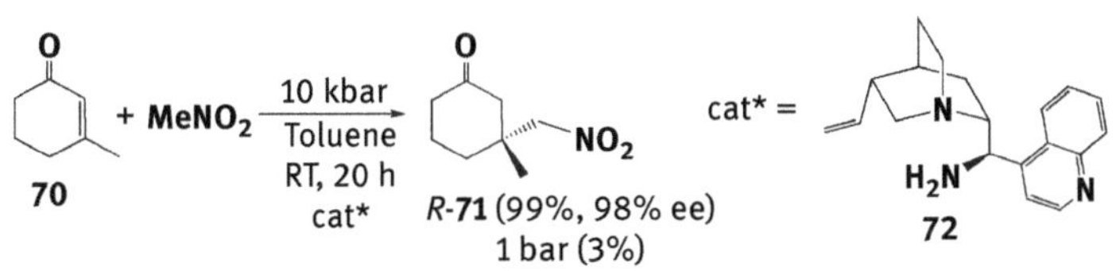

Scheme 4.19: Stereoselective Michael addition of nitromethane to 3-methylcyclohexenone.

Similarly, a significant increase in yields was obtained by pressure increase from ambient pressure to 8 kbar (while the enantioselectivity was maintained) in Michael reaction of 4,4-disubstituted cyclohexadienones with diethyl malonate employing chiral thiourea catalyst and 4-pyrrolidinopyridine as an additive [25]. Same effect of pressure (yields and high enantiomeric excess) was reported for Michael reaction of *α*-substituted cyclic ketones with acrylates using chiral amine catalysts at 10 kbar [26].

Chiral induction in the asymmetric Michael addition to alkyl and aryl crotonates was also achieved by using chiral imines [27]. The imines acting as Michael donors possess a chiral auxiliary group ((*R*)-1-phenylethylamine), which could be removed at the later stage of synthesis (Scheme 4.20) [28]. For example, the reaction of (*R*)-**73** with methylcrotonate at 12 kbar provided adduct (3*S*,1′*S*,1′*R*)-**75**, which was hydrolyzed to ketoester (3*S*,1′*S*)-**76** with a high yield and enantioselectivity.

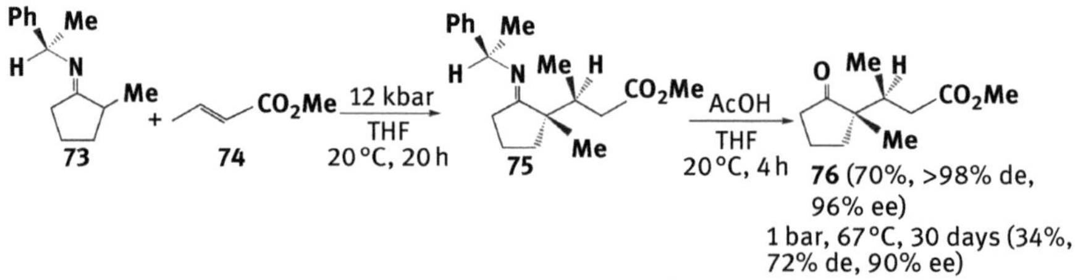

Scheme 4.20: Chiral induction of Michael addition to alkyl and aryl crotonates.

The alkylation of indoles with cyclopropane-1,1-dicarboxylic acid esters such as **78** proceeds smoothly via homo-Michael type of addition to afford 4-indolyl carbonyl compounds **79** (Scheme 4.21) [29]. The nucleophilic ring opening of an activated cyclopropane **78** in hyperbaric conditions with the aid of $Yb(OTf)_3$ (5 mol%) was optimal in acetonitrile, whereas the yields in other solvents were low or even suppressed the reaction. At atmospheric pressure, minute amounts of product **79** were obtained (3%). Similarly, Kotsuki employed cyclopropane **78** and $Yb(OTf)_3$ catalyst to homo-Michael high-pressure addition to *β*-ketoesters [30].

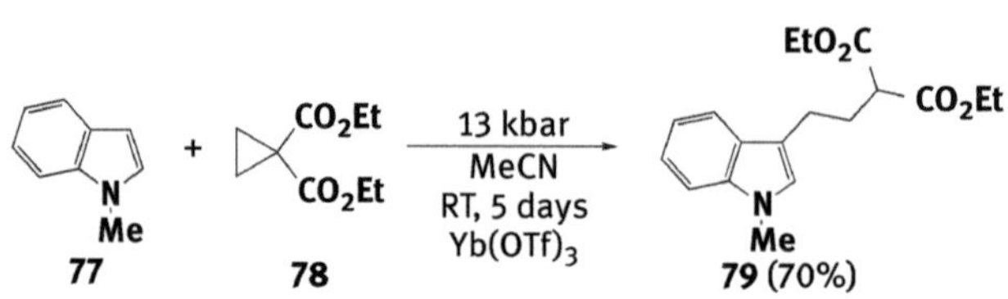

Scheme 4.21: Homo-Michael addition of cyclopropane-1,1-diester to indole.

The alkylations of indoles with *α*,*β*-unsaturated ketones catalyzed by $Yb(OTf)_3$ were studied by Harrington toward synthesis of alkaloid hapalindole C. The difficulties in obtaining reaction and acceptable yields were overcome by applying hyperbaric conditions. At 13 kbar pressure, a notable reduction in reaction time and an increase in the yields of adducts were achieved, especially with sterically congested Michael acceptors. For instance, the reaction of methylindole with 5-methyl-3-

hexen-2-one **80** as the Michael acceptor afforded **81** in high yield (Scheme 4.22) [31]. Other electrophiles such as *β*-nitrostyrene, diethyl benzalmalonate and methyl acrylate reacted with indole **77** to afford the respective products in 8–89% yields.

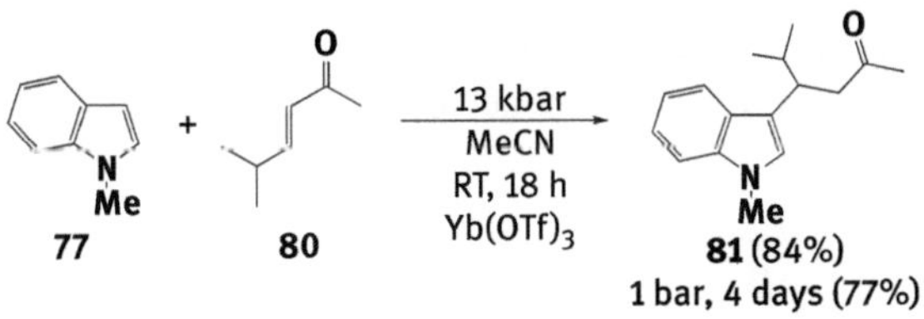

Scheme 4.22: Alkylations of indoles with *α*,*β*-unsaturated ketones.

4.5 Conjugate additions

Heathcock and Dauben reported that elevated pressures (15 kbar) do provide an alternative means to induce ketene acetal addition to sensitive activated enones with steric and conformational constraints. In such conditions, the use of Lewis acids could be avoided. This is demonstrated by applying conjugate additions of O-silylated ketene acetals to activated enones in the synthesis of the acid-labile precursor **84** of antitumor diterpenoid bruceantin (Scheme 4.23) [32]. The 1-methoxy and 1-*t*-butoxy-1-(*t*-butyl-dimethylsiloxy)ethylene enolsilanes **83** in reaction with **82** provided adducts **84** in high yield, and diastereoselectivity in the case of more sterically demanding enolsilane. Lewis catalysis with $TiCl_4$ at ambient pressure resulted in the decomposition of the substrate **82.**

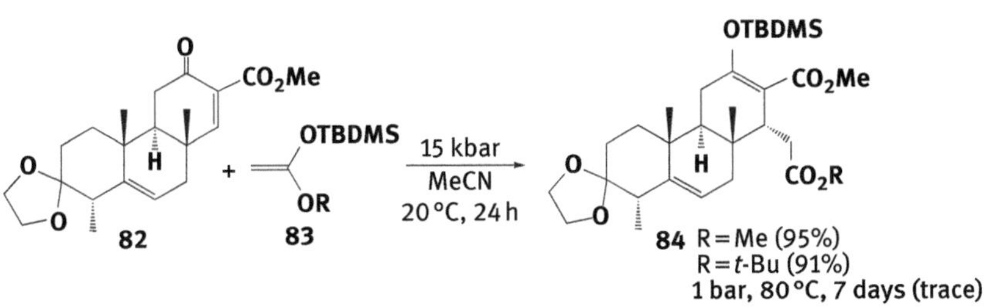

Scheme 4.23: Conjugate addition of O-silylated ketene acetals to activated enone.

The Stetter reaction is an important C–C-forming reaction by adding an *α*,*β*-unsaturated compound. Depending on the reaction conditions, the reaction of an aldehyde (acyl anion precursor) with Michael acceptor utilizing a nucleophilic catalyst

(3-benzyl-5-(2-hydroxyethyl)-4-methylthiazolium chloride **89**) could lead to the *γ*-ketonitriles (Stetter product) or acyloin products (Scheme 4.24) [33]. High pressure had a profound effect on the reaction acceleration, as well as on the dominant product in the case of sterically more crowded substrates, whereas reactions of unhindered substrates are unaffected. In the case of acrylonitrile **86** and isobutyl aldehyde, at 3 kbar the *γ*-ketonitrile product **87** dominates, whereas the reaction of isopropyl aldehyde with crotononitrile favored acyloin product **88**.

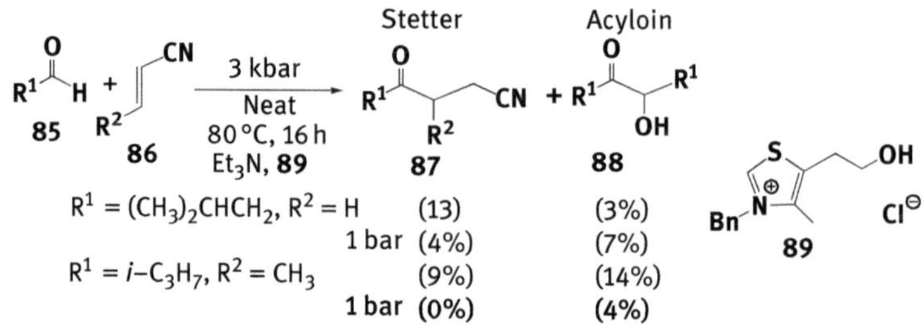

Scheme 4.24: The Stetter reaction of aldehyde with acrylonitrile.

4.6 Morita–Baylis–Hillman reaction

Because of highly negative volumes of activation ($\Delta V^{\neq}$ from −50 to −80 cm^3 mol^{-1}, acetone), Morita–Baylis–Hillman reactions are amply facilitated by pressure [34]. This ionogenic reaction proceeds by the formation of anionic species and the magnitude of $\Delta V^{\neq}$ reflects a multistep process and significant charge buildup during the activation process, which is accompanied by substantial electrostriction.

The Morita–Baylis–Hillman reaction is initiated by nucleophilic attack of bases to activated *α*,*β*-unsaturated vinyl system; tertiary 1,4-diazabicyclo[2.2.2]octane (DABCO) is often employed and the stereocontrol could be achieved by pressure. For instance, the pressure effects on the *E*/*Z*-ratio by the addition of crotononitrile (mixture of *cis*/*trans*-isomers) **90** to benzaldehyde **91** were studied by Scheeren. When reaction was carried out in aprotic solvents, the preference for the *E*-isomer increases with pressure, whereas in methanol having the smallest electrostriction, very small effect on the *E*/*Z*-ratio was found (Scheme 4.25) [35]. Triethylamine and series of tertirary alkyl

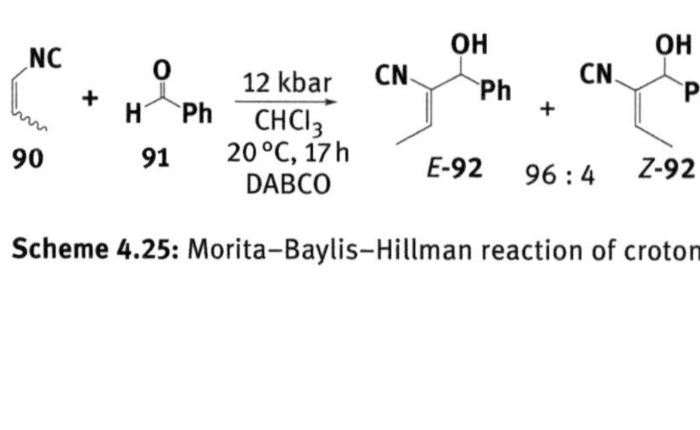

Scheme 4.25: Morita–Baylis–Hillman reaction of crotononitrile with benzaldehyde.

amines also catalyze the reaction of **90** with **91**, while the synthetic azacrown receptor functionalized with chiral *β*-amino alcohols showed good conversion but without enantioselectivity (15 kbar, 19h, 50 °C, $CDCl_3$, yield 95%, ee <5%) [36].

The solvent effects on the Baylis-Hillman reactions of aldehydes or ketones and acrylic compounds showed important pressure acceleration (Scheme 4.26) [37]. In water, better results were obtained than that in methanol (reactions in methanol and water work only for certain reagents), but the best yields (quantitative) were obtained in neat conditions (DABCO, 3 kbar, 20 °C 24 h). At atmospheric pressure, many of the reactions do not proceed. Interestingly, the addition of acrylonitrile to cyclohexanone at 3 kbar in methanol (at 20 °C) provided much better yield (12%) when the hydroxylated analog of DABCO a 3-quinuclidinol was applied instead of DABCO (8% yield). Under high-pressure conditions, vinyl sulfones are prone to polymerization and the inhibitor needs to be added. The base-catalyzed (DABCO) reaction of acrylamide and acetone (used as reactant and solvent) also led to polymerization at 9 kbar, whereas at 3 kbar low yields of Baylis–Hilmann products were obtained [38].

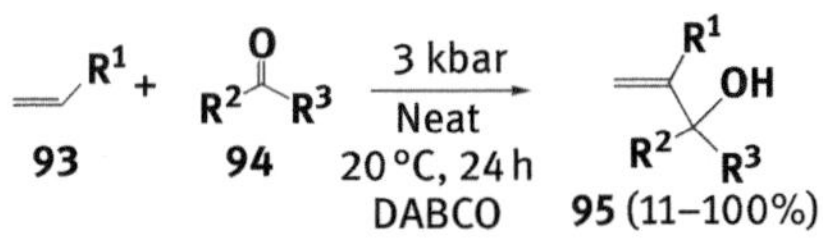

R^1 = CN, CO_2Me, SO_2Ph
R^2 = H, Ph, alkyl R^3 = Ph, alkyl

Scheme 4.26: The Morita–Baylis–Hillman reaction of aldehydes and ketones with acrylates.

While DABCO is traditionally used base in Morita–Baylis–Hillman reactions, Jenner has reported that phosphines are bases of choice for MBH dimerization of crotonitrile (3 kbar, neat 50 °C, 24 h, yields 19–100%) [39].

Enantioselectivity and reaction rate of the Baylis–Hillman reaction of 4-nitrobenzaldehyde **96** and methyl vinyl ketone were greatly enhanced under hyperbaric conditions (Scheme 4.27) [40]. The highest ee was obtained by employing 15 mol% of enantiomerically pure (*S*,*S*)-2,3-disubstituted DABCO catalyst **99**. To stimulate the reaction of the less reactive benzaldehyde **100** and methyl acrylate, 5 kbar was not high enough. However, under 10 kbar, the coupling product **102** was obtained with 14% yield and 10% ee. By the addition of methanol, the reaction was significantly improved to obtain **102** in 72% yield, whereas the enantioselectivity was not changed.

Among various chiral *β*-hydroxy amines that were tested in the Baylis–Hillman reaction between methyl vinyl ketone **97** and cyclohexyl carboxaldehyde, the *cinchona* alkaloids quinidine and cinchonine provided the highest ee's (CH_2Cl_2, 10–11 kbar) [41]. When variously substituted aldehydes were condensed with **97** at pressures ranging from 3 to 18 kbar, an increase in pressure resulted in a decrease in selectivity, with the best ee's obtained at 3 kbar. The decrease in enantioselectivity with

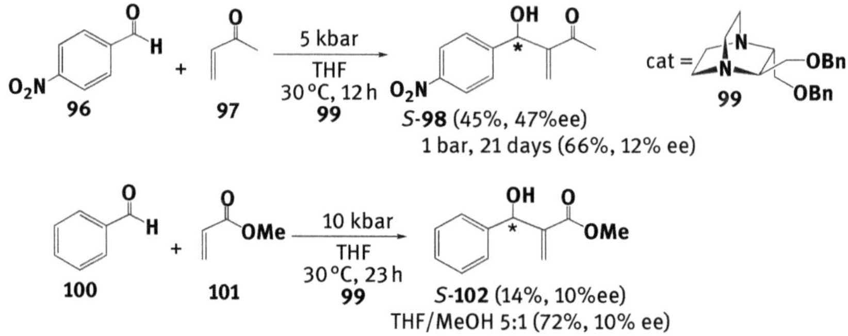

Scheme 4.27: Enantioselective Baylis–Hillman reaction.

pressure was explained by a lesser degree of selection between two transition states at high pressure where the subtle interactions are annulled.

4.7 Friedel–Crafts reaction

Maddaluno has shown that the increase in pressure is not always beneficial for the acceleration of reactions. Friedel–Crafts acylations of aromatic compounds were affected differently by high pressure, depending on their nature (Scheme 4.28) [42]. Acylation reactions of benzene and toluene (nonactivated and activated substrates) with benzoyl chloride catalyzed by $AlCl_3$ (in benzoyl chloride as the solvent) were constantly decelerated when atmospheric pressure was increased to 1–10 kbar, as evident from the sharp drop in yields. On the other hand, electronically deactivated fluorobenzene showed beneficial effect of pressure and increase in yield of the corresponding ketone **104** to 56% at 5 kbar. Further increase in pressure to 10 kbar had a similar effect as that for benzene and toluene, that is, a decrease in yield. The same pressure effect was observed when aromatic substrates were used as the solvent and the results could be explained by the largest volume of activation for deactivated aromatics, thus being the most susceptible to pressure changes. This is in

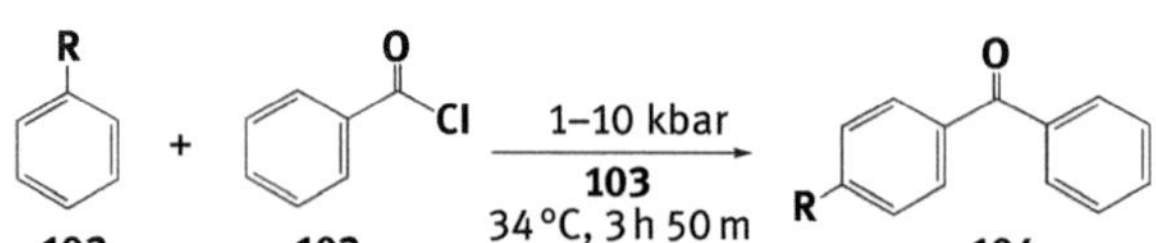

R=H, Me, F

R	1 bar	1 kbar	5 kbar	10 kbar
H	34	46	34	23
Me	73	59	59	41
F	5	12	56	24

Scheme 4.28: Friedel–Crafts acylations.

accordance with the activation volumes obtained for the electrophilic nitration of aromatics by Coillet ($\Delta V^{\neq}$ = −10, −22 and −23.5 cm^3 mol^{-1} for toluene, benzene and chlorobenzene, respectively) [43].

In variance to Friedel–Crafts acylation reaction described earlier, profitable influence of pressure was reported for the stereoselective Friedel–Crafts reactions. The reaction of 2-methylfuran with *n*-butyl glyoxylate catalyzed by catalyst(salen) cobalt(II) showed an increase in yield from 15% to 70% and ee from 44% to 72%, by the pressure rise from 1 bar to 10 kbar (25 °C, 20 h) [44, 45]. Similar effect was noted in asymmetric Friedel–Crafts alkylation of indoles with enones with chiral amine catalyst (9-amino-9-deoxy-*epi*-cinchonine) in the presence of acid additive such as benzoic acid (1 bar–10 kbar, 6–95%, 83% ee) [46].

Among the other electrophilic aromatic substitution reactions that were carried out in hyperbaric conditions, epoxide addition to indoles could be highlighted. On the basis of results obtained for the achiral substrates [47], this reaction was employed as the key step in an enantioselective synthesis of (+)-diolmycin A2. The optimal conditions for electrophilic aromatic substitution reaction of indole and epoxides was found to be 10 kbar, in dichloromethane, with $Yb(OTf)_3$ (5 mol%) catalyst. When these conditions were applied to indole reaction with chiral epoxyalcohol intermediate **106**, adduct **107** was obtained enantioselectively and in high yield (Scheme 4.29) [48]. Catalytic debenzylation of **107** provided (*S*,*S*)-diolmycin A2, an anticoccidial agent.

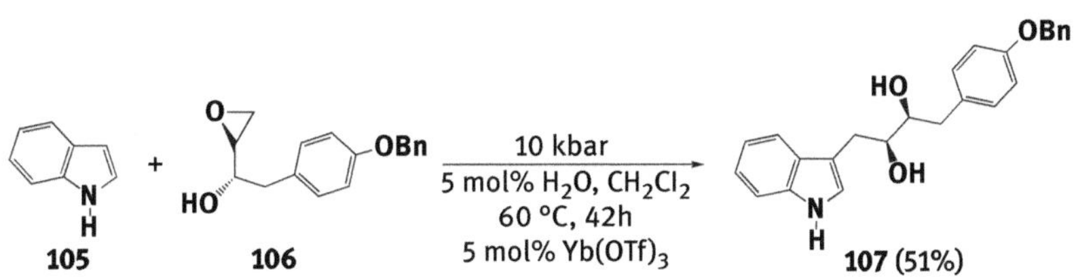

Scheme 4.29: Epoxide addition to indole.

Competition between aromatic substitution and substitution at the methyl group arises in the alkylation of 2-methylfuran with diethyl mesoxalate (Scheme 4.30) [49]. The ratio of products **110** and **111** depends on the choice of solvent and pressure. On the other hand, the reaction of 2-methyl furan with dimethyl mesoxalate took place regioselectively, providing a single product **115** [50]. In the case of 2,5-dimethylfuran, alkylation takes place at the one of the methyl groups, leading to **115**. An ene-type concerted mechanism with transition state such as **111** is proposed for the reactions of 2,5-dimethylfuran [51]. When a chiral carbonyl compound (*R*)-2,3-*O*-isopropylidene-D-glyceraldehyde was employed in the reaction with **114** (20 kbar, 50 °C, CH_2Cl_2), alcohols corresponding to **115** were obtained as a mixture of diastereoisomers (de 60%) [52, 53]. The addition of (*R*)-2,3-*O*-isopropylidene-D-glyceraldehyde to 2-methylfuran (10 kbar, 20 °C, 12h, CH_2Cl_2) produced a chiral variant of product **113**, in

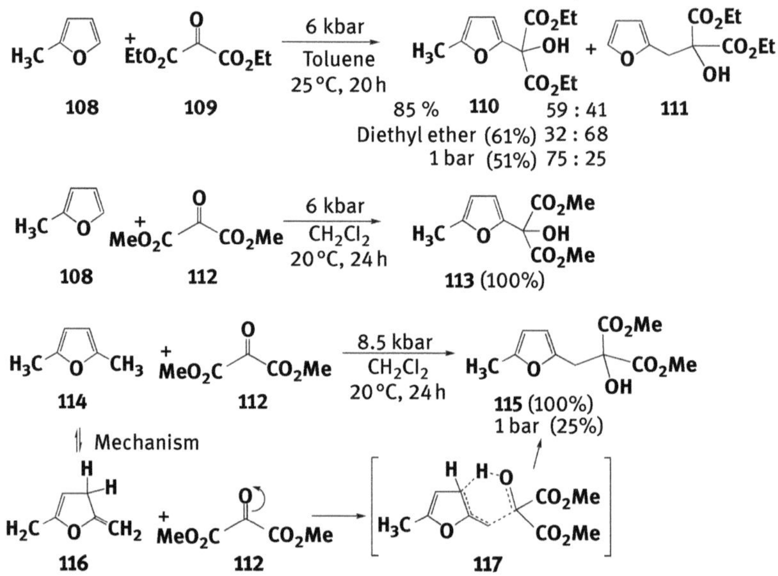

Scheme 4.30: Akylations of 2-methylfuran with diethyl mesoxalate.

which the major isomer is the *anti-* alcohol (*anti/syn* 76:24) [54]. The *anti*-selectivity was further enhanced by the use of $ZnCl_2$ catalyst to *anti/syn*-ratio of 82:18.

Related are the enantioselective hydroxyalkylation of indoles at 3-position, which were carried out at 9 kbar in the presence of chiral Brønsted acid catalysts (72–88% yield, ee 86–94%) [55] and *Cinchona* alkaloids (60–85%, ee 59–89%) [56].

4.8 Palladium catalyzed coupling reactions

Several papers describe palladium-catalyzed coupling reactions in hyperbaric conditions. Sugihara considered that initial steps in the Heck reaction, the oxidative addition of the Pd(0) species to alkyl or aryl halide and palladium reagent, and the complexation of the Pd(0) species on the double bond of olefins, are accelerated by pressure due to the negative reaction volumes [57]. Detailed kinetic studies by de Meijere revealed negative activation volumes for Heck reactions of iodobenzene and 4-nitrophenyl triflate with acrylates ($V^{\neq}$ = −5 to −37 $cm^3\ mol^{-1}$, MeCN) [58] where several mechanistic steps in the catalytic cycle are accelerated by pressure.

The Pd-catalyzed intramolecular cyclization of ethers **118** led to the formation of isochromanes. The major products of the Heck reaction are *trans*-1,2-disubstituted isochromanes **119**, whereas the minor products were *cis*-disubstituted isochromanes **120**, along with the double-bond isomers **121** and **122** (Scheme 4.31) [59]. The observed diastereoselectivity of coupling at ambient pressure is contradictory from the

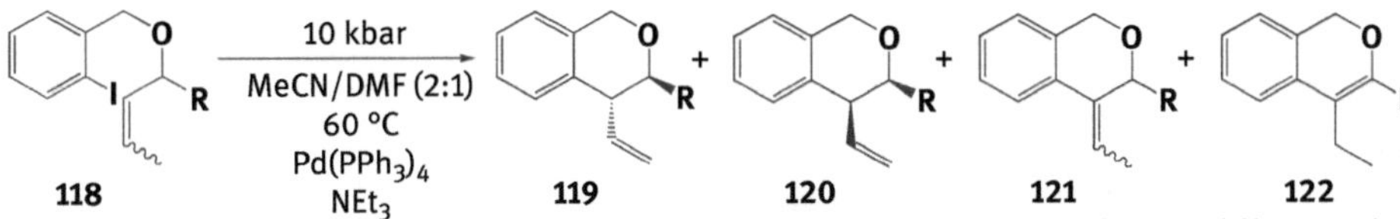

118 a) R = Me b) R = *i*-Pr *E* or *Z* → **119** + **120** + **121** + **122**

	Yield	d.r. **119/120**	Regioisom. r. **(119+120)/(121+122)**
E-**118a**	45%	6.1:1	1:2.2
1 bar	88%	15:1	1.6:1
Z-**118b**	55%	3.2:1	2.0:1
1 bar	83%	1:1	6.0:1

Scheme 4.31: Intramolecular Heck reaction of ethers **118**.

expected increase in the bulkiness of the substituents from methyl to isopropyl and decreases with the increased size of the substituents. Under high pressure at 10 kbar for substrate **118b**, the diastereoselectivity increases and the regioselectivity decreases. On the other hand, for **118a** the regioselectivity is opposite and the diastereoselectivity decreases. These differences are in accord with the larger pressure effect for the sterically more congested substrates, and the yield decrease from 1 bar to 10 kbar could be explained by the different reagents used in the Heck reaction at 1 bar ($Pd(OAc)_2$, PPh_3, TBAB, KOAc, DMF, 80 °C). In contrast, the intramolecular Heck reaction of the amine structurally corresponding to **118** is highly promoted by pressure. Isoquinolinone was obtained in 70% yield at 10 kbar (16:1 d.r. and 7:1 regioselectivity), whereas at ambient pressure, only traces of compounds were present [60].

Elevated pressures have a promoting influence on the Suzuki–Miyaura cross-coupling of aryl halides and aryl boronic acid. Remarkably, Young has shown that the reaction can be carried out in hyperbaric conditions without using palladium salts, which can be replaced by Ni (II), Co (II) and Fe(III) salts (Scheme 4.32) [61]. No reaction was observed with $NiCl_2$, $CoCl_2$ and $FeCl_3$ catalysts at ambient pressure. The high-pressure promotion of this S_NAr reaction is likely obtained by the influence on the acceleration of turnover-limiting oxidative addition of the aryl halide. The principal influence of pressure on the reactions catalyzed by Ni, Co and Fe salts may lay in the acceleration of the reduction to a catalytically active oxidation state of the metal.

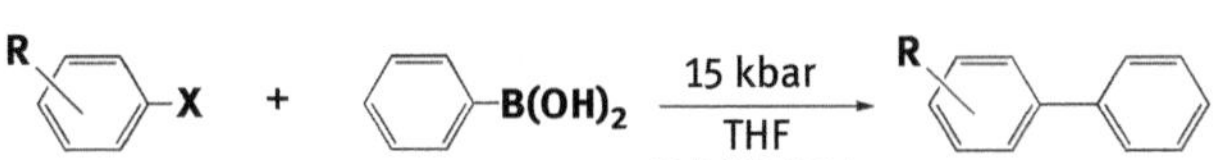

123a R = 4-NO_2, X = Cl
123b R = H, X = Br
124

125a	$Pd(OAc)_2$	99%
	$NiCl_2$	54%
	1 bar	1%
125b	$NiCl_2$	99%
	$CoCl_2$	99%
	$FeCl_3$	97%

Scheme 4.32: Suzuki–Miyaura cross-coupling of aryl halides and aryl boronic acid.

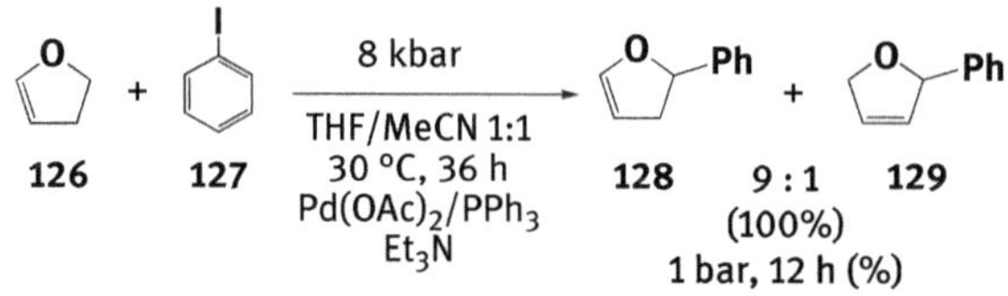

Scheme 4.33: Heck arylation of 2,3-dihydrofuran and iodobenzene.

Heck arylation of 2,3-dihydrofuran and iodobenzene shows a 23-fold rate increase from 1 bar to 8 kbar and a significant increase in turnover numbers (Scheme 4.33) [62, 63]. In this reaction, two regioisomeric products (**128** and **129**) were formed, and the determined activation volume for this reaction is -12 cm^3 mol^{-1}. When using the Pd-(*R*)-BINAP complex as the chiral catalyst at 10 kbar, no significant chiral induction (lower than 12% ee) was obtained [64]. However, the replacement of **127** with phenylnonaflate afforded **128** with significant increase in ee to 89% as compared to 47% ee at ambient pressure.

4.9 Cycloaddition reactions

In this section, cycloaddition reactions, other than Diels–Alder and dipolar cycloaddition reactions that are covered in separate chapters, are described.

4.9.1 [2+2] Cycloaddition reactions

Experimental studies on the kinetics of [2 + 2] cycloaddition reactions have revealed that ketene [2 + 2] cycloadditions follow a concerted, but nonsynchronous, mechanism with some solvent dependency ascribed to dipolar nature of transition state. Some [2 + 2] reactions exhibiting concerted, nearly synchronous bond formation are without significant charge separation in the transition state, for which the observed rate constant is practically independent of the solvent polarity.

Huisgen determined very large activation volumes for thermal reactions of different enol ethers with TCNE-producing cyclobutane rings ($\Delta V^{\neq} = -35$ to -37 cm^3 mol^{-1}, 25 °C, DCM) [65]. Similar large negative activation volume was determined for the reaction of TCNE with cyclopropanone diethyl acetal (-20.9 and -24.5 cm^3 mol^{-1}, in DCM and acetonitrile) [66]. Smaller $\Delta V^{\neq}$ values are found in the case of reactions of alkynes (Fischer carbene complexes) with 3,4-dihydro-2*H*-pyran to form cyclobutenes (-14.9 to -17.8 cm^3 mol^{-1}, 25 °C, neat) [67]. These small $\Delta V^{\neq}$ values indicate that the cycloaddition process does not involve the formation of a charged intermediate or a very dipolar transition state. In addition, rate constants do not exhibit a significant solvent dependency, which points to the isopolar transition state reaction and concerted mechanism.

Cycloadditions involving TCNE proceed via dipolar, zwitterionic intermediate (very negative $\Delta V^{\neq}$, strong solvent dependency), where the overall value $\Delta V^{\neq}$ is the sum of effects due to the intrinsic volume changes and to solvational effects (electrostriction). Concerted reaction of diphenylketen with vinyl ethers exhibits large activation volumes ($\Delta V^{\neq}$ = −22 to −32 cm^3 mol^{-1}, 25 °C, solvent) without much solvent dependency determined. Anomalously high $\Delta V^{\neq}$ values are found in benzene and toluene (−44 to −52) [68] and − 50.7 cm^3 mol^{-1} (toluene) [69].

For the reverse process – a cycloreversion by a cyclobutane ring opening, le Noble determined activation volume ($\Delta V^{\neq}$ = −16.7 cm^3 mol^{-1}, 25 °C, methanol) [70], which suggests charge development, that is, the zwitterionic character of the intermediate and the stabilization of charged TS in polar solvents. This $\Delta V^{\neq}$ is significantly larger than −2 to −3 cm^3 mol^{-1} (CH_2Cl_2) obtained for pericyclic conrotatory ring opening of cyclobutenes [71].

In line with the reactivity of TCNE, the [2 + 2] cycloaddition reactions of 1,1-dicyanoalkenes **131** with enol ethers **130** are facilitated by high pressure and afford 1-alkoxy-2,2-dicyanobutanes **132** in moderate-to-high yields (Scheme 4.34) [72]. A dipolar mechanism was suggested.

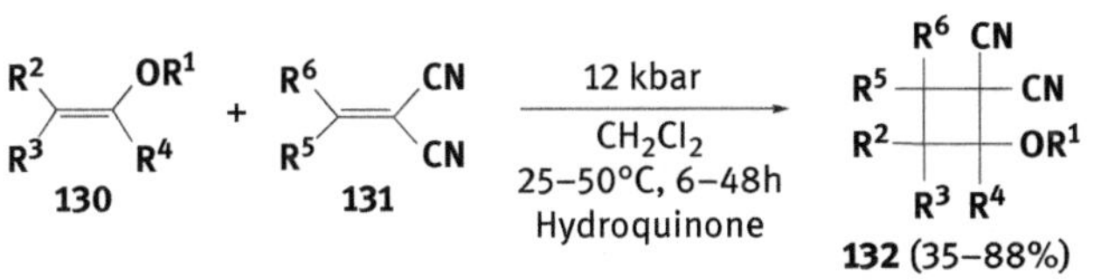

Scheme 4.34: [2 + 2] Cycloaddition reactions of 1,1-dicyanoalkenes with enol ethers.

The [2 + 2] cycloadditions of dicyanoalkenes have synthetic applications, for instance, in the functionalization of electron-deficient alkenes dehydroepiandrosterone derivatives **133** (Scheme 4.35) [73]. In this system, the steric hindrance from the C-ring and the methyl substituent at the C–D ring junction reduces the reactivity. Besides the expected cyclobutane cycloadducts **135** and also products **136** arising by an ene reaction of 1,1-dimethoxyprop-1-ene and **133** were obtained.

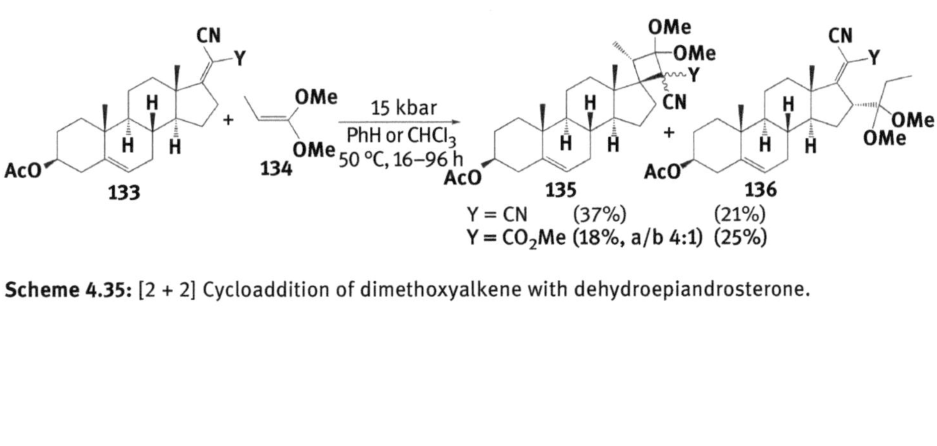

Scheme 4.35: [2 + 2] Cycloaddition of dimethoxyalkene with dehydroepiandrosterone.

The [2 + 2] cycloadditions of 4-methylphenyl 1,2-propadienyl sulfone and enol ethers at 15 kbar provided cyclobutanes ((3-alkoxycyclobutylidene)methyl 4-methylphenyl sulfones) **139** as single regioisomer, while the cycloaddition of allene **137** with enamine **140** gave a mixture of two regioisomers (Scheme 4.36) [74].

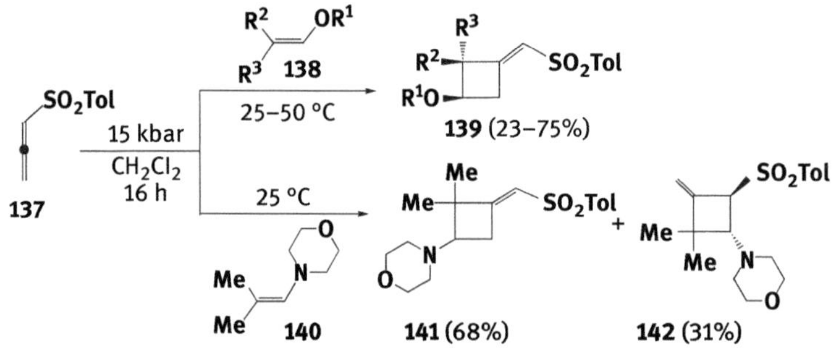

Scheme 4.36: [2 + 2] Cycloadditions of propadienyl sulfone with enol ethers.

Interestingly, high pressure does not have a promoting effect on the [2 + 2] cycloaddition of tetrachlorobenzyne on norbornadiene, and the **145/146** product ratio is not much changed (Scheme 4.37) [75]. By applying pressure, the product **146** is slightly favored ($\Delta\Delta V^{\neq}$ = −0.7 cm^3 mol^{-1}, RT, diethyl ether), which was ascribed to solvation of a zwitterionic intermediate **147**.

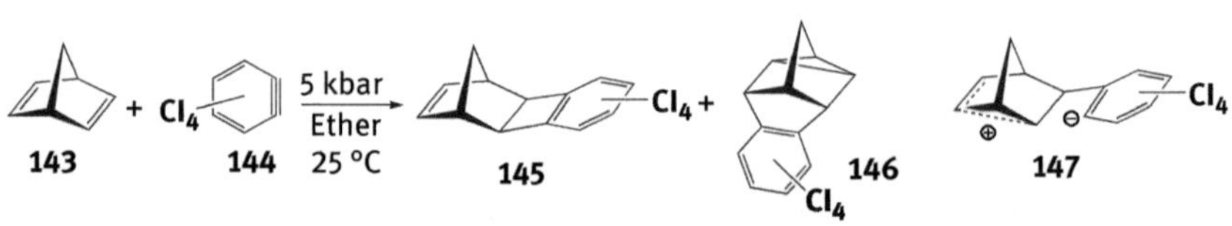

Scheme 4.37: [2 + 2] Cycloaddition of tetrachlorobenzyne with norbornadiene.

The photochemical [2 + 2] cycloaddition reactions require using high-pressure apparatuses equipped with the sample chamber with an optical window for the irradiation by light source. In the study of the photodimerization of methyl 3-methoxy-2-naphthoate, Sasse found that the rate acceleration at high pressure is of lesser extent than that for the Diels–Alder cycloadditions (Scheme 4.38) [76]. Activation volumes are accordingly smaller, $\Delta V^{\neq}$ = −9 and −10 cm^3 mol^{-1} (25 °C, toluene and methanol, respectively). Pressure has some influence on the diastereoselectivity ratio of the photochemical [2 + 2] cycloaddition reaction of cyclopentenones with 3,3-dimethyl-l-buten (3 kbar) [77], or the reaction of (*E*)-stilbene to chiral fumarate (4 kbar) [78]. In variance, enantiospecificity of the irradiation of chiral 2-pyridone to afford bicyclo

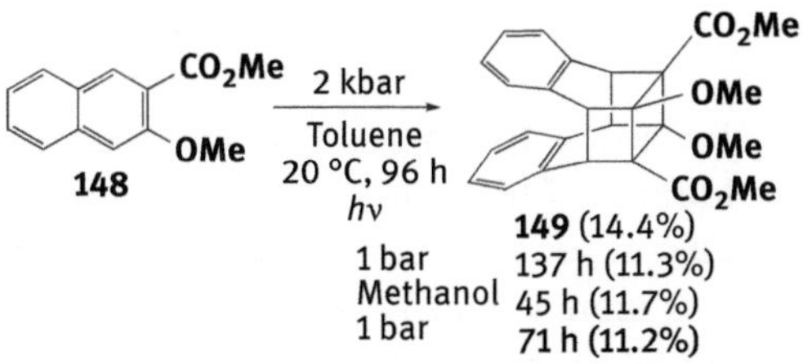

Scheme 4.38: Photochemical [2 + 2] cycloaddition of naphthalene derivative.

[2.2.0] photoproduct rises from 4% to 27% ee by the increase in pressure from 1 bar to 1 kbar [79]. Curiously, the [4 + 4] photodimerization of anthracene in hexane solution was found to be retarded by an increase in pressure, which was explained by pressure-induced decrease in the fluidity of the solvent [80].

4.9.2 [4+4] Cycloadditions

Examples of hyperbaric [4 + 4] cycloadditions in the literature are limited to the cyclodimerization of butadiene derivatives, 1,1,2,2,3,3-hexamethyl-4,5-bis(methylene)cyclopentane **150** and *o*-quinodimethane **151**, where the [4+4] cyclodimerization process is in competition with [4 + 2] cycloaddition (Scheme 4.39) [81]. In the case of the diene **150**, Klärner obtained $\Delta V^{\neq}$ for products ($\Delta V^{\neq} = -15.8$ and -15.5 cm^3 mol^{-1} for **152** and **153**, respectively) [82]. A constant ratio of 3:4 at various pressures suggests that the $\Delta V^{\neq}$ of the [4 + 2] and the [4 + 4] processes are similar ($\Delta\Delta V^{\neq} = -0.3$ cm^3 mol^{-1}), which is associated with alike geometry of the transition states and analogous mechanisms. Since the formation of **153** via thermal [$\pi 4_s + \pi 4_s$] process is forbidden by orbital symmetry rules, it could originate from the diradical intermediate **154** in a stepwise process. From diradical **154**, product **152** is formed via a nonconcerted Diels–Alder reaction. On the basis of the small $\Delta\Delta V^{\neq}$ value for the *o*-quinodimethane [4 + 2] and the [4 + 4] dimerizations (−5 cm^3 mol^{-1}), similar mechanistic conclusion on two parallel stepwise processes could be drawn [83].

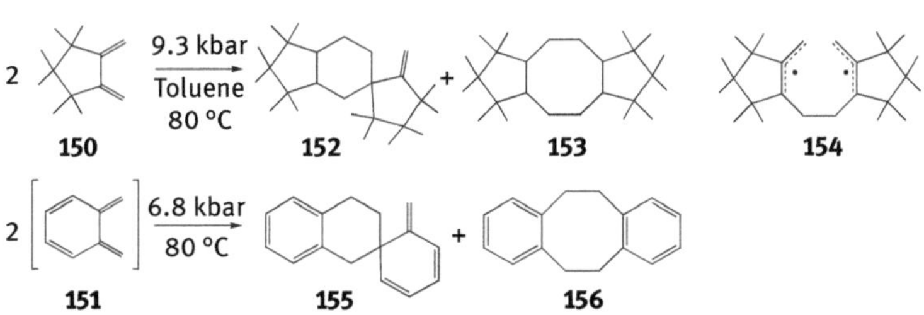

Scheme 4.39: [4 + 4] Cycloadditions of 1,1,2,2,3,3-hexamethyl-4,5-bis(methylene)cyclopentane and *o*-quinodimethane.

4.9.3 [6+4] and [8+2] Cycloadditions

Higher-order cycloadditions, that is, [6 + 4] and [8 + 2], are mostly associated with tropone chemistry. Thermal [6 + 4] cycloadditions of tropones could be readily accelerated applying pressure. Takeshita studies of kinetics tropone and cyclohexadiene addition revealed highly negative activation volumes ($\Delta V^{\neq}$ = −37.6, −28.2 and −32.6 $cm^3\ mol^{-1}$) in cumene, dioxane and DMF (Scheme 4.40) [84]. No significant solvent effect was observed, and all data are consistent with concerted mechanism as well as with the results for the cycloaddition of tropone with 2,3-dimethyl-1,3-butadiene ($\Delta V^{\neq}$ = −33.1, −30.5 and −30.1 $cm^3\ mol^{-1}$) in cumene, dioxane and DMF, respectively [85]. Obtained $\Delta V^{\neq}$ values are significantly larger than those reported in an earlier study on cycloaddition of tropone with cyclopentadiene ($\Delta V^{\neq}$ = −7.5 $cm^3\ mol^{-1}$) in dioxane [86].

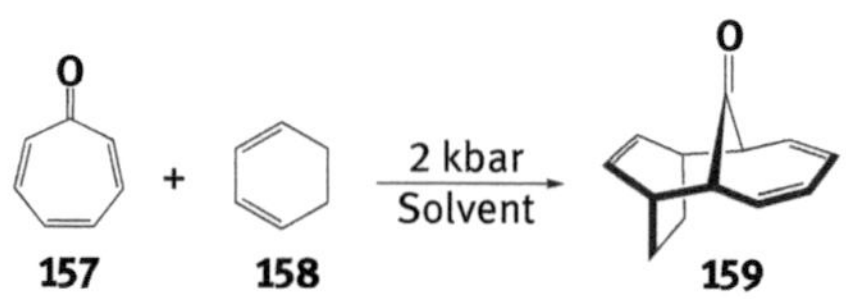

Scheme 4.40: [6 + 4] Cycloaddition of tropone with cyclohexadiene.

In the case of palladium-catalyzed [6 + 4] cycloadditions of tropone with functionalized alkene **161**, Trost observed that the ratio of kinetic and thermodynamic products changed by applying pressure. At 1 bar, thermodynamic products **162** and **163** were favored (1:6.3, 54% yield, $PdOAc_2$), while at 15 kbar kinetic products **164** and **165** dominate (ratio 2.3:1) (Scheme 4.41) [87].

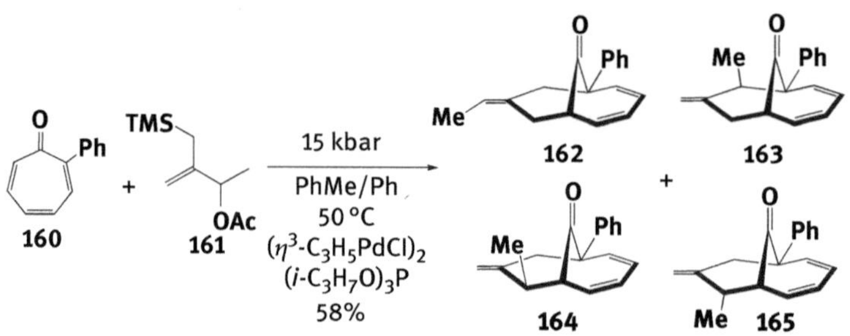

Scheme 4.41: [6 + 4] Cycloaddition of tropone with alkene.

The high-pressure reactions of 2,3-bis(methoxycarbonyl)-7-oxanorbornadiene **166** with 1-acetyl-cyclohepta[*b*]pyrrol-2(1*H*)-one **168** and **167** afford complex mixtures of cycloaddition products. In the reaction of **166** with 7-bromo **167**, the formation of

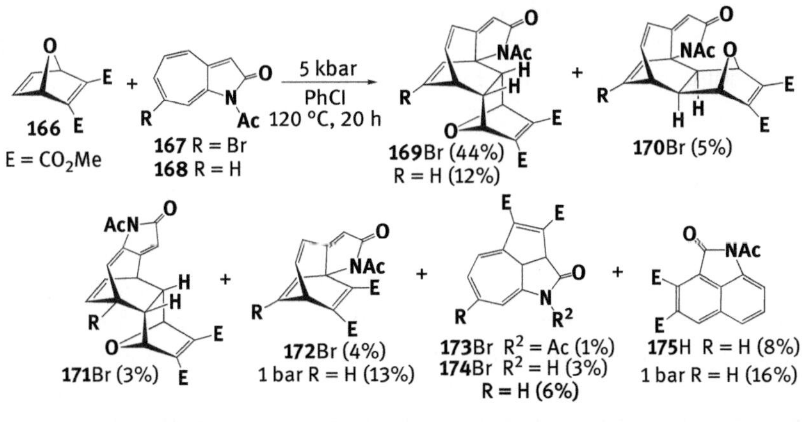

Scheme 4.42: Cycloaddition reactions of 2,3-bis(methoxycarbonyl)-7-oxanorbornadiene with 1-acetyl-cyclohepta[*b*]pyrrol-2(1*H*)-one.

the [4 + 2] adducts **171** and **172** dominates over [8 + 2] adducts **173** and **174** (ratio 56/4), while the ratio in the reaction of **166** with **168** (R=H) is 14/16 (Scheme 4.42) [88]. When the reaction of **168** was carried out at 1 bar, only cycloadducts **172** and **175** were isolated, without a decrease in the overall yield [89]. Products **173** and **174** arise from the thermal cycloreversion of initial [8 + 2] cycloadducts and loss of furan, whereas further oxidation led to **175**. Structurally related to **168**, methyl 2*H*-cyclohepta[*b*]furan-2-one-3-carboxylate in the reaction with ethoxyethene at 3 kbar provided mixtures of [4 + 2] and [8 + 2] adducts [90].

Cycloadditions of 11-methylene-l,6-methano[10]annulene **176** with dicyanoacetylene are very sluggish at atmospheric pressure. At 60 °C, in pentane, after 24 h, only low yield (3–4%) of 1:1 adduct **178** was obtained as the equilibrating methylenenorcaradiene-heptafulvene valence tautomers in equal amounts (Scheme 4.43) [91]. By

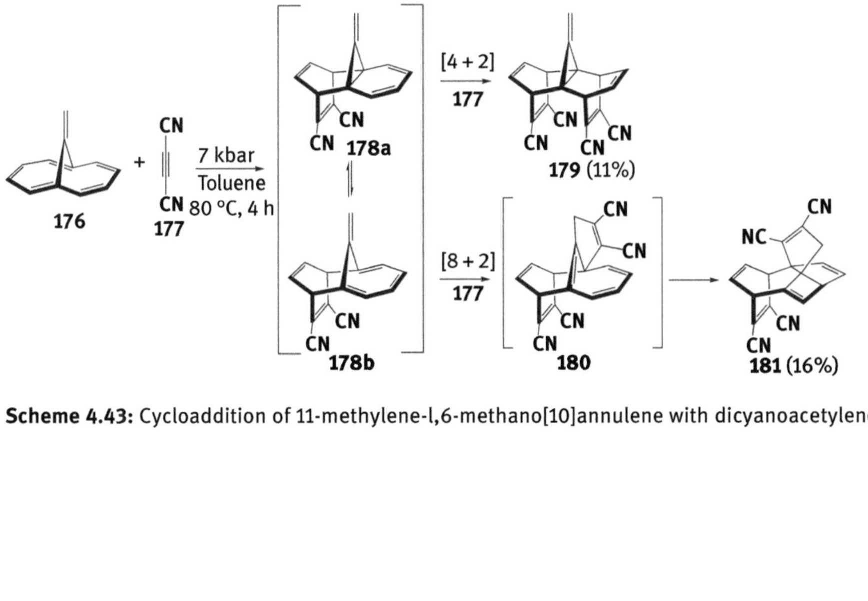

Scheme 4.43: Cycloaddition of 11-methylene-l,6-methano[10]annulene with dicyanoacetylene.

applying 7 kbar pressure, instead of an increase in 1:1 adduct **178**, two 2:1 cycloadducts **179** and **181** in 1:4 ratio were obtained, which arise from the consecutive [4 + 2] and [8 + 2] cycloadditions on the initially formed **178.**

4.10 Wittig reaction

Synthetically powerful way of the preparation of C=C bond by the Wittig reaction is also covered in the chapter dealing with the C–P bond formation reactions. The large negative volumes of activation that were determined by Isaacs [92] for the reaction of triphenyl-*p*-nitrobenzylidene phosphorane and *p*-benzaldehydes ($\Delta V^{\neq}$ = −21, −29 and −19 $cm^3\ mol^{-1}$, in dioxane, acetonitrile and dichloromethane, 1 kbar) indicate that the high pressure will be advantageous in forcing less reactive partners in this type of reaction. Indeed, at 10 kbar, the Wittig reaction of aromatic, indol and pyrrole aldehydes with triphenylphosporanylidene acetone afforded the corresponding olefins in 30–85% yield [93]. These yields are much higher than those obtained at normal pressure (0–12%) in the same solvent (dichloromethane). The utility of high pressure on the Wittig reaction could be exemplified by the olefination of sterically hindered cyclic ketones such as camphor **182** with *n*-butylidenetriphenylphosphorane (Scheme 4.44) [94].

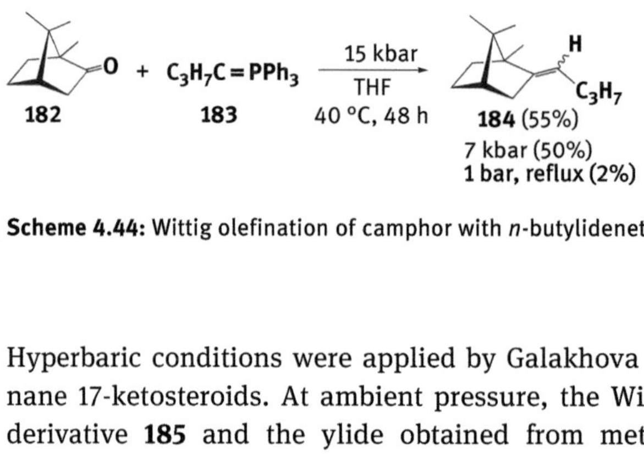

Scheme 4.44: Wittig olefination of camphor with *n*-butylidenetriphenylphosphorane.

Hyperbaric conditions were applied by Galakhova to the olefination of the pregnane 17-ketosteroids. At ambient pressure, the Witting reaction of androsterone derivative **185** and the ylide obtained from methoxymethyltriphenylphosphonium chloride afforded the product in low yield (10%). When 5 kbar of pressure was applied, the stereoisomeric methoxyolefins **187** and **188** were obtained in 97:3 ratio and in high yield (Scheme 4.45) [95]. These results are in good accord with the account on the Wittig reactions of series of steroidal ketones at 9 kbar (in benzene) where in all cases hyperbaric conditions were the only route to the product [96].

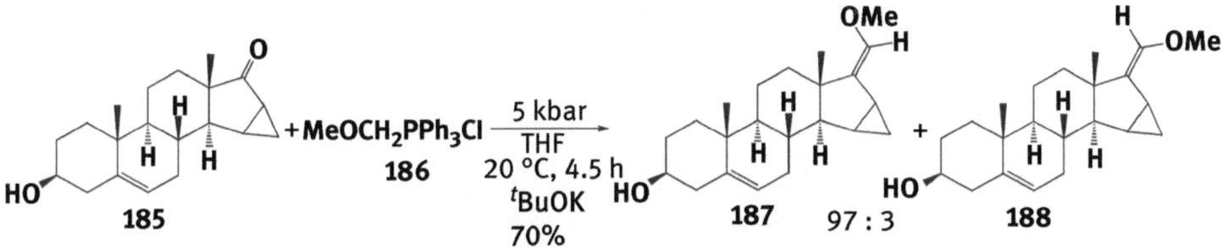

Scheme 4.45: The Witting reaction of androsterone with the ylide.

4.11 Ene reaction

Besides cycloadditions, ene reactions are pericyclic reactions that were studied in hyperbaric conditions. Detailed kinetic study of various types of uncatalyzed ene reactions (1 bar–1.3 kbar) has helped to establish activation volumes within the range of −30.9 to −42.1 cm^3 mol^{-1} (25 °C, DCM) [97]. The large negative $\Delta V^{\neq}$ values indicate that the ene reactions are greatly facilitated by pressure. Kinetic results suggest that the transition state for this process is less concerted than that for the cycloaddition reactions. Hence, the ene reactions of bicyclic substrates with congested reaction centers such as unreactive *α*-pinene proceed readily at 1 kbar with diethyl mesoxalate and DMAD (Scheme 4.46) [98]. Similar promoting effect pressure has on the gold-catalyzed intramolecular Conia-ene reaction of sterically congested ε-acetylenic-*β*-ketoesters[99] and on the silica gel-catalyzed cyclizations of unsaturated carbonyl compounds (15 kbar) [100].

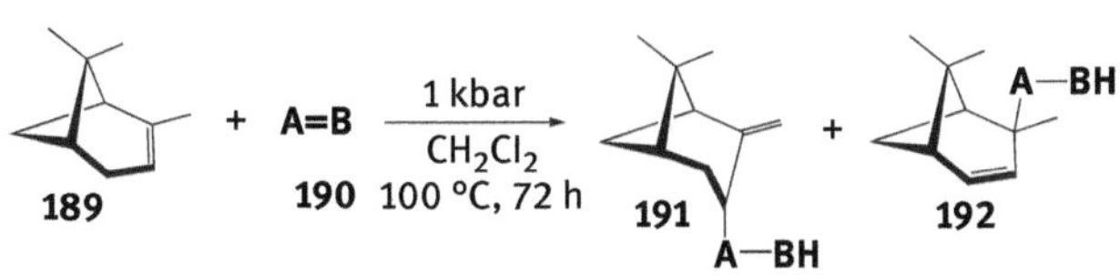

a A-BH = HO-C($CO_2Et)_2$
b A-BH = MeO_2C-C=$CHCO_2Me$

a 1 kbar (54%) mixt. **191+192**
1 kbar (0%)
b 1 kbar (78%) **191** *Z/E* 32:68
1 bar (0%)

Scheme 4.46: The ene reactions *α*-pinene with diethyl mesoxalate and DMAD.

Jenner observed that the $AlCl_3$-catalyzed reaction of cyclohexene and methyl propynoate leads to two products **194** and **195**, originating from competing [2 + 2] cycloaddition and ene reactions (Scheme 4.47) [101]. Although the rate is greatly accelerated, the chemoselectivity of this reaction is not affected by pressure, which suggests that the activation volumes of two processes are very similar, and likely proceeds through a dipolar acyclic transition state with a common intermediate.

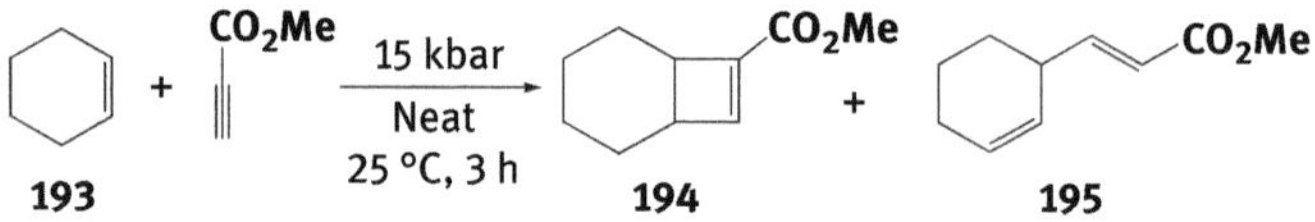

1 bar, 7 days (12%) 71 : 29
3 kbar (42%) 72 : 28
15 kbar (93%) 70 : 30

Scheme 4.47: Competing [2 + 2] cycloaddition and ene reactions.

4.12 Condensation and polymerization reactions

The C–C bond-forming condensation reactions are also facilitated by pressure. Radical copolymerizations of maleic anhydride with alkene monomers showed very negative activation volumes ($\Delta V^{\neq}$ = −16.5 to −55 cm^3 mol^{-1}) with the largest values are found for sterically hindered monomers (1.5 kbar, 70 °C, $CHCl_3$) [102, 103]. The promotive effects were reported for polymerization of *α*,*α*-dimethyl-*γ*-butyrolactone (20–24 kbar, 200 °C)[104], copolymerization of 2,3-epoxy-l-propyl methacrylate with 2-vinyl-5-ethylpyridine (AIBN, 15 kbar, THF 60 °C)[105] and polymerization phenylacetylene (9 kbar, 220 °C) [106]. Triarylmethanes were obtained by hyperbaric condensations from aldehydes (8 kbar, 160 °C, 20 h, neat) [107], as well as by the condensation of acidic phenol with aldehyde (3 kbar, 60 °C, 24 h, EtOH) [108]. Dimeric and trimeric condensation products were formed at high pressure from thiolactones (silica, 17 kbar, 170 °C) [109], and without acidic catalyst in more forcing conditions (12–25 kbar, 170–210 °C) [110].

4.13 Organotin reactions

The formation of the C–C bonds with the aid of stannanes has been widely explored under high-pressure conditions. The procedure is mild technique for these relative unstable substrates. The examples of direct C–Sn bond-forming reactions are covered in Chapter 6. For instance, allylstannes were used in regioselective Wurtz coupling with allyl halides (10 kbar) to obtain products in head-to-tail manner [111], and palladium-catalyzed cross-coupling (alkoxycarbonylation) of chloroformate with vinyl and arylorganotins was performed at 10 kbar [112].

Investigation of the catalytic version of high-pressure allylation of aldehydes led to the selection of the most efficient chromium salen complexes [113]. Reactions carried out at 10 kbar markedly outperformed allylations in classical conditions. These complexes were applied in efficient asymmetric allylation of aromatic and aliphatic aldehydes (Scheme 4.48). A high-pressure reaction of aldehydes and allyltributyltin was catalyzed by chiral (1*R*,2*R*)-chromium complex **199** (1 or 2 mol%) [114]. In optimal conditions (10 kbar, 20 °C, 24 h), chiral homoallyl alcohols were afforded in

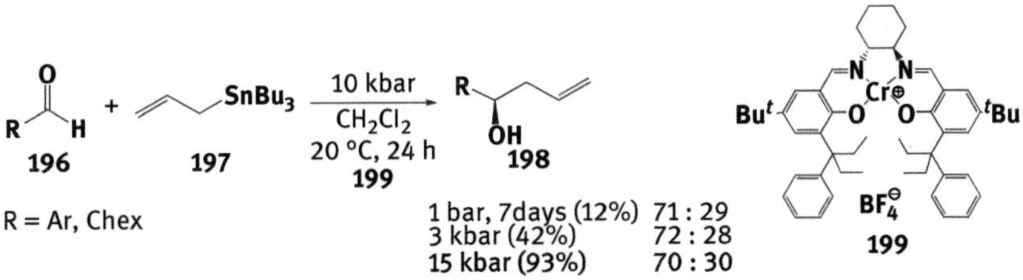

Scheme 4.48: Allylation of aldehydes and allyltributyltin.

good yields (84–92%) and high level of asymmetric induction (83–92% ee). These catalyzed reactions are faster than a noncatalytic ones at room temperature performed by Matsumoto, which were thought to proceed via a six-membered cyclic transition state [115]. The alternative way to induce diastereoselectivity in noncatalyzed homoallylation was by employing chiral *α*-amino aldehydes. Moderate *syn*-diastereoselectivity (de 20–76%) was achieved at 9 kbar [116].

Yamamoto also showed that high-pressure addition of allylic stannanes to aldehydes could result in the reverse stereochemistry to that achieved with Lewis acids [117]. For the reaction of octanal with crotyltin, $\Delta V^{\neq}$ values for *E*- and *Z*-isomers are determined to be –30.3 and –32.1 $cm^3\ mol^{-1}$ (30 °C, DCM, 7.5 kbar), which cause an increase of the *erythro* product. The ability of high pressure to invoke the preference for the reverse stereochemistry than obtained by Lewis acid catalysts is well illustrated in the reaction of ethyl glyoxylate with *α*-substituted *γ*-silyloxyallyltributyltins (Scheme 4.49) [118].

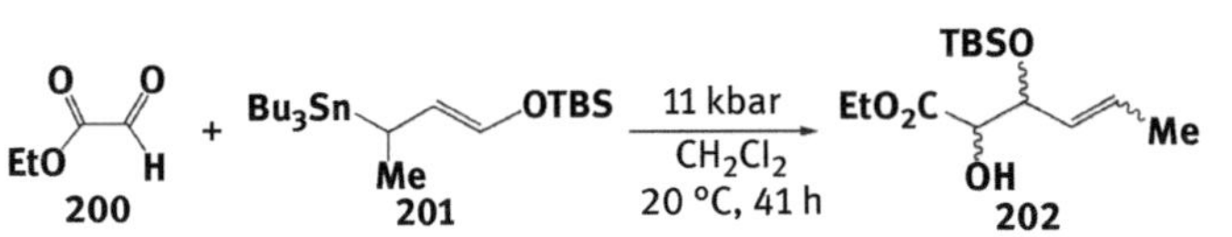

	Yield (%)	*Syn E*	*Syn Z*	*Anti E*	*Anti Z*
	77	6	6	0	88
Cr-salen **199**	73	92	4	3	1
1 bar, 100 °C 12 h	73	0	6	0	94
1 bar, $MgBr_2$	88	78	9	10	3
1 bar, $InCl_3$	87	78	5	14	3

Scheme 4.49: The addition of allylic stannanes to ethyl glyoxylate.

Other C–C bond-forming reactions were successfully carried out in hyperbaric conditions. Trimethylsilyl cyanide was employed by Matsumoto in asymmetric trimethylsilylcyanation of benzaldehyde by chiral titanium catalysts to obtain cyanohydrin [119, 120], which were also prepared by Kotsuki via noncatalytic (8 kbar, $MeNO_2$) cyanation of acetals with TMSCN [121]. Various cyclizations were also promoted by pressure in conjunction with light or catalyst stimuli such as photo-induced electron transfer-oxidative cyclization of unsaturated silyl enol ethers (1.5 kbar) [122] and gold-catalyzed intramolecular cycloisomerization of 1,6- and 1,7-enynes [123].

References

[1] Dauben, WG., & Bunce, RA. Organic reactions at high pressure. A Robinson annulation sequence initiated by Michael addition of activated cycloalkanones with hindered enones. J Org Chem 1983,48, 4642–4648.

[2] Desmaële, D. Unexpected alkylation reaction of amines, acids and phenols by alkyl (triphenylphosphoranylidene)acetates. Tetrahedron Lett 1996, 37, 1233–1236.

[3] Sekiguchi, Y., Sasaoka, A., Shimomoto, A., Fujioka, S., & Kotsuki, H. High-pressure-promoted asymmetric aldol reactions of ketones with aldehydes catalyzed by L-Proline. Synlett 2003, 1655–1658.

[4] Misumi, Y., & Matsumoto, K. Diastereoselective asymmetric nitro-aldol reaction of α-amino aldehydes under high pressure without catalyst. Angew Chem Int Ed Engl 2002, 41, 1031–1033.

[5] Yamamoto, Y., Maruyama, K., & Matsumoto, K. Aldol reactions, conjugate additions, and alkylation reactions of O-silylated ketene acetals under neutral conditions. Tetrahedron Lett 1984, 25, 1075–1078.

[6] Bellassoued, M., Reboul, E., & Dumas, F. High pressure induced mukaiyama type aldol reaction of bis trimethylsilyl ketene acetals. Tetrahedron Lett 1997, 32, 5631–5634.

[7] Matsumoto, K. High-pressure syntheses: the henry reaction. Angew Chem Int Ed Engl 1984, 23, 617–618.

[8] Jenner, G. Effect of high pressure on Michael and Henry reactions between ketones and nitroalkanes. New J Chem 1999, 23, 525–529.

[9] Misumi, Y., Bulman, RA., & Matsumoto, K. High pressure mediated asymmetric henry reaction of nitromethane with carbonyl compounds catalyzed by cinchona alkaloids. Heterocycles 2002, 56, 599–605.

[10] Agafonov, NE., Sedishev, IP., Dudin, AV., Kutin, AA., Stashina, GA., & Zhulin, VM. Henry condensation under high pressure 2. Effect of aromatic aldehyde type and pressure on the yield of ω-nitrostyrenes and secondary processes. Russ Chem Bull 1991, 40, 366–372, (Izv. Akad. Nauk SSR, Ser. Khim. 1991, 426–7.).

[11] Newitt, DM., Linstead, RP., Sapiro, RH., & Boorman, EJ. Liquid-phase reactions at high pressures. Part I. Hydrolysis of esters, and the Knoevenagel reaction. J Chem Soc 1937, 876–883.

[12] Jenner, G. Steric effects in high pressure Knoevenagel reactions. Tetrahedron Lett 2001, 42, 243–245.

[13] Jenner, G. The pressure dependence of sterically hindered organic reactions of ketones. High Press Res 2002, 22, 511–514.

[14] Diedrich, MK., Hochstrate, D., Klärner, F-G., & Zimny, B. The effect of pressure on pericyclic rearrangements (intramolecular Diels-Alder reaction, Cope rearrangement, and electrocyclic reaction). High Press Res 1994, 13, 7–11.

[15] Diedrich, MK., Hochstrate, D., Klärner, F-G., & Zimny, B. The pressure effect in pericyclic rearrangements: the cope rearrangement, electrocyclization, and the intramolecular Diels-Alder Reaction. Angew Chem Int Ed 1994, 33, 1079–1081.

[16] von, E., Doering, W., Birladeanu, L., Sarma, K., Teles, JH., Klärner, F-G., & Gehrke, J-S. Perturbation of the degenerate, concerted cope rearrangement by two phenyl groups in

"Active" positions of (*e*)-1,4-diphenylhexa-1,5-diene. acceleration by high pressure as criterion of cyclic transition states. J Am Chem Soc 1994, 116, 4289–4297.

[17] Matsumoto, K. Synthesis under high pressure: Michael Addition. Angew Chem Int Ed Engl 1980, 19, 1013–1014.

[18] Matsumoto, K. High pressure Michael addition catalyzed by fluoride ions. Angew Chem Int Ed Engl 1981, 20, 770–771.

[19] Dauben, WG., & Gerdes, JM. Organic reactions at high pressure. Michael addition of activated acyclic donors with β,β-disubstituted enone receptors. Tetrahedron Lett 1983, 36, 3841–3844.

[20] Hashimoto, S., Matsumoto, K., & Otani, S. Synthesis under high pressure. Michael additions to a sterically hindered steroidal 4-En-3-one system. J Org Chem 1984, 49, 4543–4544.

[21] Kotsuki, H., & Arimura, K. $Yb(OTf)_3$-Catalzed Michael addition reactions of β-ketoesters on silica gel supports and at high pressure. Tetrahedron Lett 1997, 38, 7583–7586.

[22] Sera, A., Takagi, K., & Yamada, H. High-Pressure asymmetric Michael additions of Thiols, Nitromethane, and Methyl Oxoindancarboxylate to Enones. J Org Chem 1988, 53, 1157–1161.

[23] Kwiatkowski, P., Dudziński, K., & Łyżwa, D. Effect of High pressure on the organocatalytic asymmetric Michael reaction: highly enantioselective synthesis of γ-nitroketones with quaternary stereogenic centers. Org Lett 2011, 13, 3624–3627.

[24] Kwiatkowski, P., Cholewiak, A., & Kasztelan, A. Efficient and highly enantioselective construction of trifluoromethylated quaternary stereogenic centers via high-pressure mediated organocatalytic conjugate addition of nitromethane to β,β-disubstituted enones. Org Lett 2014, 16, 5930–5933.

[25] Miyamae, N., Watanabe, N., Moritaka, M., Nakano, K., Ichikawa, Y., & Kotsuki, H. Asymmetric organocatalytic desymmetrization of 4,4-disubstituted cyclohexadienones at high pressure: a new powerful strategy for the synthesis of highly congested chiral cyclohexenones. Org Biomol Chem 2014, 12, 5847–5855.

[26] Horinouchi, R., Kamei, K., Watanabe, R., Hieda, N., Tatsumi, N., Nakano, K., Ichikawa, Y., & Kotsuki, H. Enantioselective synthesis of quaternary carbon stereogenic centers through the primary amine-catalyzed Michael addition reaction of α-substituted cyclic ketones at high pressure. Eur J Org Chem 2015, 4457–4463.

[27] Camara, C., Keller, L., Jean-Charles, K., Joseph, D., & Dumas, F. A comparative study of high pressure versus other activation modes in the asymmetric Michael reaction of chiral imines. High Press Res 2004, 24, 149–162.

[28] Camara, C., Joseph, D., Dumas, F., d'Angelo, J., & Chiaroni, A. High pressure activation in the asymmetric Michael addition of chiral imines to alkyl and aryl crotonates. Tetrahedron Lett 2002, 43, 1445–1448.

[29] Harrington, P., & Kerr, MA. The high pressure reaction of cyclopropanes with indoles catalyzed by ytterbium triflate. Tetrahedron Lett 1997, 34, 5949–5952.

[30] Kotsuki, H., Arimura, K., Maruzawa, R., & Ohshima, R. Ytterbium(III) trifluoromethanesulfonate-catalyzed homoconjugate addition reactions of β-ketoesters with activated cyclopropane derivatives at high pressure. Synlett 1999, 650–652.

[31] Harrington, P., & Kerr, MA. The ultra high pressure conjugate addition of indoles to electron-deficient olefins. Can J Chem 1998, 76, 1256–1265.

[32] Bunce, RA., Schlecht, MF., Dauben, WG., & Heathcock, CH. Organic reactions at high pressure. conjugate additions of o-silylated ketene acetals to activated enones. Tetrahedron Lett 1983, 24, 4943–4946.

[33] Gacem, B., & Jenner, G. Effect of pressure on stetter reactions: synthesis of hindered aliphatic acyloins and γ-ketonitriles. High Press Res 2004, 24, 233–237.

[34] Hill, JS., & Isaacs, NS. Functionalisation of the α-position of acrylate systems by the addition of carbonyl compounds: highly pressure-dependent reactions. Tetrahedron Lett 1986, 16, 5007–5010.

[35] van Rozendaal, ELM., Voss, BMW., & Sheeren, HW. Effect of solvent, pressure and catalyst on the *E/Z*-selectivity in the Baylis-Hillman reaction between crotononitriie and benzaldehyde. Tetrahedron 1993, 49, 6931–6936.

[36] Schuurman, RJW., V D Linden, A., Grimbergen, RPF., Nolte, RJM., & Sheeren, HW. Effect of branching in alkyl groups of tertiary amines on their performance as catalysts in the high pressure promoted Baylis-Hillman reaction. Tetrahedron 1996, 52, 8307–8314.

[37] Jenner, G. Comparative high pressure and water effect in the Baylis-Hillman reaction. High Press Res 1999, 16, 243–252.

[38] Jenner, G., & Ben Salem, R. High pressure base-catalyzed oligomerization of ketones. High Press Res 2005, 25, 1–9.

[39] Jenner, G. Phosphine-catalyzed dimerization of activated alkenes under ambient and high pressure conditions. Tetrahedron Lett 2000, 41, 3091–3094.

[40] Oishi, T., Oguri, H., & Hirama, M. asymmetric Baylis-Hillman reactions using chiral 2,3-disubstituted 1,4-diazabicyclo[2.2.2]octanes catalysts under high pressure conditions. Tetrahedron Asymm 1995, 6, 1241–1244.

[41] Markó, IE., Giles, PR., & Hindley, NJ. Catalytic enantioselective Baylis-Hillman reactions. correlation between pressure and enantiomeric excess. Tetrahedron 1998, 53, 1015–1024.

[42] Mugnier, F., Harrison-Marchand, A., Igersheim, F., & Maddaluno, J. Effects of high pressure on Friedel-Crafts benzoylation. J Phys Org Chem 2005, 18, 268–271.

[43] Coillet, DW., & Hamann, SD. Chemical effects of pressure. Part 7. – The rates of electrophilic aromatic nitrations in solution. Trans Faraday Soc 1961, 57, 2231–2238.

[44] Kwiatkowski, P., Wojaczyńska, E., & Jurczak, J. (Salen)Co(II) complex-an efficient catalyst for the high-pressure Friedel-Crafts reaction of 2-methylfuran with alkyl glyoxylates. Tetrahedron Asymm 2003, 14, 3643–3645.

[45] Kwiatkowski, P., Wojaczyńska, E., & Jurczak, J. Asymmetric Friedel-Crafts reaction of furans with alkyl glyoxylates catalyzed by (salen)Co(II) complexes. J Mol Catal A 2006, 257, 124–131.

[46] Łyźwa, D., Dudziński, K., & Kwiatkowski, P. High-Pressure accelerated asymmetric organocatalytic Friedel Crafts alkylation of indoles with enones: application to quaternary stereogenic centers construction. Org Lett 2012, 14, 1540–1543.

[47] Kotsuki, H., Nishiuchi, M., Kobayashi, S., & Nishizawa, H. Novel neutral alkylations of indoles and pyrroles with vinyl epoxides at high pressure. J Org Chem 1990, 55, 2969–2972.

[48] Kotsuki, H., Teraguchi, M., Shimomoto, N., & Ochi, M. Ytterbium(III) trifluoromethanesulfonate catalyzed high pressure reaction of epoxides with indole. an enantioseleetive synthesis of (+)-Diolmycin A2. Tetrahedron Lett 1996, 37, 3727–3730.

[49] Jurczak, J., & Belniak, S. Effect of pressure on the reaction of 2-akylfurans with diethyl mesoxalate. High Press Res 1992, 11, 119–123.

[50] Jenner, G., Papadopoulos, M., Jurczak, J., & Koźluk, T. Reactions of furans, thiophenes and pyrroles with activated carbonyl compounds under thermal and high pressure conditions. Tetrahedron Lett 1984, 25, 5747–5750.

[51] Jurczak, J., Koźluk, T., Pikul, S., & Sałański, P. An unusual reaction of 2,5-dimethylfuran with carbonyl compounds: high-pressure ene reaction? Chem Commun 1983, 1447–1448.

[52] Jurczak, J., & Pikul, S. Asymmetric induction in the high-pressure reaction between 2,3-*O*-isopropylidene-*D*-glyceraldehyde and 2.5-dimethylfuran. Tetrahedron Lett 1984, 25, 3107–3110.

[53] Jurczak, J., & Pikul, S. The high-pressure reaction of 2,5-dimethylfuran with 2,3-*O*-isopropylidene-*D*-glyceraldehyde. Tetrahedron 1988, 44, 4569–4573.

[54] Pikul, S., & Jurczak, J. Stereochemical consequences of nucleophilic additions of 2,3-O-isopropylideneglyceraldehyde. High-pressure approach versus the use of organometallic reagent. Tetrahedron Lett 1985, 26, 4145–4148.
[55] Biedrzycki, M., Kasztelan, A., & Kwiatkowski, P. High-Pressure accelerated enantioselective addition of indoles to trifluoromethyl ketones with low-loading of chiral BINOL derived phosphoric acid. ChemCatChem 2017, 9, 2453–2456.
[56] Kasztelan, A., Biedrzycki, M., & Kwiatkowski, P. High-Pressure-Mediated asymmetric organocatalytic hydroxyalkylation of indoles with trifluoromethyl ketones. Adv Synth Catal 2016, 358, 2962–2969.
[57] Sugihara, T., Takebayashi, M., & Kaneko, C. Effects of high pressure on the heck reaction. Is it possible to control dehydropalladation of alkylpalladium intermediates having β-Hydrogens? Tetrahedron Lett 1995, 36, 5547–5550.
[58] Buback, M., Perković, T., Redlich, S., & de Meijere, A. The effect of high pressure on the heck reaction – A contribution to a deeper understanding of the mechanism. Eur J Org Chem 2003, 2375–2382.
[59] Tietze, LF., Burkhardt, O., & Henrich, M. Intramolecular heck reaction for the synthesis of isochromanes under ambient and high pressure. Liebigs Ann 1997, 1407–1413.
[60] Tietze, LF., Burkhardt, O., & Henrich, M. Diastereo- and regioselective intramolecular heck reaction of α-amino alcohol derivatives for the synthesis of enantiomerically pure isoquinolines and benzazepines at ambient and high pressure. Liebigs Ann 1997, 887–891.
[61] Guo, Y., Young, DJ., & Andy Hor, TS. Palladium-free Suzuki–Miyaura cross-coupling at elevated pressures. Tetrahedron Lett 2008, 49, 5620–5621.
[62] Hillers, S., & Reiser, O. First determination of the rate acceleration of a palladium-catalysed coupling reaction by use of high pressure. Chem Commun 1996, 2197–2198.
[63] Hillers, S., & Reiser, O. Dramatic increase of turnover numbers in palladium-catalyzed coupling reactions using high-pressure conditions. J Am Chem Soc 1996, 118, 2087–2088.
[64] Hillers, S., & Reiser, O. Palladium-catalyzed coupling reactions of 2,3-dihydrofuran using high-pressure conditions. Tetrahedron Lett 1993, 34, 5265–5268.
[65] von Jouanne, J., Kelm, H., & Huisgen, R. Effect of pressure on the [2+2→4] cycloaddition reaction of tetracyanoethylene to enol ethers. J Am Chem Soc 1979, 101, 151–158.
[66] Wiering, PG., & Steinberg, H. Effect of pressure on the (2+2) cycloaddition of cyclopropanone diethyl acetal to tetracyanoethylene. Rec Trav Chim 2010, 101, 267–269.
[67] Pipoh, R., van Eldik, R., Wang, SLB., & Wulff, WD. Pressure acceleration of [2+2] cycloaddition reactions on the coordinated ligand of chromium and tungsten pentacarbonyl carbene complexes. Organomet 1992, 11, 490–492.
[68] Swieton, G., von Jouanne, J., Kelm, H., & Huisgen, R. The effect of pressure on the [2+2→4] cycloaddition diphenylketen to enol ethers. J Chem Soc Perkin Trans 2 1983, 37–43.
[69] Isaacs, NS., & Rannala, E. Reaction studies at high pressure. Part I. activation volumes of some [2+2] and dipolar cycloadditions. J Chem Soc Perkin Trans 2 1975, 1555–1560.
[70] le Noble, WJ., & Mukhtar, R. Nature of the intermediates involved in the [2+2] cycloaddition of tetracyanoethylene and enol ethers. J Am Chem Soc 1975, 97, 5398–5399.
[71] Jenner, G. Effect of pressure on the conrotatory opening of cyclobutenes. An unexpected result? Tetrahedron 1998, 54, 2771–2776.
[72] Aben, RWM., Goudriaan, J., Smits, JMM., & Scheeren, HW. Synthesis of 1-Alkoxy-2,2-dicyanocyclobutanes by high pressure promoted polar [2+2] cycloadditions of enol ethers with 1,1-dicyanoalkenes. Synthesis 1993, 37–39.
[73] Blanco-Ania, D., Aben, RWM., van Berkom, LWA., Scheeren, HW., & Rutjes, FPJT. High-pressure-mediated extension of the privileged steroid scaffold. Eur J Org Chem 2014, 1438–1444.

[74] Aben, RWM., Braverman, S., & Scheeren, HW. High pressure-promoted [2+2] cycloaddition reactions of 4-methylphenyl 1,2-propadienyl sulfone with enol ethers. Eur J Org Chem 2003, 894–897.
[75] le Noble, WJ., & Mukhtar, R. effect of pressure on the competing [2+2] and [2+2+2] cycloadditions of tetrachlorobenzy ne and norbornadiene. J Am Chem Soc 1974, 96, 6191–6192.
[76] Hamann, SD., Linton, M., & Sasse, WHF. High-pressure photodimerization of a naphthalene derivative. Aust J Chem 1980, 33, 1419–1423.
[77] Buback, M., Bünger, J., & Tietze, LF. Pressure-induced diastereoselectivity in photochemical [2+2] cycloaddition reactions. Chem Ber 1992, 125, 2577–2582.
[78] Saito, H., Mori, T., Wada, T., & Inoue, Y. Pressure control of diastereodifferentiating [2+2] photocycloaddition of (*E*)-stilbene to chiral fumarate upon direct and charge-transfer excitation. Chem Comm 2004, 1652–1653.
[79] Ayitou, AJ-L., Fukuhara, G., Kumarasamy, E., Inoue, Y., & Sivaguru, J. Enantiospecific photochemical transformations under elevated pressure. Chem Eur J 2013, 19, 4327–4334.
[80] Tanaka, F., & Osugi, J. Pressure effects on fluorescence and photodimerization of anthracene and 9-methylanthracene in solution. Rev Phys Chem Jpn 1972, 42, 85–93.
[81] Klärner, F-G., Krawczyk, B., Ruster, V., & Deiters, UK. Evidence for pericyclic and stepwise processes in the cyclodimerization of chloroprene and 1,3-butadiene from pressure dependence and stereochemistry. Experimental and theoretical volumes of activation and reaction. J Am Chem Soc 1994, 116, 7646–7657.
[82] Baran, J., Mayr, H., Ruster, V., & Klärner, F-G. Stepwise [4+2]- and [4+4]-cyclodimerizations of 1,1,2,2,3,3-hexamethyl-4,5-bis(methylene)cyclopentane. J Org Chem 1989, 54, 5016–5019.
[83] Roth, WR., & Scholz, BP. das energieprofil des *o*-chinodimethan ⇄ benzocyclobuten – gleichgewichtes, ii (energy profile of the *o*-quinidimethane ⇄ benzocyclobutene equilibrium II). Chem Ber 1981, 114, 3741–3750.
[84] Takeshita, H., Sugiyama, S., & Hatsui, T. kinetic studies of the high-pressure [6+4]π cycloaddition of tropone to cyclohexa-1,3-diene. J Chem Soc Perkin 2 1986, 1491–1493.
[85] Sugiyama, S., & Takeshita, H. High-pressure kinetics of the (6+4)π cycloaddiotion of tropone with 2,3-dimethyl-1,3-butadiene, an acyclic diene. The temperature and solvent effects. Bull Chem Soc Jpn 1987, 60, 977–980.
[86] le Noble, WJ., & Ojosipe, BA. Effect of pressure on the [4+6] cycloaddition of tropone to cyclopentadiene. J Am Chem Soc 1975, 97, 5398–5399.
[87] Trost, BM., Parquette, JR., & Marquart, AL. Effect of high pressure on a transition-metal-catalyzed cycloaddition. J Am Chem Soc 1995,117, 3284–3285.
[88] Tian, GR., Mori, A., & Takeshita, H. Synthetic photochemistry. LII: substituent effects on a bromine atom on the high-pressure cycloaddition between 1-acetyl-7-bromocyclohepta[*b*] pyrrol-2(1*H*)-one and 2,3-Bis(methoxycarbonyl)-7-oxanorbornadiene and the photorrarangement of 2-Acety-9-bromo-2-azatricyclo[6.2.2.0^{1,5}] dodeca-4,6,9,11-tetraen-3-one. Bull Chem Soc Jpn 1989, 62, 3727–3729.
[89] Mori, A., Takeshita, H., Higashi, M., & Yamaguchi, H. Synthetic photochemistry. XLIX: synthesis and photoisomerisation of 2-acetyl-2-azatricyclo[6.2.2.0^{1,5}] dodeca-4,6,9,11-tetraen-3-one. An MO explanation of the cycloaddition of cyclohepta[*b*]pyrrol-2(1*H*)-ones. Bull Chem Soc Jpn 1989, 62, 1567–1571.
[90] Mori, A., Nukii, Y., Takeshita, H., & Nozoe, T. High-pressure cycloaddition reactions between methyl 2*H*-cyclohepta[*b*]furan-2-one-3-carboxylate and ethoxyethene. Pressure effects on the [8+2] and [4+2] cycloadditions. Heterocycles 1993, 35, 863–868.

[91] Klärner, F-G., Dogan, BMJ., Weider, R., Ginsburg, D., & Vogel, E. Organic reactions at high pressure: cycloaddition reactions of 11-methylene-1,6-methano[10]annulene. Angew Chem Int Ed 1986, 25, 346–348.

[92] Isaacs, NS., & Abed, OH. Reaction at high pressure, part 18; volumes of activation and of reaction for the Wittig reaction. Tetrahedron Lett 1986, 27, 995–996.

[93] Nonnenmacher, A., Mayer, R., & Plieninger, H. Hochdruckversuche, XII. Über die anwendung von hohem druck bei wittig -Reaktionen mit resonanzstabilisierten yliden. liebigs Ann Chem 1983, 12, 2135–2140.

[94] Dauben, WG., & Takasugi, JJ. Organic reactions at high pressure. Wittig reaction of hindered ketones with nonstabilized ylides. Tetrahedron Lett 1987, 28, 4377–4380.

[95] Galakhova, TN., Stashina, GA., Zhulin, VM., Kamernitskii, AV., & Levina, IS. Wittig methoxymethylenation of 15β,16β-methyleno-5-androsten-3β-ol-17-one induced by high pressure. Russ Chem Bull 1990, 2613.

[96] Isaacs, NS., & El-Din, GN. The application of pressure to some difficult Wittig reaction. Tetrahedron Lett 1987, 28, 2191–2192.

[97] El'yanov, B., Gonikberg, EM., & Jenner, G. Linear free energy relationship for the pressure dependence of rate constants in ene reactions. J Chem Soc Perkin 1992, 2, 137–142.

[98] Ben Salem, R., Bernard, A., & Jenner, G. High pressure synthesis of ene adducts derived from bicyclic terpenes. High Press Res 1991, 7, 147–149.

[99] Chaładaj, W., Kołodziejczyk, A., & Domański, S. Gold(I)-catalyzed conia-ene cyclization of internal ε-Acetylenic β-ketoesters under high pressure. ChemCatChem 2017, 9, 4334–4339.

[100] Dauben, WG., & Hendricks, RT. The acid-catalyzed cyclizations of unsaturated carbonyl compounds utilizing silica gel at high pressure. Tetrahedron Lett 1992, 33, 603–606.

[101] Jenner, G., & Papadopoulos, M. Influence de la pression dans les réactions entre les cycloalcènes et le propynoate de méthyle catalysées par $AlC1_3$. Etude Mécanistique. Tetrahedron Lett 1996, 37, 1417–1420.

[102] Jenner, G., & Kellou, M. Investigation of strained transition states in free radical copolymerizations by means of high -pressure kinetics. Chem Comm 1979, 851–852.

[103] Jenner, G., & Kellou, M. Etude de l'effet stérique en chimie organique moyen du volume d'activation. Etude de l'influence de la pression dans les polymérisations de l'anhydride maléique avec des éthyleniques substitués. Tetrahedron 1981, 37, 1153–1160.

[104] Glet, W., & Korte, F. Verhalten von lactonen unter dücken von 20–24 kb, eine modellreaktion zur entstehung des erdöls. Angew Chem 1967, 79, 990–991.

[105] Tanaka, Y., Atsukawa, M., Shimura, Y., Okada, A., Sakuraba, H., & Sakata, T. Self-crosslinkable polymers, 2. High pressure copolymerization of 2,3-epoxy-l-propyl methacrylate with 2-vinyl-5-ethylpyridine. Makromol Chem 1976, 177, 3265–3283.

[106] Jarre, W., Bieniek, D., & Korte, F. Verhalten yon phenylacetylen unter hohem druck hochdruckreaktionen, XI. Naturwiss 1975, 62, 391–392.

[107] Yasumoto, M., Yanagiya, K., Shibuya, I., & Sekiguchi, T. Synthesis of triarylmethanes under high pressure. Nippon Kagaku Kaishi 1989, 7, 1099–1102.

[108] Ohishi, T., Kojima, T., Matsuoka, T., Shiro, M., & Kotsuki, H. High-yielding TfOH-catalyzed condensation of phenols with aromatic aldehydes at high pressure. A model synthesis of the benzylidene biphenol key skeleton of blepharismins. Tetrahedron Lett 2001, 42, 2493–2496.

[109] Proetzsch, R., Bieniek, D., & Korte, F. Hochdruckreaktionen IV. Verhalten von thiolactonen unter hohen drucken. Tetrahedron Lett 1972, 6, 543–544.

[110] Proetzsch, R., Bieniek, D., & Korte, F. Hochdruckreaktionen XII. Verhalten von thiolactonen und ausgewählten schwefelhaltigen heterocyclen unter hohen drucken. Zeit Naturforsch 1976, 31b, 529–530.

[111] Yamamoto, Y., Maruyama, K., & Matsumoto, K. Regioselective head-to-tail coupling of allylic trialkylstannanes with allylic halides under high pressure. Chem Comm 1984, 548–549.

[112] Jousseaume, B., Jurczak, J., Sałański, P., & Verlhac, J-B. Palladium-catalyzed alkoxycarbonylation of vinyl and arylorganotins under high pressure. High Press Res 1991, 7, 162–164.

[113] Kwiatkowski, P., & Jurczak, J. High-pressure enantioselective allylation of aldehydes catalyzed by (salen)chromium(iii) complexes. Synlett 2005, 227–230.

[114] Kwiatkowski, P., Chaładaj, W., & Jurczak, J. A sterically modified (salen)chromium(iii) complex – an efficient catalyst for high-pressure asymmetric allylation of aldehydes. Synlett 2005, 2301–2304.

[115] Yamamoto, Y., Maruyama, K., & Matsumoto, K. Organometallic high pressure reactions. Allylation of aldehydes with allylic stannanes under mild conditions. J Chem Soc Chem Commun 1983, 489–490.

[116] Cokley, TM., Isaacs, NS., McCluskey, A., & Young, DJ. High pressure allylstannane addition to α-amino aldehydes: *Syn* diastereoselectivity in the absence of chelation control. Main Gr Met Chem 1997, 20, 581–582.

[117] Yamamoto, Y., & Saito, K. Intervention of the boat transition state in the allylic tin-aldehyde condensation at high pressure. Chem Commun 1989, 1676–1678.

[118] Lumbroso, A., Kwiatkowski, P., Blonska, A., Le Grognec, E., Beaudet, I., Jurczak, J., Jarosz, S., & Quintard, J-P. Addition of γ-silyloxyallyltins on ethyl glyoxylate: evaluation of the influence of the experimental conditions on the stereochemical course of the reaction. Tetrahedron 2010, 66, 1570–1580.

[119] Choi, MCK., Chan, SS., & Matsumoto, K. Catalytic asymmetric synthesis of (*s*)-acetophenone cyanohydrin under high pressure. Tetrahedron Lett 1997, 38, 6669–6672.

[120] Choi, MCK., Chan, S-S., Chan, M-K., Kim, JC., Iida, H., & Matsumoto, K. Tri-tiol catalysed trimethylsilylcyanation of acetophenones under high pressure. Heterocycles 2004, 62, 643–653.

[121] Kumamoto, K., Nakano, K., Ichikawa, Y., & Kotsuki, H. High-pressure-promoted uncatalyzed cyanation of acetals using trimethylsilyl cyanide as a cyanide source in nitromethane. Synlett 2006, 1968–1970.

[122] Ackermann, L., Heidbreder, A., Wurche, F., Klärner, F-G., & Mattay, J. Mechanistic studies on PET-oxidative cyclization of unsaturated silyl enol ethers: dependence of the regioselectivity on alcohol addition and pressure effects. J Chem Soc Perkin 1999, 2, 863–869.

[123] Kocój, M., Jarosińska, E., & Chaładaj, W. High-pressure au-catalyzed cycloisomerization of internal 1,6-and 1,7-enynes. Synlett 2018, 29, 1319–1323.

5 C–N bonds

The carbon–nitrogen bond formation reactions are also described in chapters that cover hetero Diels–Alder, 1,3-dipolar and ketene [2 + 2] cycloadditions, as well as the examples of the Menshutkin reaction applied in supramolecular chemistry. This chapter presents the reactions that are not pericyclic. These include important reactions such as the Menshutkin reaction, nitrogen alkylation, aza-Michael and peptide bond forming reactions.

5.1 Menshutkin reaction

The early accounts of the formation of C–N bonds in the Menshutkin reaction (nitrogen quarternization) with the aid of high pressure were given by Fawcett [1]. For the reaction between pyridine and ethyl iodide in acetone at 5–8.5 kbar and 15–30 °C, it has been demonstrated the beneficial increase in the reaction rates going from ambient pressure to 3 kbar. Furthermore, Fawcett [2] and Perrin [3] have found that the rates of formation of pyridinium halides under high pressure are significantly accelerated.

Notable pressure acceleration of Menshutkin reactions by the use of high pressure is the consequence of the very large negative volumes of activations associated with the quaternization. Additional volume contraction due to electrostriction is obtained since this reaction involves charged species and highly polar transition states [4]. The kinetic studies by le Noble revealed that activation volumes $\Delta V^{\neq}$ for the Menshutkin reaction of 2,6-substituted pyridines with alkyl iodides at 4–9 kbar in acetone vary from −21.9 to −35 cm^3 mol^{-1} [5]. The $\Delta V^{\neq}$ values increase with the raise in sterical hindrance (substitution), in the order H<Me<Et<iPr, approximately by 2–3 cm^3 mol^{-1} for each additional carbon atom in substituents. This behavior was explained by the fact that the sterically crowded compounds have to some extent larger densities in comparison to their less crowded isomers, which implies that transition states for hindered isomers have smaller volume requirements than the unhindered substrates, and consequently,

https://doi.org/10.1515/9783110556841-005

hindered reactions are more accelerated. Acceleration of rates of sterically hindered Menshutkin reaction can be related to the Hammond postulate, with the transition states of the more hindered reactions are more product-like [6]. Similar pressure effects on 2,6-buttressed pyridines were found by le Noble, with $\Delta V^{\neq}$ within the range of −21.4 to −50 $cm^3\ mol^{-1}$ in acetonitrile [7]. The pressure influence on sterically hindered Menshutkin reaction was also reported by Harris and Weale [8]. Alkylations of *N,N*-dimethylaniline with methyl,ethyl and *i*-propyl iodides in methanol have $\Delta V^{\neq}$ of −26, −34 and −47 $cm^3\ mol^{-1}$, respectively.

Sterically hindered 2,6-di-*t*-butyl pyridine reacted with methyl iodide under 3 kbar pressure to obtain l-methyl-2,6-di-*t*-butyl pyridinium iodide, and this reaction does not occur at ambient pressure [9]. When more forcing conditions were applied – higher pressures and temperatures (5–6 kbar, 90 °C, 10–15 h) [10], in dioxane, two products were obtained: 2,6-di-*tert*-butyl-*N*-methylpyridinium iodide and 2,6-di-*tert*-butylpyridinium hydrogen iodide in a 2:8 ratio. In pressurization reaction of 2,6-di-*t*-butyl pyridine or 4,5-dimethylacridine with methyl fluorosulfonate, mixtures of *N*-methyl and protonated compounds were obtained, with the inversed 8:2 ratio. Similar mixture of *N*-methyl and protonated 2,6-di-*t*-butyl pyridine was obtained with methyl fluorosulfonate at 4.5 kbar and 60 °C. The ratio strongly depended on the dryness of the reaction mixture. In fully dry reaction conditions, *N*-methyl product is predominant, whereas in wet solvent, protonated product was formed in quantitative yield [11].

Interesting example is the synthesis of 2-pyridones by Hilbert–Johnson reaction under high pressure (Scheme 5.1). The first mechanistic step of the Hilbert–Johnson reaction is nitrogen quaternization (the Menshutkin reaction) of 2,6-dimethoxypyridine **1** by the reaction with iodoalkanes. Matsumoto has reported that two methoxy groups at *ortho*-positions make the steric bulk, and the reaction is sensitive to the size of

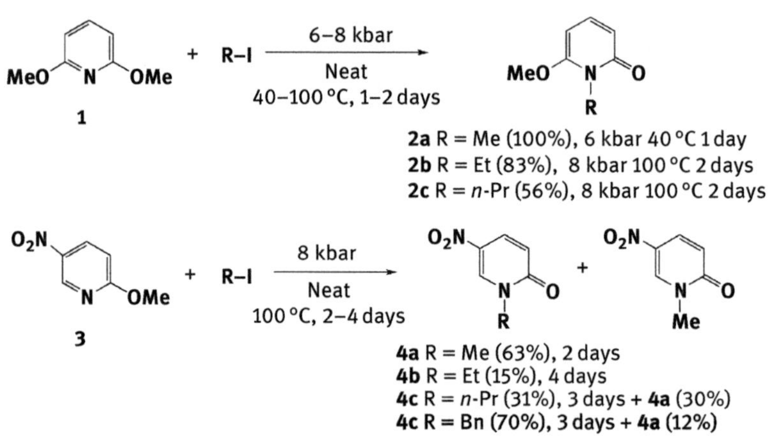

Scheme 5.1: Nitrogen quaternization of pyridines.

iodoalkanes [12]. While methyl iodide was fairly reactive, ethyl iodide required higher temperature and pressure to achieve high yield of **2b**. Iodopropane produced moderate yield (56%), even at 8 kbar and 100 °C. In the case of sterically less hindered 2-methoxy-5-nitropyridine **3**, Hilbert–Johnson reaction provided 2-pyridones **4** in lower yields. With the increase in bulkiness of alkyl halide from methyl, ethyl, *n*-propyl iodide to benzyl bromide, yield is gradually rising. In the case of ethyl iodide and benzyl bromide, **4a** was also obtained. In the postulated reaction mechanism, after nitrogen quaternization step, a lactim–lactam tautomerization assisted by halide takes place, with the elimination of MeX group, which in the case of bulkier groups (*n*-Pr and Bn) and long time reacts to form **4a**.

When tetrathiafulvalene (TTF)/2,2-bipyridine-cyclophane *cis*-**5** was subjected to quaternization via the reaction with 1,2-dibromoethane at 10 kbar, the TTF-diquat cyclophane *cis*-**6** product was obtained in 9% yield (Scheme 5.2) [13]. Albeit only the low yield of bis-Menshutkin reaction was obtained, the replacement of thermal conditions [reflux MeCN/DMF (*N*,*N*-dimethyl formamide)] to room temperature and high pressure prevented the degradation of substrate and formation of many side-products. Mechanism of this reaction likely takes place in two steps: initial intermolecular Menshutkin reaction, followed by intramolecular reaction.

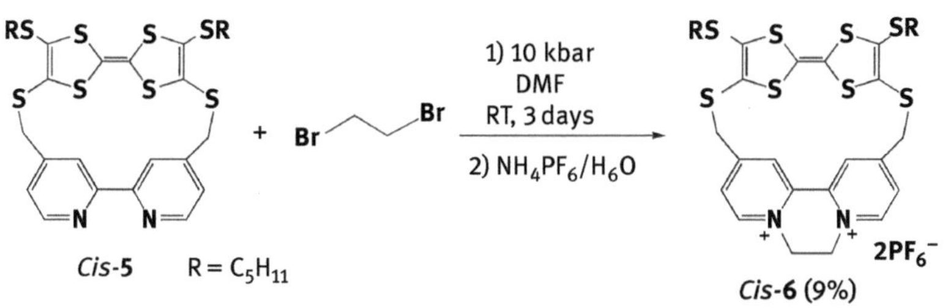

Scheme 5.2: Bis-quaternization of tetrathiafulvalene/2,2-bipyridine-cyclophane.

Quarternization of triethylamine with dibromomethane at 5 kbar produces as a major product bis-salt **8**, whereas alkylation at atmospheric pressure gives exclusively mono-salt **7** (Scheme 5.3) [14]. Two equivalents of amine and one equivalent of CH_2Br_2 were employed. When dichloromethane was used, only mono-salt of triethylamine was obtained in quantitative yield (25 °C, 10 kbar, 18 h) similarly to tri-*n*-butylamine, quinuclidine, diethylethylamine and *N*-methylmorpholine. On the other hand, *N*-quarternization of tertiary amines such as trimethylamine, pyridine and 4-diazabicyclo[2.2.2]octane (DABCO) with excess dichloromethane produced significant amounts of bis-quarternized amines **10** and **11**. Diiodomethane was also used as quaternization reagent at 5 kbar, in 30% methanol.

While sparteine **12** is usually unreactive toward alkyl halides, when subjected to pressures in excess of 20 kbar in the presence of diiodomethane, a methylene-

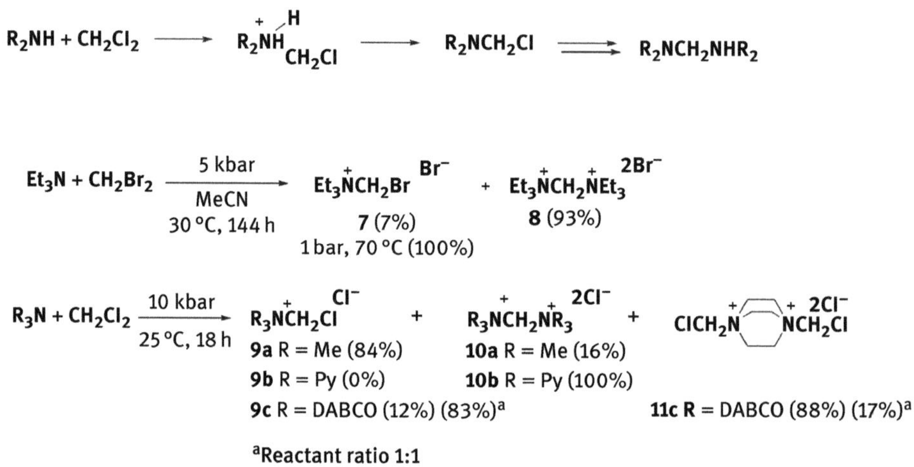

Scheme 5.3: Quarternization of amines with dibromomethane and dichloromethane.

bridged bis-ammonium salt **13** was formed as a single product in quantitative yield [15]. Structurally related sterically hindered sparteine alkaloids α-isolupanine **14** and 15-oxosparteine **16** could be quaternized with methyl iodide at high pressure and this is demonstrated to be the most practical method (Scheme 5.4) [16].

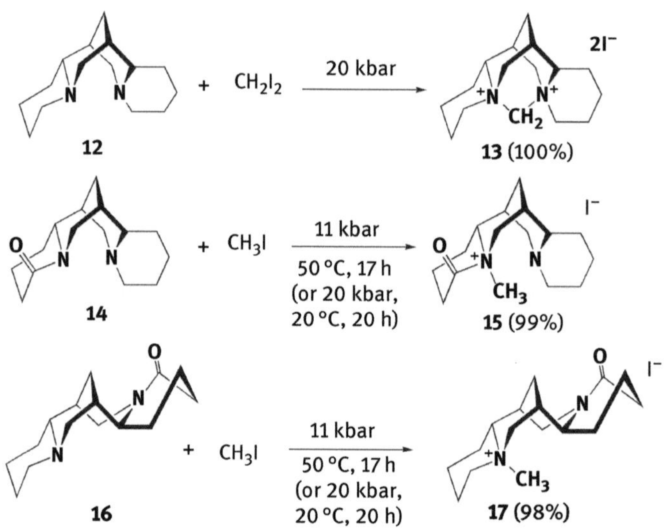

Scheme 5.4: Quaternization of sparteine alkaloids.

The Menshutkin reaction at high pressures is often employed in the synthesis of macrocycles such as cryptands and azacrown ethers [15, 17, 18]. Double quaternization reaction of diamines with bis-halides to afford the bis-quaternary salt is followed by demethylation by treatment with triphenylphosphine in DMF. Formation of the bis-quaternary salt consists of two pressure-accelerated processes: slow intermolecular quaternization, followed by fast intramolecular quaternization. The formation of cyclic products over linear bis-quaternary salts is favored by high-pressure conditions. Hence, high-pressure reaction of *N,N'*-dimethyl diazacoronand **18** with bis-(2-iodoethyl) ether **19** in acetone produced only cryptand **21** (Scheme 5.5). Additional examples are given by the same authors [19, 20]. Jurczak and coworkers have established a strong dependence on aprotic solvent polarity and the leaving group in **19**: double quaternization is accelerated in polar solvents such as DMF and acetonitrile, and the reactivity order is the following: I > Br > Tos > Cl [21].

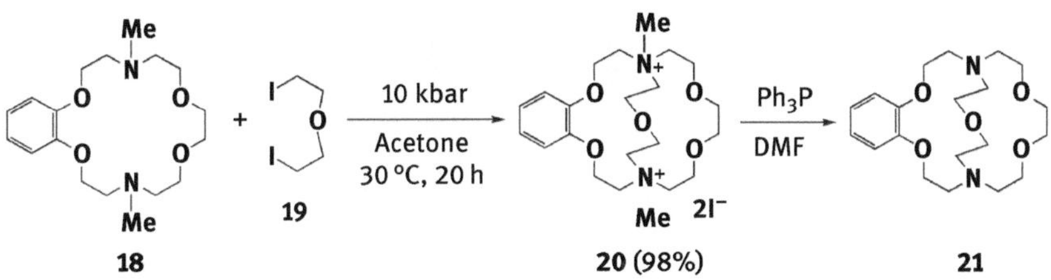

Scheme 5.5: Double quaternization reaction in synthesis of cryptand.

In a similar manner, chiral diaza-crown cryptand **24** derived from methyl 4,6-*O*-benzylidene-*α*-D-mannopyranoside was prepared in a stereocontrolled manner by double quaternization reaction under high pressure (Scheme 5.6) [22]. Another synthetic strategy to chiral cryptands was developed, in which the substrates consist of achiral

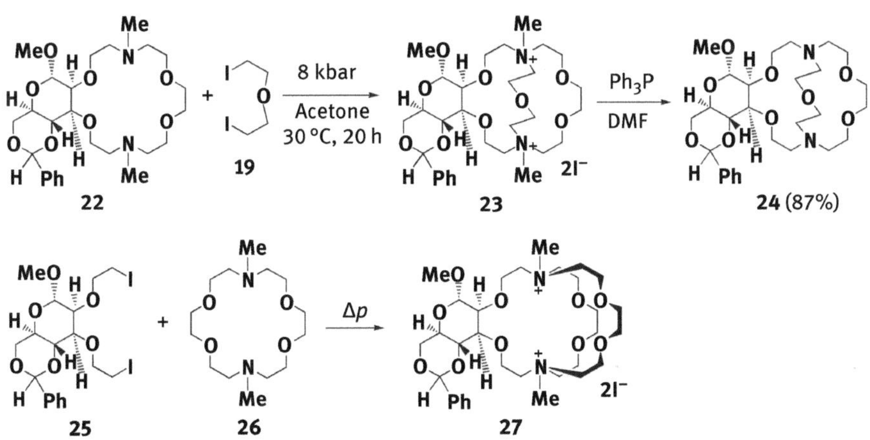

Scheme 5.6: Double quaternization of chiral diaza-crown ether.

N,N'-dimethyldiazacoronands and chiral bridging components such as 1,4-diiodo compound **25** [15].

The cyclic bis-quaternary salts of *N,N'*-dimethyl diazacoronands **31** were also prepared in almost quantitative yields by the reaction of *α,ω*-di-iodo compounds **29** or **30** with diamines **28** (Scheme 5.7) [23, 24].

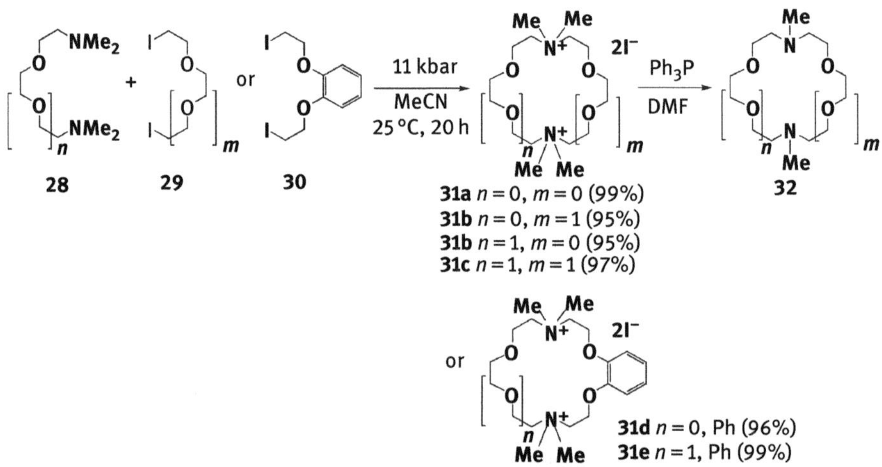

Scheme 5.7: Synthesis of bis-quaternary salts of *N,N'*-dimethyl diazacoronands.

Chiral *N,N'*-dimethyl diazacoronands **35** and **36** were prepared by different synthetic approach than diaza-crown cryptand **27** (Scheme 5.6). Synthesis starts with the reaction of diiodoethers and chiral (*R,R*)- and (*S,S*)-2,3-dimethoxy-1,4-bis(dimethylamino)butane fragments **32**, which were derived from L- and D-tartaric acid (Scheme 5.8) [25]. Both enantiomeric forms of bis-quaternary salts **35** and **36** were obtained in high yield. At ambient pressure, these reactions gave only polymeric products.

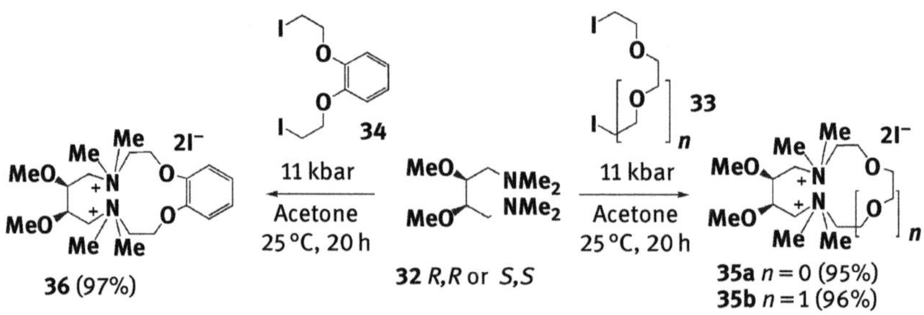

Scheme 5.8: Synthesis of chiral *N,N'*-dimethyl diazacoronands.

Okamoto has found that the Menshutkin reaction of 2,4,6-tri-*t*-butyl-*N*-methylaniline with methyl iodide under high pressure produced 2,4-di-*t*-butyl-*N*,*N*-dimethylanilinium iodide **38** and isobutylene instead of the expected anilinium salt, 2,4,6-tri-*t*-butyl-*N*,*N*-dimethylanilinium iodide (Scheme 5.9) [26, 27]. The same result was obtained in dioxane solution or without solvent, using a large excess of methyl iodide. The de-*t*-butylation reaction from the *ortho* position also occurs in the reaction with ethyl iodide under the identical conditions yielding 2,4-di-*t*-butyl-*N*,*N*-diethylanilinium iodide **41**. When the reaction was pressurized for a shorter time (8–10 h), 2,4-di-*t*-butyl-*N*-ethyl-*N*-methylanilinium iodide **39** was produced, indicating that **39** is likely initially produced from **38**, which is followed by the exchange of methyl by ethyl group. Experimental results suggest that the mechanism of the de-*t*-butylation from **38** under high pressure is a concerted process promoted by the relief of steric hindrance.

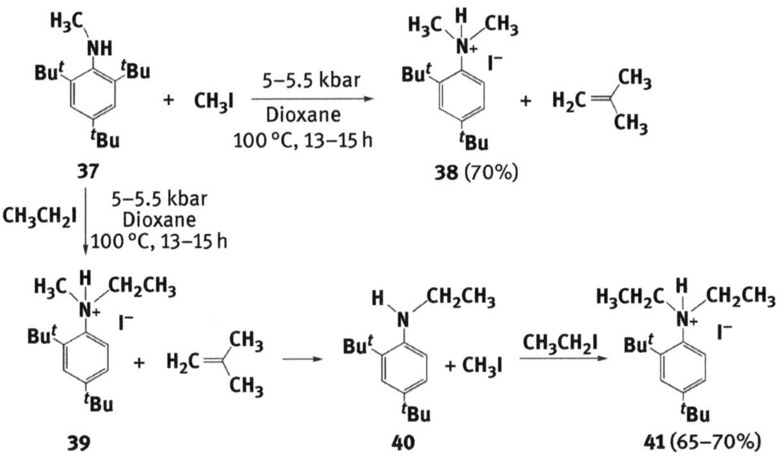

Scheme 5.9: The Menshutkin reaction of 2,4,6-tri-*t*-butyl-*N*-methylaniline with alkyl iodides.

Matsumoto has shown that high-pressure conditions enhance nitrogen crown ether functionalization. When diaza-crown ethers **42** were subjected to S_NAr reaction with the haloheteroaromatics at 8 kbar, high yields of 15-, 18- and 21-diaza-crown ethers **43** were obtained (Scheme 5.10) [28]. During the workup procedure, after depressurization,

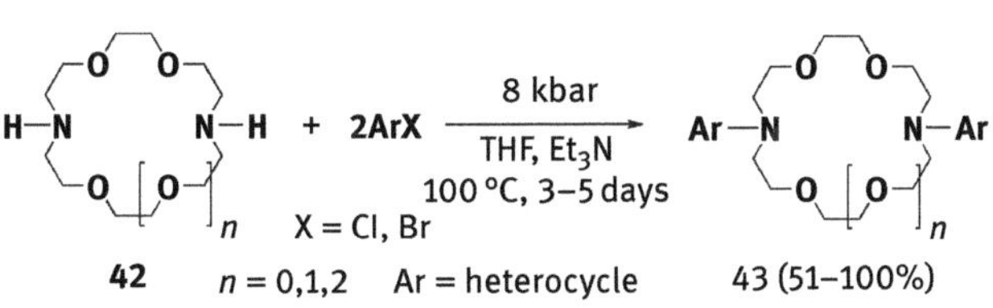

Scheme 5.10: *N*-Alkylation of diaza-crown ethers.

the quaternary salts (formed in Menshutkin reaction) were removed by filtration, and the products **43** remained in the filtrate. Comparison of the high pressure results with reaction under atmospheric pressure (2-bromothiazole, 100 °C, 4 days) showed that only 2% yield of product was obtained, whereas 51% yield was obtained at 8 kbar. Diaza-crown thioethers undergo the analogous S_NAr reaction under identical high-pressure conditions [29].

5.1.1 *N*-Alkylation

Identical reaction conditions were applied by Matsumoto in the preparation of lariat ethers possessing heteroaromatic substituents on the sidearms **45** from aza-crown ethers **44** and heteroaromatic chlorides (Scheme 5.11) [30].

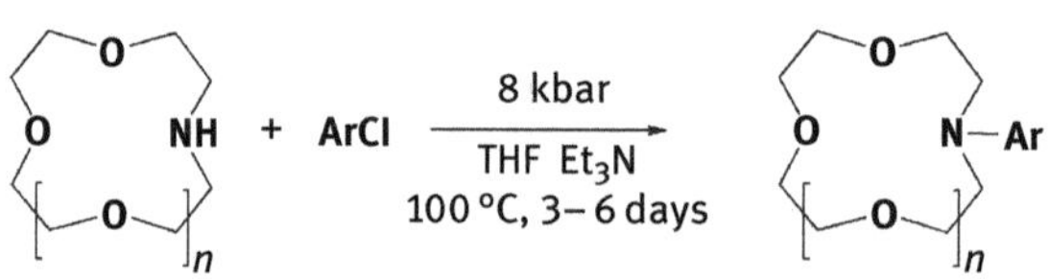

44 *n* = 1,2,3 Ar = heterocycle, Bn **45** (44–99%)

Scheme 5.11: Alkylation of aza-crown ethers.

High-pressure S_NAr reactions of homo piperazine **46** with heteroaromatic halides in high yields were achieved by the employment of the above-mentioned reaction conditions (Scheme 5.12) [31].

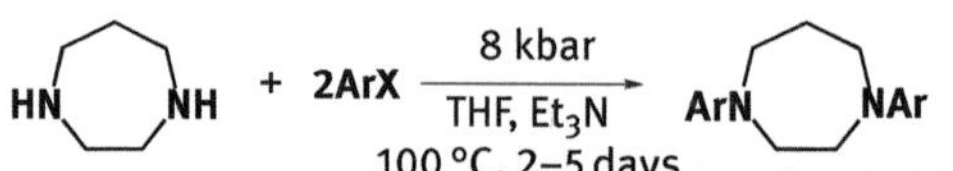

46 X = Cl, Br **47** (44–99%)

Ar = heterocycle

Scheme 5.12: Homo piperazine alkylation.

4-Mono and dialkylamino pyridines were prepared under high pressure by nucleophilic aromatic substitution (S_NAr) of 4-chloropyridine with primary and secondary amines (Scheme 5.13) [32]. Aromatic amination is clearly promoted by high-pressure technique, and products are generally obtained in moderate to high yields. This is an advantageous protocol for the transformation of **48** compared to the conventional protocol at ambient pressure.

Instead of hetero-Diels–Alder cycloaddition of Danishefsky's diene **50** across carbonyl bond of δ-valerolactam **49**, the *N*-alkenylation reaction of amide nitrogen with formation of product **52** was obtained by Jenner (Scheme 5.14) [33]. The addition of Lewis acid catalyst such as $BiCl_3$ slightly improved the yield to 46%,

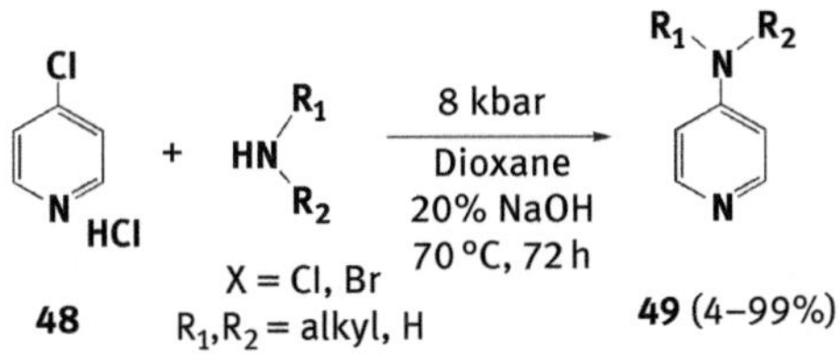

Scheme 5.13: Reaction of 4-chloropyridine with primary and secondary amines.

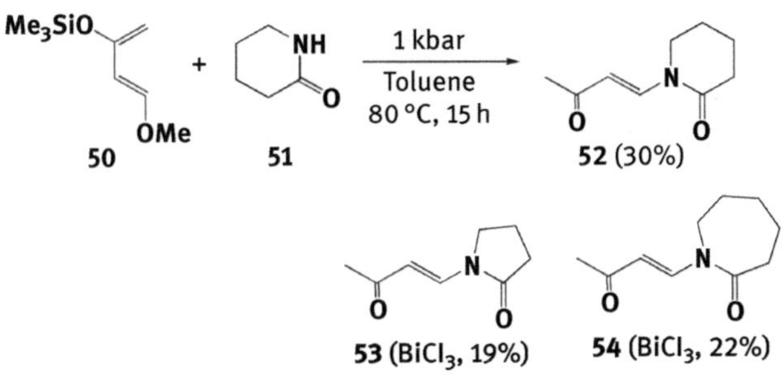

Scheme 5.14: *N*-Alkenylation reaction of Danishefsky's diene.

whereas the increase of pressure led to polymerization. The change in the ring size of lactams to 2-pyrrolidinone and ε-caprolactam led to formation of the same type of products (**53** and **54**, respectively, with the aid of $BiCl_3$ catalyst) in regioselective *trans*-manner. It is suggested that the reaction takes place via Michael reaction, where addition of amide nitrogen occurs on the diene carbon possessing a methoxy group.

Several pyrazole or imidazole nitrogen heterocycles were successfully used for chiral epoxide-opening reactions and formation of chiral diazoles **58**–**59** under high pressure (Scheme 5.15) [34]. Epoxides such as (*R*)-(+)- and (*S*)-(−)-styrene oxide, or (*R*)-(+)-2,3,4,5,6-pentafluorostyrene oxide and also (+)-camphopyrazole as a chiral pyrazole component were employed. This high-pressure-promoted *N*-alkylation of pyrazoles or imidazoles and construction of optically active derivatives does not require any catalyst or additive.

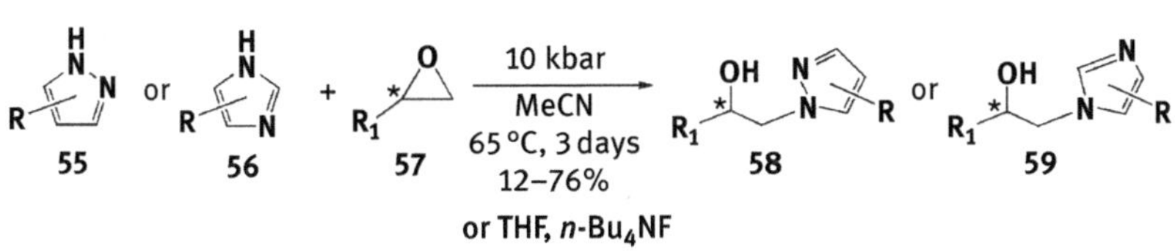

Scheme 5.15: *N*-Alkylation of pyrazoles and imidazoles.

5.2 Formation of peptide/amide bonds

5.2.1 Peptide coupling

Remarkably poor reactivity of the amino bond of *N*-carboxymethylamino acids in various conventional methods of peptide synthesis is likely due to the adverse combination of a steric hindrance and negative inductive effect of substituents. Yamada has found that the peptide bond formation in coupling reactions of *N*-(benzyloxycarbonyl)amino acid *N*-hydroxysuccinimide esters **60** with *N*-(carboxymethyl) amino acid diesters **61** was greatly enhanced at 10 kbar (Scheme 5.16) [35]. Control experiments at 1 bar demonstrated that these reactions proceed in low yield (<10%), or are completely unreactive. Reaction rate enhancement was particularly significant for sterically bulkier substrates.

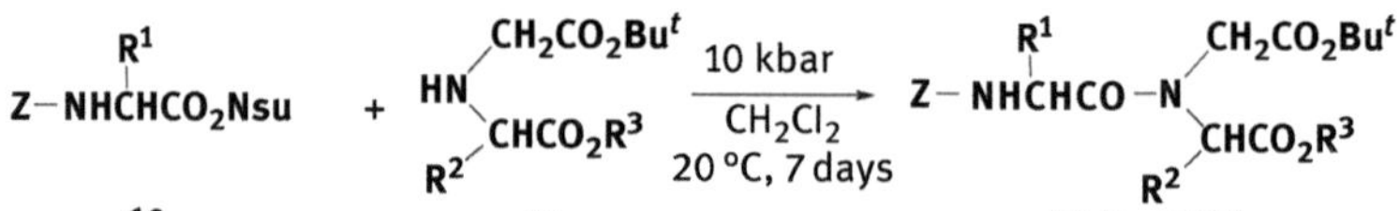

60 **61** **62** (2–82%)

R^1 = Bn, Me, ^{i}Bu, ^{i}Pr   R^2 = Me, ^{i}Bu, ^{i}Pr
R^3 = Et, ^{t}Bu

Scheme 5.16: Peptide coupling reaction.

Similarly, Nakahara reported high-pressure coupling reaction of Fmoc-aminoacyl fluorides **63** (Fmoc-Leu-F, Fmoc-Asp(OBut)-F and Fmoc-Gln(Trt)-F**)** with *N*-ethyl-*S*-triphenylmethyl-cysteine allyl ester **64** in the presence of DIPEA (diisopropylethyl amine), which gave high yields of the coupling products **65** (Scheme 5.17) [36]. The coupling products were obtained as mixtures of major and minor diastereomers in which epimerization occurred at both chiral atoms.

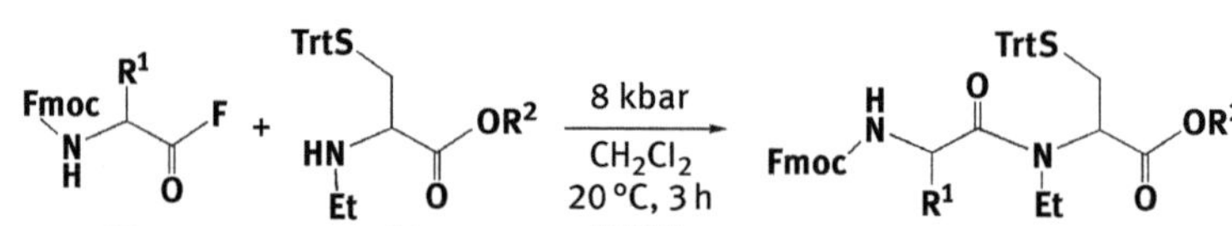

63 **64** **65** (main product 67–74% +minor product 14–25%)

R^1 = CH$_2$CH(CH$_3$)$_2$, R^2 = -CH$_2$CH=CH$_2$
CH$_2$CO$_2$Bu^t,
CH$_2$CH$_2$CONHTrt

Scheme 5.17: Peptide coupling of Fmoc-aminoacyl fluorides.

5.2.2 Aminolysis of esters

Ester aminolysis (amidation of esters) is the method described to be facilitated by high pressure in several papers. Matsumoto has shown that aminolysis near the room temperature could be achieved in reaction of nonactivated esters **66** and primary amines at 8 kbar without catalysts (Scheme 5.18) [37]. High yields were obtained at 45 °C for some reactions, whose ambient pressure requires heating at 190–230 °C leading to side-products. The advantage of high pressure was illustrated by successful application of optimal reaction conditions for the amidation of the relatively heat-sensitive alkyl 2-arylsulfinylacetates **69**.

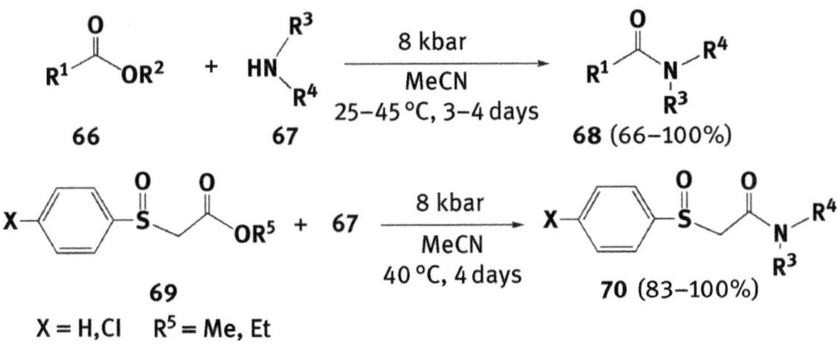

Scheme 5.18: Aminolysis of esters with primary amines.

Klärner and coworkers have applied higher pressures (10–11 kbar) and methanol as solvent to obtain dipeptides from *N*-protected amino acid methyl esters and L-amino acid sodium salts at room temperature [38]. Detailed kinetic study of the amidation reaction of *N*-benzoylalanine methyl ester **69** with the sodium salt of glycine **70** in methanol forming the dipeptide **71** was carried out by Klärner (Scheme 5.19) [39]. Strongly negative activation volume $\Delta V^{\neq} = -19.3\ \mathrm{cm^3\ mol^{-1}}$ was determined, which

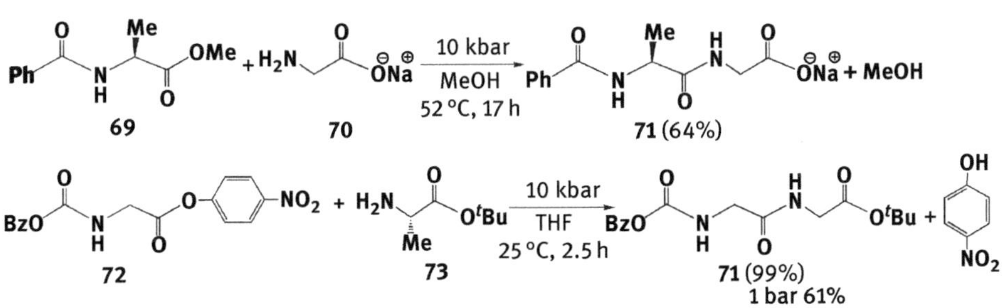

Scheme 5.19: Aminolysis of *N*-protected amino acid esters.

provides a good evidence for the formation of zwitterionic intermediate in the rate-determining step. Even the reaction in tetrahydrofuran (an aprotic solvent) affording the dipeptide **74** (from *N*-benzyloxycarbonyl-glycine *p*-nitrophenyl ester **72** and alanine *t*-butyl ester **73**) is also accelerated by increasing the pressure from 1 bar to 10 kbar. Further studies showed a positive Hammett value (ρ = +6.07) for the aminolysis of *p*-substituted phenyl esters (*p*-X-C_6H_4-O-C(O)-Ar). This indicates that in the rate-determining step, the C–O bond cleavage occurs via S_N2-type substitution mechanism.

Jurczak performed double aminolysis of oligoglycolic dimethyl esters, with oligoglycolic diamines in similar hyperbaric reaction conditions (10 kbar, 48 h) to obtain macrocyclic crown-ether diamides in improved yields [40]. Analogous double amidation reactions were carried out on diazacoronands such as **76** with oligoglycolic α,ω-dicarboxylic *p*-nitrophenyl and pentafluorophenyl esters. Here, nitromethane and acetonitrile were used as solvents, and macrocycles were obtained in 12–63% yields (Scheme 5.20) [41].

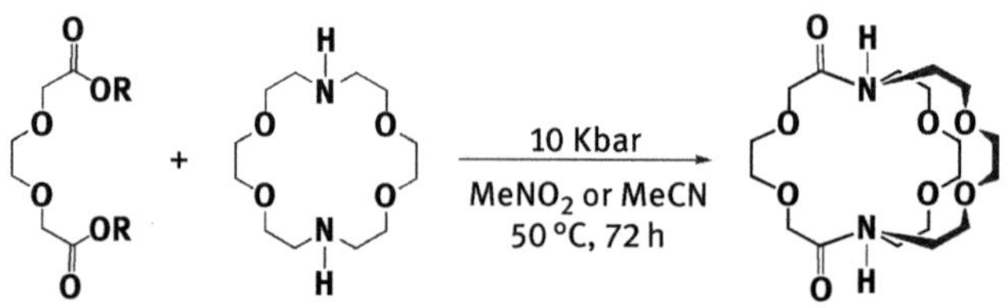

75 **76** **77** (63%) R = *p*-$NO_2C_6H_4$ (41%) R = C_6F_5

Scheme 5.20: Double aminolysis of oligoglycolic dimethyl esters.

Aminolysis of 2,2,2-trichloroethyl carbamates (Troc-carbamates) **78** with amines was also promoted by high pressure, and unsymmetrical ureas are obtained in high yields (Scheme 5.21) [42]. The process is entirely chemoselective on the ester side, and transamidation does not take place. Superiority of high-pressure conditions are illustrated in the reaction of Troc-protected aniline with benzylamine, which does not occur in refluxing THF after 30 h. Similarly to reaction of carbamates, dimethyl-carbonate with aliphatic amines at 8 kbar (RT, 3–16 h) provided carbamates in high yields [43].

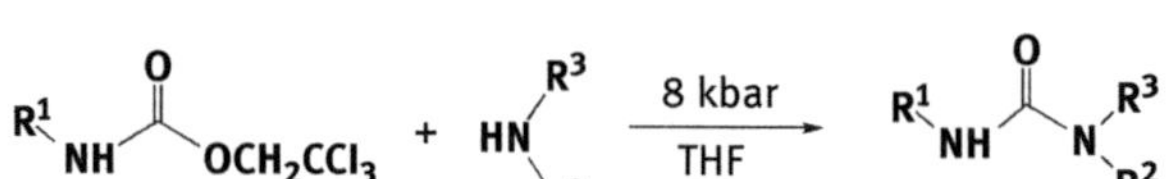

78 **79** **80** (41–99%)

R^1, R^2,R^3 = Alkyl, Ar

Scheme 5.21: Aminolysis of Troc-carbamates with amines.

5.2.3 Transamidation

Transamidation reaction offers another entry to amides and Kotsuki et al. have shown that lactams **81** could be more easily converted to ω-amino-carboxamides **83** (Scheme 5.22) [44]. Test reaction of lactam **81** (R = H, n = 1) with stoichiometric amount of benzylamine provided corresponding amide in 97% yield, whereas in reflux conditions (77 h) lactam was fully recovered. High-pressure conditions led to the chemoselective formation of amides, and ester functionalities are not affected.

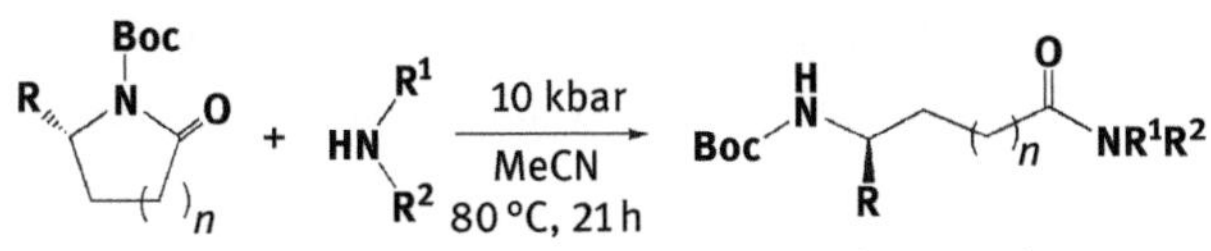

Scheme 5.22: Transamidation of lactams to ω-amino-carboxamides.

5.2.4 [2 + 2] cycloaddition

Isocyanate [2 + 2] cycloaddition on alkenes presents a convenient synthetic route to β-lactams. Positive effect of pressure on the addition of phenyl isocyanate to 2,3-dihydrofuran was evident from very large activation volume $\Delta V^{\neq} = -28\ \text{cm}^3\ \text{mol}^{-1}$ (neat) [45]. The feasibility of conducting this reaction with the aid of pressure was demonstrated by Jurczak. Toluene-4-sulfonyl isocyanate readily reacted at 10 kbar with glycals, a 3,4-dihydro-2*H*-pyran derivatives to obtain the corresponding β-lactams in regio-stereospecific manner (Scheme 5.23) [46]. The isocyanate addition occurs exclusively at the *trans*-position in relation to C-3 acetoxy group. In classical thermal conditions, glycals are unreactive in such [2 + 2] cycloadditions (due to retro-addition) and isocyanate causes dimerization of sugar substrate. By heating, β-lactam, adducts rearrange to the corresponding α,β-unsaturated amides, which is possible for some adducts only under high pressure, whereas at atmospheric conditions, the retro-process is favored over rearrangement [47] .

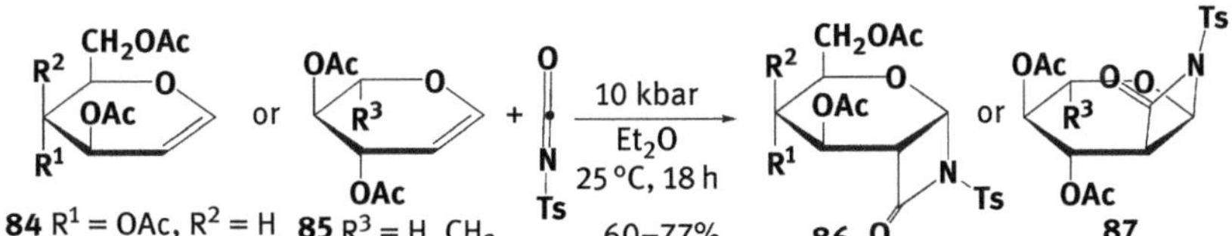

Scheme 5.23: Lactam formation by isocyanate [2 + 2] cycloaddition with alkenes.

The acyl isocyanates are relatively less reactive than sulfonyl isocyanates, which require mild heating (50 °C) and led to the parallel formation of *α,β*-unsaturated amides by the decomposition of *β*-lactam ring. The content of the reaction mixture highly depends on the temperature and time. Interestingly, pressurizations at room temperature in the case of the [2 + 2] cycloadditions of benzoyl and 3,5-dinitrobenzoyl isocyanate on di-*O*-benzyl-D-arabinal **88** were accompanied by the competitive [4 + 2] cycloadditions and formation of *α,β*-unsaturated amides **91** (Scheme 5.24) [48]. The presence and isolation of thermally unstable cycloadducts **90** are due to the applied mild reaction conditions. In the case of more reactive trichloroacetyl isocyanate, cycloadditions with di-*O*-acetyl-L-rhamnal, both at 1 bar and 10 kbar favor the [4 + 2] product over [2 + 2], and the content of amide in the reaction mixture increases with time and temperature [49].

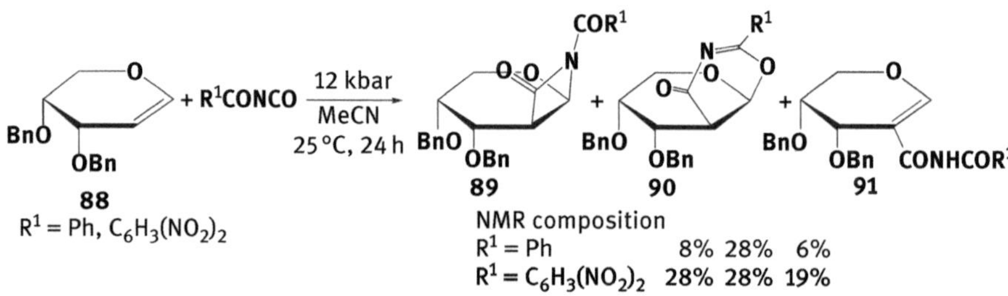

Scheme 5.24: Side reactions in the synthesis of β-lactams.

Another competitive side-reaction in isocyanate additions to less reactive alkenes is dimerization (trimerization) of phenyl isocyanate to **88** and **89** (Scheme 5.25) [45]. Taguchi has found that additions of phenyl isocyanate to more reactive enol ethers provided the expected *β*-lactams (8 kbar, 100 °C, 20 h, neat) in high yields.

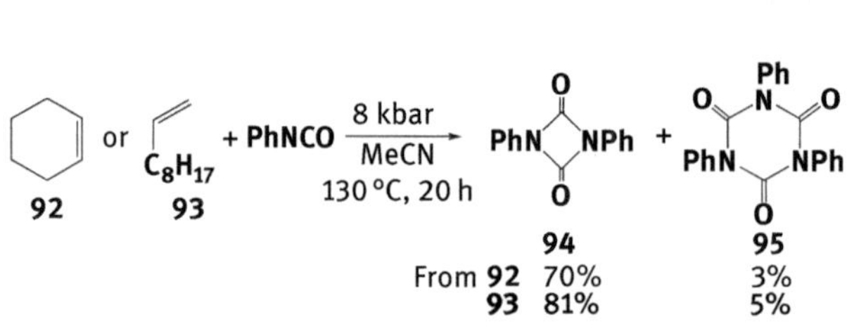

Scheme 5.25: Dimerization of phenyl isocyanate.

The reactivity of isocyanates in the high-pressure additions of isocyanates to enol ethers (12 kbar, MeCN, 50 °C) could be also increased by the addition of $ZnCl_2$ catalyst. Scheeren and coworkers have prepared the corresponding *β*-lactams in 35–78%

yields [50]. On the other hand, isothiocyanates were found to be reactive enough to be condensed with amines under high pressure affording thioureas, without the use of catalyst (6 kbar, 40 °C, 18–44 h, THF, 47–95%) [51]. Formation of thioureas by thermal condensation is troubled with the formation of undesired side-product, symmetrical thiourea (originated from the decomposition of isothiocyanate, THF reflux). Under pressure, this side-reaction is suppressed, and only minute amounts of side-product were obtained, with the increase in yield of unsymmetrical thiourea.

The reverse process, amide C–N bond cleavage under pressure, was much less studied. For base-induced decomposition of chloroacetylhydrazide to nitrogen, hydrazine, chloride and acetate ions, le Noble and Chang determined apparent positive activation volume ($\Delta V^{\neq}_{app}$ = +3 cm^3 mol^{-1}), at 4 kbar in bicarbonate buffer, which needs to be corrected by the change in pH of buffer by pressure ($\Delta V^{\neq}$ = –5 cm^3 mol^{-1}), thus becoming slightly negative ($\Delta V^{\neq}$ =–2 cm^3 mol^{-1}) [52]. The whole process was found not to be concerted and the first intermediate is an anion.

Amide bond formation is also the crucial mechanistic step in several high-pressure-promoted reactions, such as condensation of carboxylic acids with *o*-phenylenediamines to synthesize benzimidazoles [53].

5.3 Aza-Michael reaction

The importance of aza-Michael reaction in synthetic organic chemistry is evidenced in large number of publications dealing with the pressure effects on the process. Synthetic studies have shown significant acceleration of reaction, which are supported by the kinetic measurements by Jenner. The extraordinary large values of activation volumes were measured for the addition of isopropylamine and *t*-butylamine to acrylonitrile leading to *β*-aminonitriles in MeCN ($\Delta V^{\neq}$ = –65 and –60 cm^3 mol^{-1}, respectively, 3 kbar) [54] and perfluorohexane (C_6F_{14}, –65 cm^3 mol^{-1}) [55]. This indicates the reaction mechanism proceeding via a nucleophilic attack on the double bond with formation of a zwitterionic species. The size of $\Delta V^{\neq}$ implies sizeable electrostriction due to the emergence of the zwitterion and a highly ordered, late transition state. Large activation volumes were also found for the reaction of acrylonitrile and *tert*-butylamine in various solvents ($\Delta V^{\neq}$ from –21 to –55 cm^3 mol^{-1}) (Scheme 5.26) [56]. These values decrease with the increase in solvent polarity: diethyl ether (-55), chloroform (-54), acetonitrile (-56), methanol (-35), ethylene glycol

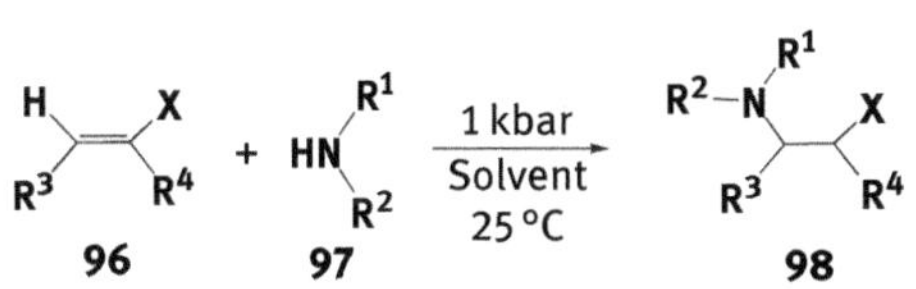

Scheme 5.26: Synthesis of β-aminonitriles by aza-Michael reaction.

(-33), formamide (-21) and water (-25), which means a diminishing sensitivity of the reaction to pressure. This observation is consistent with difficulties in formation of zwitterions in less polar solvents, and assisted by pressure. In polar solvents, the introduction of an ionic charge cannot magnify pressure effect much further, that is, electrostriction is much diminished.

A wide range of solvents could be applied in hyperbaric aza-Michael reactions. To the above-mentioned ones, the list is extended by Maddaluno and coworkers with DCM, THF, iPrOH, EtOH, polyhalogenated alcohols such as TCE (trichloroethanol), TFE (trifluoroethanol) and HFIP (hexafluoroisopropanol) [57]. Water was employed as a solvent of choice by Jenner in aza-Michael reactions (3 kbar, 30–50 °C, 24 h, 45–100% yield) [58] and by Kotsuki in reactions of *α*, *β*-unsaturated enones with nitrogen heterocycles (6 kbar, 60 °C, 20 h, yield up to 99%) [59]. However, β-aminoesters in aqueous solution were found to quickly reverse to substrates, while β-aminonitriles and β-aminoamides are more stable [60].

The reactivity was also enhanced by the combined effects of high pressure and Lewis acid catalysts. Jenner have found that $Eu(fod)_3$, $Yb(OTf)_3$ and $ZrCl_4$ catalysts improved the reaction yield, and ytterbium catalyst is the most efficient (in DCM, at 3 kbar, 36 °C, 15 h) [61]. For the reaction of *t*-butylamine with acrylonitrile, $LiClO_4$ in diethyl ether was the catalyst of choice (9 kbar, 30–50 °C, 24 h) [62]. Furthermore, Kotsuki reported that for addition of amides and ureas onto α,β-unsaturated enones (6 kbar, 60 °C, MeCN) the *p*-toluenesulfonic acid catalyst was the most efficient, and less efficient are 2,4-dinitrobenzenesulfonic acid, (+)-(*S*)-camphorsulfonic acid and diphenyl phosphate [63].

Conjugate addition of amines represents an attractive synthetic way to aziridines. Maddaluno has reported that the optimal conditions for the reaction of 4-*tert*-butylcyclohexylidene bromoacetate **99** with benzyl amine are at 11 kbar in methanol (Scheme 5.27) [64]. High yield of aziridines **102/103** was obtained in 4 days, whereas at atmospheric pressure and room temperature condensation

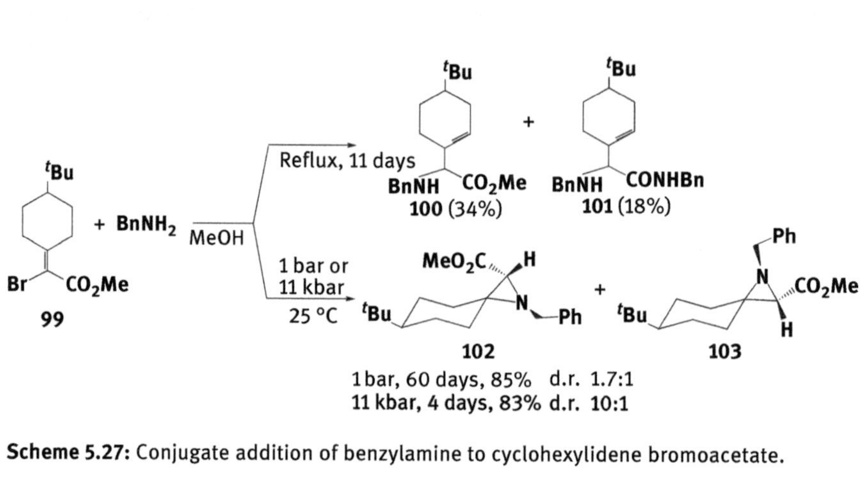

Scheme 5.27: Conjugate addition of benzylamine to cyclohexylidene bromoacetate.

required 60 days to afford similar yields. Such a long time is an important drawback to its synthetic utility at classical conditions. In addition to the increase in the reaction rate, the application of pressure significantly improved stereoselectivity. The diastereomeric ratio increases from 1.7:1 to 10:1 at atmospheric pressure and at 11 kbar, respectively. Differences could arise from variations in steric congestion in the transition states or by the kinetic control of reaction under pressure, and the thermodynamic control in thermal conditions. When the reaction was carried in refluxing methanol, aziridine was not isolated, but instead a mixture of the α-amino β,γ-unsaturated ester **100** and the corresponding amide **101** arising from transformation of initial aziridine. The employment of less nucleophilic amines such as aniline led to preferable formation of products analogous to **100** and **101**, whereas the bis-spiroaziridination was achieved by using dibromodiester as substrate [65].

The side-reactions leading to **100** and **101** were used as an advantage to synthesize cyclic amide and ester products using *N,O*- and *N,N*-binucleophiles (Scheme 5.28) [66]. The condensation of ethyl cyclopentylidenebromoacetate **102** with ethylenediamine at reflux provided a mixture of cyclopentenylpiperazinone **103** and cyclopentyldihydropyrazinone **104**. These products arose from the initial aza-Michael addition, followed by the amidation of the ester functionality. Under high pressure, at room temperature, piperazinone **103** was obtained as a single product in better yield. When **102** was condensed with *N*-benzylaminoethanol, aza-Michael reaction was followed by transesterification, leading to various mixtures of 1,3-dihydro-[1,4]oxazinone **107**, aminoalcohols **105** and **106**.

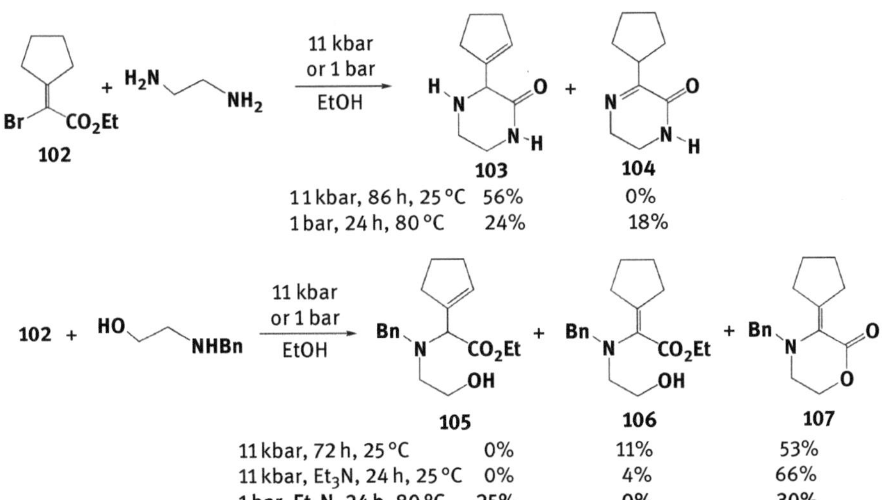

Scheme 5.28: Aza-Michael addition/amidation and /transesterification reactions.

Double aza-Michael condensation of cyclohexa-1,4-diene diester **108** with primary amines provided 7-azabicyclo[2.2.1] skeleton. This is an unprecedented reaction, in which the low reactivity of secondary amines in conjugate additions is enhanced by high pressure. The combination of binucleophiles and bielectrophiles provided aza-bridged structures such as **110**, and tricyclic products **111** by further amidation reaction of bridgehead esters in high yields (Scheme 5.29) [67]. In these conditions, only intramolecular aza-Michael reaction takes place, and the products arising from the double aza-Michael reaction of two amines with two double bonds within **108** were not obtained, even with the longer and more flexible diamines such as diethylenetriamine. Instead, the dimeric bridged structure **113** was obtained by two separate double Michael additions of diethylenetriamine to **108**, with subsequent lactam ring formation. In the case of secondary amines, double aza-Michael addition of two amine molecules (*N,N′*-dimethylethylenediamine and morpholine) to **108** furnished dispirocyclic bis-(lactam) **116** and diamino diester **117** [68].

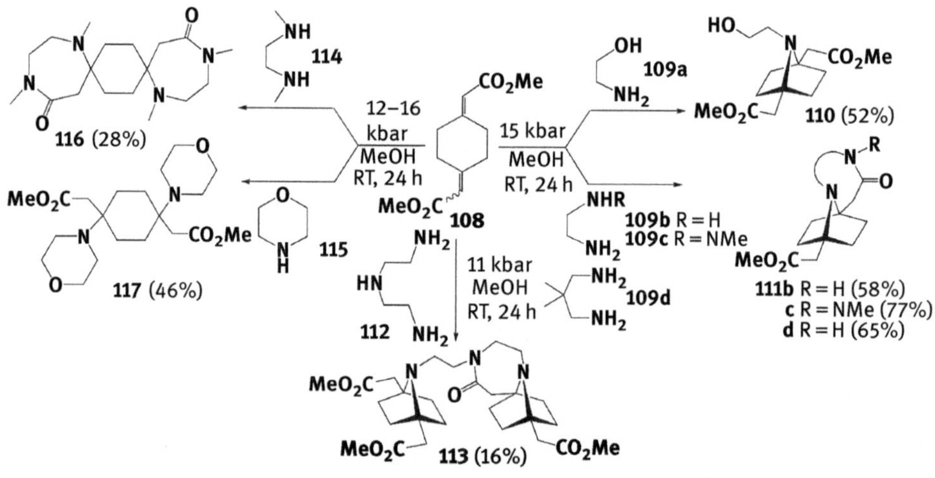

Scheme 5.29: Double aza-Michael condensations of cyclohexa-1,4-diene diester.

Tsypysheva and coworkers have shown that the low nucleophilic reactivity of urea functionality could be increased to act as donor in hyperbaric aza-Michael reactions. Cyclic carboxamide derivative of natural product (–)-cytisine **118** readily reacted at 6 kbar (in the presence of 10 mol% of *p*-TsOH) with β-nitrostyrene 3-buten-2-one and 2-cyclohexen-1-one to obtain 12-*N*-adducts **119**–**121** in high yields (Scheme 5.30) [69]. In the case of DMAD (dimethylacetylene dicarboxylate), the obtained yield of Michael adduct is much lower; however, no cycloaddition product was noted. However, *N*-phenylmaleimide participated exclusively in Diels–Alder reaction affording the *endo*-cycloadduct **123**.

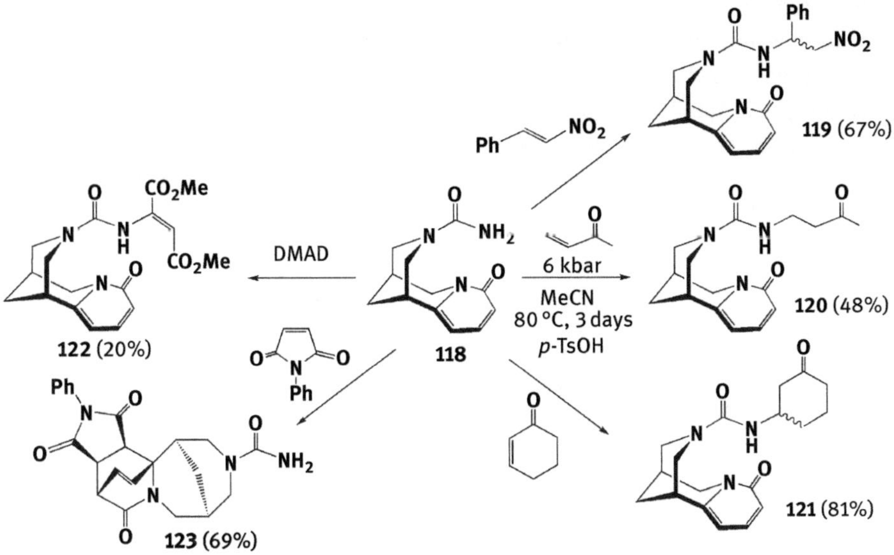

Scheme 5.30: aza-Michael reactions of (–)-cytisine carboxamide.

The choice of amine plays an important role in the reaction outcome, as shown by Kotsuki and coworkers (Scheme 5.31) [70]. Whereas the addition of primary amines onto low-activated double bond of ester **124** led to the formation of aza-Michael adducts **125**, secondary amines in many cases afforded two products **125** and **126**. The later one is the elimination product of **125**. The ratio of products is related to the nucleophilicity of amines, acidity of the NH proton of amine and the preorganization of the reagents via hydrogen bonding. The advantageous use of pressure conditions could be illustrated by the addition of sterically congested diisopropylamine, which does not occur in atmospheric conditions.

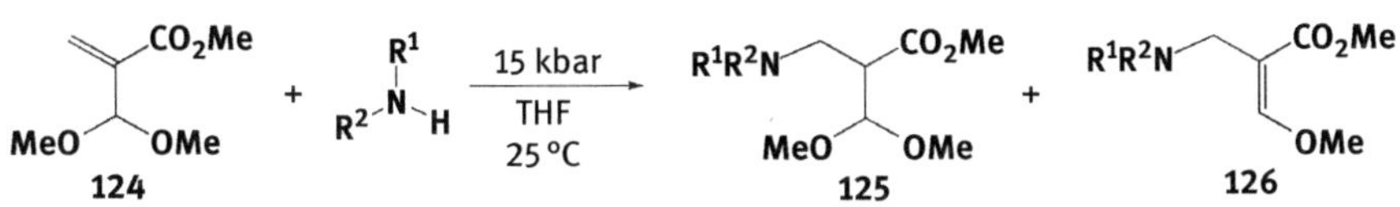

	125	126
$PhNH_2$ 48 h	34%	0%
$BnNH_2$ 24 h	73%	0%
$(i\text{-}Pr)_2NH$ 5 h	66%	0%
1 bar, Rfx, 24 h	0%	0%
Bn_2NH 5 h	32%	56%
Bn(Et)NH 48 h	73%	21%
1 bar, Rfx, 48 h	55%	22%

Scheme 5.31: Amine influence on the aza-Michael reaction.

High enantioselectivity could be induced in asymmetric aza-Michael reactions with the aid of high pressure. The additions of primary amines to alkyl crotonates proceed sluggishly in thermal conditions, and this reaction is very efficient at 15 kbar in methanol at room temperature. Maddaluno obtained the highest de's for the reaction of benzyl amine with chiral 8-phenylmenthyl crotonate derivatives **127** possessing *p*-*t*-butylphenyl, *p*-phenoxyphenyl and β-naphthyl substituents on menthyl core (Scheme 5.32) [71]. The resulting enantioselectivity is the effect of the "π-stacking" in which the aryl substituents shield one face of the C=C bond, directing the approach of amine from the other face.

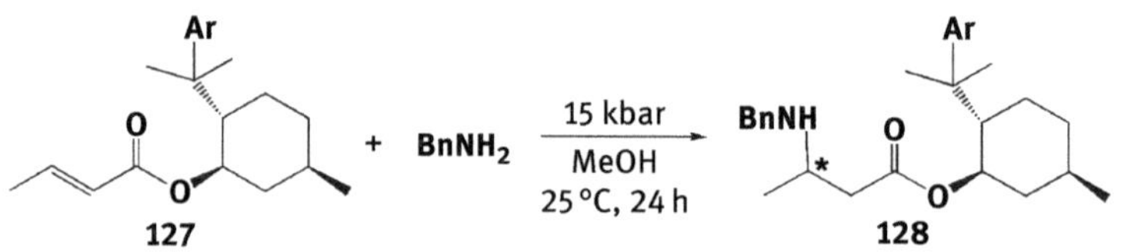

	Yield	de
Ar = *p*-*t*-BuPh	65%	75%
p-PhOPh	90%	95%
β-Naphth	50%	>99%

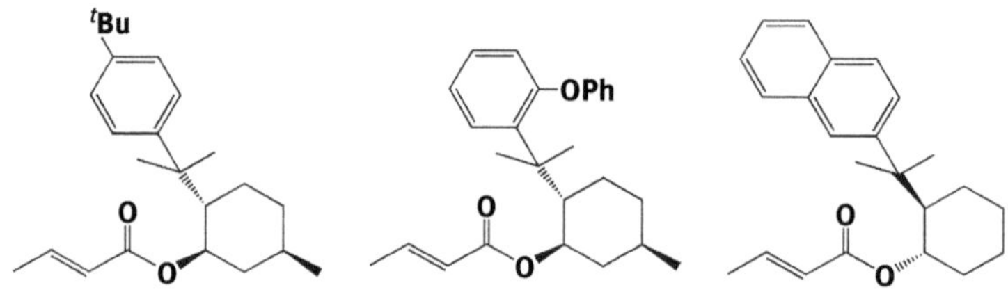

127 8-*p*-*t*-BuPhM **127** 8-*p*-PhOPhM **127** 8-*β*-NaphthM

Scheme 5.32: Asymmetric aza-Michael reaction of chiral 8-phenylmenthyl crotonate.

The removal of the methyl substituent from the cyclohexane ring of the crotonate **127** does not affect the efficacy of the chiral inducer and high-pressure-mediated (14 kbar, MeOH) addition of diphenylmethanamine led to the conjugate adduct analogous to **128** with a very high de (98%) [72]. The crucial role of the high pressure to the stereoselectivity is evidenced by ambient pressure reaction (5 days, 40 °C, MeOH), which provided the β-amino ester with a low de (10%) and 50% reaction yield. Stereoselectivity of these aza-Michael reactions was further studied computationally and explained by the formation of the reactive complexes and their conformations [73]. For this purpose, the effects of pressure on intra- and intermolecular energies were taken into account by the modification of the molecular mechanics method. The other approach to induce stereoselectivity of piezo-aza-Michael addition of diphenylmethanamine was the use of chiral sugar-based crotonates (MeOH, 12 kbar, RT) [74].

Among other related reactions, the nucleophilic addition of 2-halogenopyridines with dimethylacetylene dicarboxylate which produces 9a*H*- and 4*H*-quinolizines starts with aza-Michael addition to acetylenic bond. The process is greatly enhanced by pressure (10 kbar, Et_2O, RT, 7 days) in view of formation of dipolar intermediates and net volume contraction by conversion of three molecules into a single one [75]. Reactions of pyridines and indoles with DMAD to afford 1:2 and 1:3 adducts which start with aza-Michael reaction, followed by cycloadditions are further examples of pressure (10 kbar)-promoted condensation [76].

5.4 Functional transformations of carboxylic acid derivatives

The condensations of carboxylic acid derivatives are strongly facilitated by high pressure.

The formation of oximes from ketones is the subject of several synthetic studies, which show that hindered ketones form oximes with problems and the most hindered oximes can be prepared only by the application of high pressure, demonstrating the profitable pressure effect in reactions that involve sterically congested ketones [77]. For instance, reaction of cortisone-3,20-bis-dioxolane **133** with hydroxylamine hydrochloride to produce cortisone-3,20-bis-dioxolane-11-oxime (3,20-bis-ethylenedioxy-5-pregnene-17α-21-diol-11-oxime) **134** could be achieved with difficulties at ambient pressure in low yields providing mixtures of products. The low reactivity 11-ketosteroid functionality was significantly enhanced at 9.2 kbar affording the oxime in 74% yield (Scheme 5.33) [78]. Negative volume of activation has been observed ($\Delta V^{\neq}$ = −20 cm^3 mol^{-1}) for this transformation in ethanol–water–pyridine in 1:1.3:2.1 solvent system.

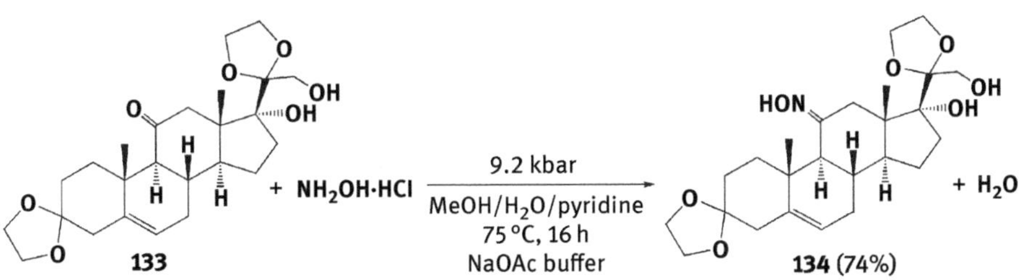

Scheme 5.33: Reaction of cortisone-3,20-bis-dioxolane with hydroxylamine.

Studies on the Maillard reaction (reaction of amino acids with sugars) in hyperbaric conditions shed some light on the reaction mechanism. Reactions between tryptophan and glucose were used as models for the early stages of the mechanism. Isaacs has reported that ring opening of the pyranoside ($\Delta V^{\neq}$ = −12 cm^3 mol^{-1}, H_2O) [79], aldehyde to amine condensation ($\Delta V^{\neq}$ = −21 cm^3 mol^{-1}, 6 kbar, water) and the formation of the Amadori product ($\Delta V^{\neq}$ = −14 cm^3 mol^{-1}) are favored by pressure, whereas the last step, decomposition of Amadori product to heterocyclic products, is retarded by pressure ($\Delta V^{\neq} = +17\ cm^3\ mol^{-1}$) [80].

Pinner condensation of *N*-monosubstituted amidine with an *α*-substituted-*β*-ketoester which affords pyrimidinones could be carried out in much better yields by the application of pressure (Scheme 5.34) [81]. The ratio of formed 3-substituted-4 (3*H*)-pyrimidinones **137** and **138** depends on the reaction conditions applied (solvent) and could be channeled toward each of two products. The initial reaction step in this transformation is the formation of C–N bond by the addition of nitrogen onto carbonyl. In thermal conditions, reaction is much less satisfactory, and amidine substrate **135** decomposes to a mixture of products.

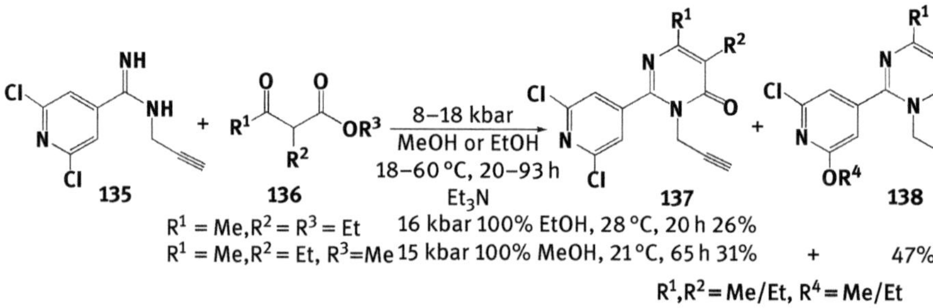

Scheme 5.34: Pinner condensation of amidine with α-substituted-β-ketoesters.

The same type of substrate, ethyl acetoacetate **140** and 2,4-pentanedione **141** were condensed with amines to form enamines at 3 kbar and near room temperature in better yields than in atmospheric conditions (Scheme 5.35) [82]. Further increase in efficiency of more difficult reactions was achieved by the catalysis with Ytterbium triflate. The rate increase by pressure was connected to the formation of zwitterionic intermediates, which exert considerable electrostriction in the transition state.

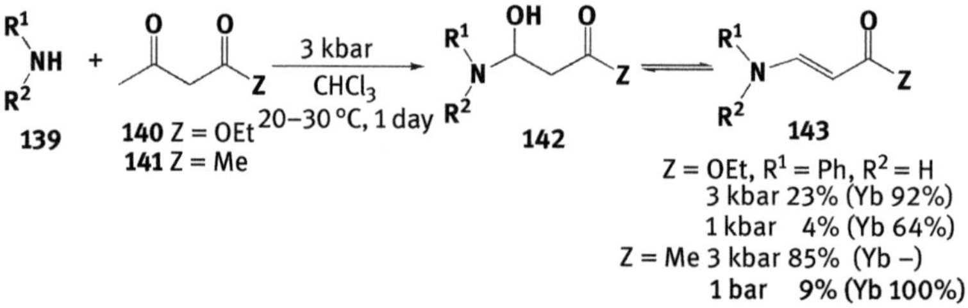

Scheme 5.35: Condensations of ethyl acetoacetate and 2,4-pentanedione with amines.

In the Biginelli tricomponent reaction, an aldehyde condenses with urea and ethyl acetoacetate to form 3,4-dihydro-2(1*H*)-pyrimidinones **150**. This process is facilitated by the application of high pressure, especially in the case of sterically demanding reactants (Scheme 5.36) [83]. The first reaction step is the nucleophilic addition of aldehyde to urea forming the new C–N bond of intermediate **146**, which is followed by water elimination. The intermediate **147** adds at the α-position of the ketoester and the ureide intermediate **149** subsequently loses water to obtain the final product. Since Jenner established that pressure does not have effect on the final steps leading to **149** and **150**, the rate-determining step has to be the formation of **146**.

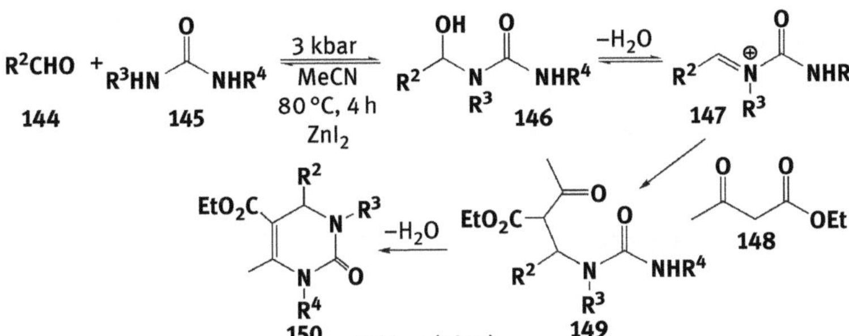

	3 kbar (1 bar)
R^2 = Et, R^3 = H, R^4 = H	73% (43%)
R^2 = ^{t}Bu	35% (2%)
R^2 = Et, R^3 = Me, R^4 = H	90% (85%)
R^2 = ^{t}Bu	15% (0%)

Scheme 5.36: Biginelli reaction of aldehyde with urea and ethyl acetoacetate.

Other reactions of functional derivatives of carboxylic acids by forming carbon–nitrogen bonds that are promoted by high pressure include anhydride transformation to imide. The reaction of the pentacyclic anhydride **151** with various aromatic amines under mild (room temperature) and forcing conditions (13 kbar) was partially successful in the production of amic acids such as **153** (Scheme 5.37) [84]. The major product that is favored by pressure was bis-amide **154**, rather than the amic acid. Although anhydride **151** is thermally stable, the corresponding amic acids **153** formed in the reaction are much more prone to retrocyclization and the use of mild temperature in

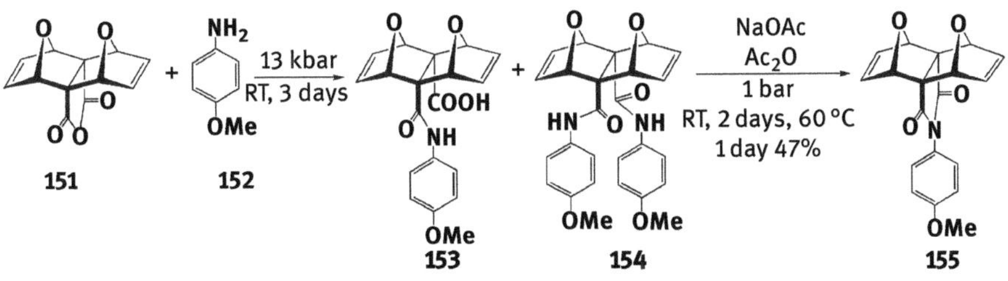

Scheme 5.37: Pentacyclic anhydride transformation to imide.

conjunction with high-pressure conditions helped to increase the yield obtained by thermal reaction.

Synthesis of 1,3,5-triazines by the nitrile condensation has been the subject of several studies, which point that high pressures and temperatures are required. Different conditions were applied by Bengelsdorf (35–50 kbar, 350–500 °C, neat, 5–7 min) [85], McKusick and coworkers (7–8.5 kbar, 70–150 °C, alcohols, 18 h) [86], Korte and coworkers (8–10 kbar, 200 °C, alcohols, sulfolan, 10–15 h) [87] and Shibuya et al. (4 kbar, 160 °C, neat, 20 h) [88]. The pressure acceleration is associated with the reaction mechanism involving dipolar intermediates produced by self-ionization under the extreme reaction conditions and a net volume contraction. Due to multiple condensation reactions, phthalonitrile was also efficiently condensed to phthalocyanine at 10 kbar in the presence of DABCO superbase [89].

5.5 Miscellaneous reactions

The *N*-derivatization of aromatic compounds via nucleophilic aromatic substitution (S_NAr) in hyperbaric conditions was exploited by several authors. For the S_NAr reactions promoted by pressure, highly negative activation volumes were determined by Brower. The $\Delta V^{\neq}$ of aryl halide reactions with piperidine vary from – 8 to –64 $cm^3\ mol^{-1}$ (1.4 kbar, water), and the highest value was obtained for bromoquinoline [90]. These significantly negative values are associated with the formation of the intermediate ionic Meisenheimer complex and significant electrostriction [91].

Besides examples that were described earlier in the chapter, Matsumoto applied this favorable pressure effect to synthetic purposes of obtaining aminopyridine derivatives from halopyridines (Scheme 5.38) [92]. Primary and secondary amines reacted readily at modest temperatures and without Lewis acid catalyst when subjected to 6–8 kbar pressure.

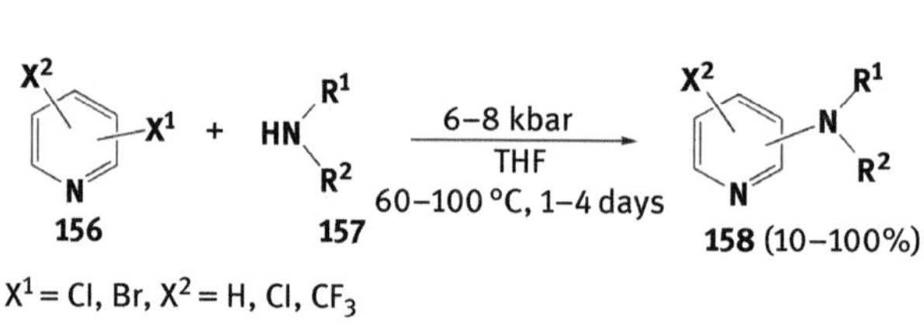

Scheme 5.38: Synthesis of aminopyridines from halopyridines.

Synthetic utility of hyperbaric conditions for the nucleophilic aromatic substitutions could be exemplified by the amination of 4-amino-6-chloro-1-phenylpyrazolo [3,4-*d*]pyrimidine **159** (Scheme 5.39) [93]. Young has found that the employment of classical synthetic conditions at ambient pressure does not lead to the product,

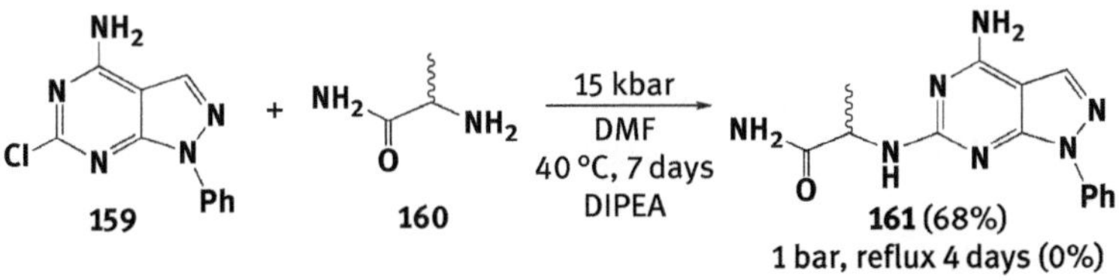

Scheme 5.39: Amination of 4-amino-6-chloro-1-phenylpyrazolo[3,4-*d*]pyrimidine.

whereas under high pressure, the C-6 substituted pyrazolo[3,4-*d*]pyrimidine product **161** was obtained in high yield.

Besides aromatic chlorides and bromides, aromatic fluorides [94] and triflates [95] were also employed as substrates in the nucleophilic aromatic substitutions. A good pressure driving toward polysubstituted products was achieved in S_NAr reactions of 5,10,15,20-tetrakis(pentafluorophenyl)porphyrin **162** with *N*-methylpiperazine (Scheme 5.40) [96]. With the increase in pressure, formation of the tetrasubstituted piperazine product **168** is being favored and higher in proportion, while mono- and bis-products are converted to **167** and **168.**

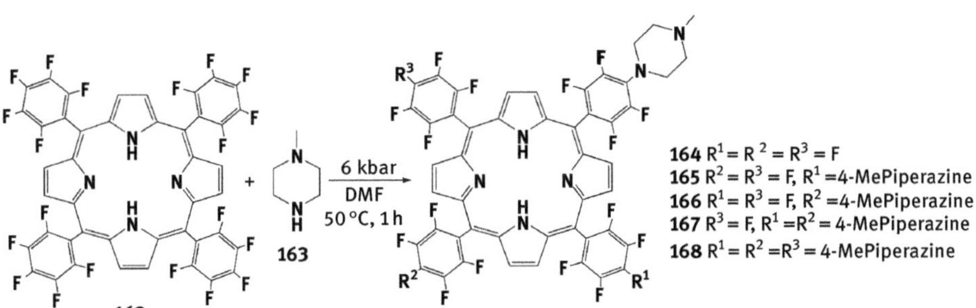

	164	165+166	167	168 (%)
1 bar, 1 h	23	0	0	0
3 kbar, 1 h	25	41	14	0
4.5 kbar, 1 h	0	21	17	53
6 kbar, 1 h	0	0	10	82

Scheme 5.40: Formation of tetrasubstituted piperazine porphyrin.

Yamamoto has found that ring opening of the epoxides with aliphatic amines to obtain β-amino alcohols could be carried out in high-pressure conditions [97]. In the presence of $Yb(OTf)_3$ catalyst (10 kbar, DCM, 18 h, RT) products were obtained in 40–90% yields, and less effective conditions were the employment of pressure only (19%). In the case of reactions of nitrogen heterocycles (imidazoles and pyrazoles) with styrene oxide (10 kbar, MeCN, 65 °C, 3 days) the *N*-(2-phenylethanol) heterocycles were prepared in high yields (53–72%), without the use of catalyst [98]. Diverse chiral imidazole and pyrazole ligands have been prepared by Kotsuki under high pressure when optically active epoxides **170** were employed under 10 kbar (Scheme 5.41) [99].

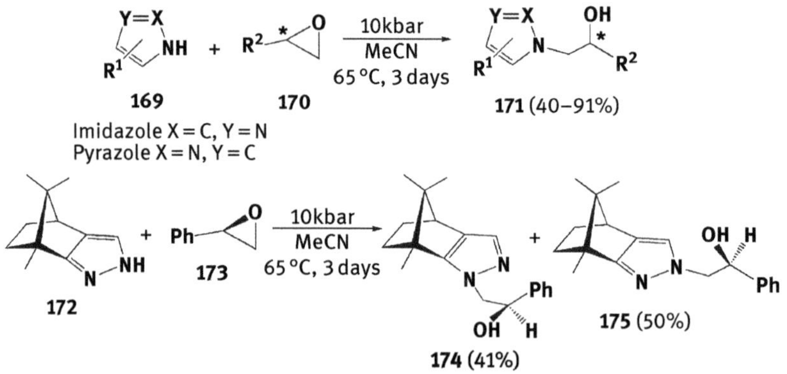

Scheme 5.41: Ring opening of epoxides with aliphatic amines.

In variance to monoheterocycles that led to single product, the reaction of (+)-camphorpyrazole **172** with (*R*)-(+)-styrene oxide **173** in the presence of TBAF (tetrabutylammonium fluoride) provided two regioisomeric products **174** and **175.**

Similarly to S_NAr reactions described earlier, hydrazinolysis of 2,2,2-trifluoro-1-phenylethyl tosylate by the ammonia, which is severely impeded S_N1 or S_N2 reaction by the presence of α-CF_3 group in tosylate, reacts smoothly at 6 kbar pressure to afford the corresponding amine [100].

Due to being condensation reactions, polymerization processes are promoted by high pressure. Pressure in the liquid phase facilitates the direct oxidative polymerization of anilines (*o*- and *m*-aminobenzenesulfonic acid, 19 kbar, with $Na_2S_2O_8$, in aqueous LiCl, $FeSO_4$, 20 °C, 18 h) [101]. Young has also reported that oligo anilines were obtained in respectable yields by high-pressure-promoted S_NAr reaction of aromatic amines with fluoronitrobenzenes (15 kbar, DMF, 30 °C, Et_3N, 3 days) [102]. On the other hand, condensation of tetramethylguanidine with carbodiimides affording biguanides was less advantageous under high pressure, in comparison to microwave irradiation [103].

The development of zwitterion intermediate in the synthesis of azetidines from imines and 2,3-dihydrofuran is critical for the reaction promotion by high pressure. Shibuya has shown that the 6,7-aryl-2-oxa-7-azabicyclo[3.2.0]heptane compounds **178** could be prepared via formal [2 + 2] process in moderate yields at 8 kbar (Scheme 5.42)

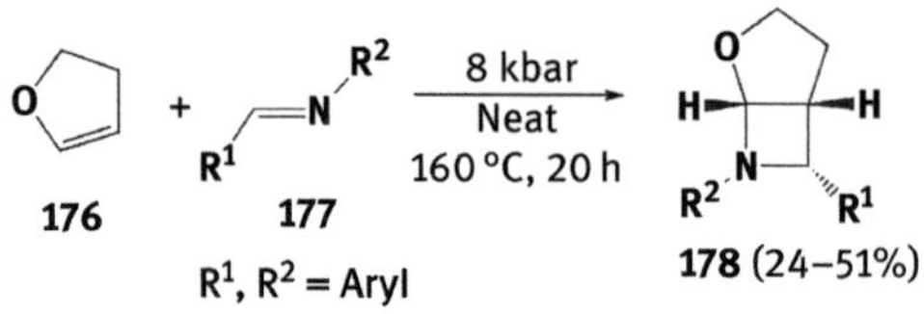

Scheme 5.42: Synthesis of azetidines from imines and 2,3-dihydrofuran.

[104]. The [2 + 2] reactions of imines with electron-rich alkenes were carried out at lower temperatures (RT to 50 °C, *tert*-butyl methyl ether) by the pressure increase to 15 kbar. Azetidines and β-amino carbonyl compounds were prepared in 45–90% yield [105].

Several C–N bond breaking processes take place in high-pressure conditions, such as the decomposition of the *N*-nitrosamide into esters (5 kbar) by le Noble et al. [106], the lactam hydrolysis of alkaloid lupanine hydrolysis (10 kbar, 24 h, 20 °C) by Barciszevski and coworkers [107] and decomposition of 3-methyl-1-*p*-tolyltriazene with benzoic acid into amine and dinitrogen (1.2 kbar) by Laila and Isaacs [108]. For this reaction in acetonitrile and chloroform, activation volumes $\Delta V^{\neq} = -4$ and $-15\ cm^3\ mol^{-1}$ were determined, and their magnitude was explained as a combined value of two processes: associative protonation and dissociative decomposition of the formed complex.

Unexpected transformation induced by high pressure of nitroaromatics with 4-(cyclohex-1-en-1-yl)morpholine **181** involves a C–N bond cleavage (Scheme 5.43) [109]. The complex sequence of processes involving C–C, C–O and C–N multiple bond formation takes place in hyperbaric conditions at 14 kbar. Polycyclic products **186** and **187** possessing nitroaromatic core, two of enamine less one of morpholine moiety were obtained in high yields, as the single diastereomers. Mechanism of this reaction starts with nucleophilic addition of enamine **181** to the nitroarene, followed by rearomatization. The attack of the oxygen atom on the electrophilic carbon atom is followed by the elimination of morpholine fragment to form the nitroso compound **185**. The reaction of a second molecule of **181** with the nitroso group is followed by double cyclization, which includes the keto group, and subsequently the iminium

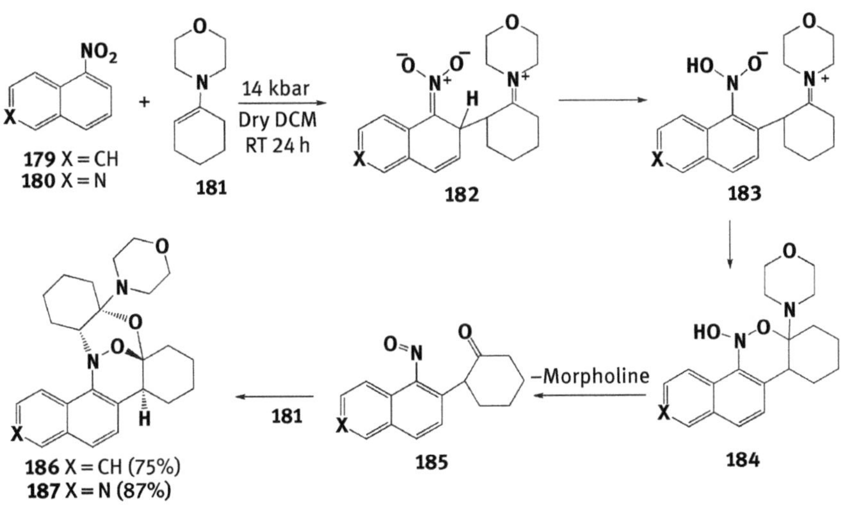

Scheme 5.43: Reaction of nitroaromatics with 4-(cyclohex-1-en-1-yl)morpholine.

species to obtain **186.** Several of the reaction steps are accelerated under high pressure, two additions and the cyclization, whereas the morpholine elimination step is disfavored. The solvation of the leaving molecule and ionic intermediates has the positive effect on the elimination process. At ambient pressure, initial addition process is not constructive and the addition of enamine requires high pressure to become efficient in the case of weak electrophiles such as nitroarenes.

References

[1] Gibson, RO., Fawcett, EW., & Perrin, MW. The effect of pressure on reactions in solution I-sodium ethoxide and ethyl iodide to 3000 kg/cm^2. II-pyridine and ethyl iodide solution to 8000 kg/cm^2. Proc. R. Soc. Lond. A 1935, 150, 223–240.

[2] Fawcett, EW., & Gibson, RO. The influence of pressure on the rate of formation of cetylpyridinium halides. J Chem Soc 1934, 396–400.

[3] Perrin, MW., & Williams, EG. The effect of pressure up to 12,000 kg/cm^2 on the reactions between amines and alkyl halides in acetone solution. Proc Roy Soc A 1937, 159, 162–170.

[4] Jenner, G. Quantitative descriptions of the pressure effect in sterically hindered reactions. Physica 1986, 139&140B, 698–700.

[5] le Noble, WJ., & Ogo, Y. The effect of pressure on sterically hindered Menshutkin reactions. Tetrahedron 1970, 26, 4119–4124.

[6] le Noble, WJ., & Asano, T. Special effect of pressure on highly hindered reactions as a possible manifestation of the Hammond postulate. J Am Chem Soc 1975, 97, 1778–1782.

[7] Cheung, CK., Weinger, RS., & le Noble, WJ. Effect of pressure on the rate of methylation of a buttressed pyridine. J Org Chem 1989, 54, 570–573.

[8] Harris, AP., & Weale, KE. Liquid-phase reactions at high pressures. Part XIII. Structural effects in the acceleration of Menshutkin reactions by pressure. J Chem Soc 1961, 146–152.

[9] Okamoto, Y., & Shimakawa, Y. On the synthesis and properties of 1-methyl 2,6-di-*t*-butyl pyridinium iodine. Tetrahedron Lett 1966, 7, 317–319.

[10] Okamoto, Y., & Lee, KI. Reactions of crowded molecules under high pressure. Reactions of 2,6-di-*tert*-butylpyridine with methyl iodide and methyl fluorosulfonate under high pressure. J Am Chem Soc 1975, 97, 4015–4018.

[11] Hou, CJ., & Okamoto, Y. Reactions of 2,6-di-*tert*-butylpyridine derivatives with methyl fluorosulfonate under high pressure. J Org Chem 1982, 47, 1977–1979.

[12] Matsumoto, K., Ikemi, Y., Suda, M., Iida, H., & Hamana, H. Hilbert-Johnson reaction under high pressure: A facile preparation of 2-pyridones. Heterocycles 2007, 72, 187–190.

[13] Ploug-Sørensen, A., Brøndsted Nielsen, M., & Becher, J. Ultra-high pressure synthesis of a tetrathiafulvalene-diquat cyclophane. Tetrahedron Lett 2003, 44, 2979–2982.

[14] Almarzoqi, A., George, AV., & Isaacs, NS. The quarternisation of tertiary amines with dihalomethane. Tetrahedron 1986, 42, 601–607.

[15] Jurczak, J., Ostaszewski, R., Pietraszkiewicz, P., & Sałański, P. High pressure approach to the synthesis of cryptands and related compounds. J Incl Phenom 1985, 5, 553–561.

[16] Jurczak, J., Tkacz, M., & Majchrzak-Kuczynski, U. Formation of quaternary salts of sterically hindered bis-quinolizidine monolactams under high pressure. Synthesis 1983, 920.

[17] Jurczak, J., Kasprzyk, S., Makosza, M., Ostaszewski, R., & Sałański, P. Recent advances in the organic synthesis at high pressure. Effect of pressure on the macrocyclization reaction. High Press Res 1990, 5, 641–643.

[18] Jurczak, J., & Ostaszewski, R. High pressure approach to the synthesis of diazacoronands and cryptands. J Coord Chem 1992, 27, 201–214.

[19] Pietraszkiewicz, M., Sałański, P., & Jurczak, J. High-pressure synthesis of cryptands with aliphatic bridging units. Heterocycles 1985, 23, 547–548.

[20] Tarnowska, A., Tarnowski, P., & Jurczak, J. The unusual course of the high-pressure reaction of 1,10-diaza-[18]-crown-6 with α,β-diiodo ethers. Tetrahedron Lett 2004, 45, 9553–9556.

[21] Pietraszkiewicz, M., Sałański, P., Ostaszewski, R., & Jurczak, J. Synthesis of cryptands under high pressure. The role of solvent and leaving group in double quaternization reactions. Heterocycles 1986, 24, 1203–1204.

[22] Pietraszkiewicz, M., Sałański, P., & Jurczak, J. High-pressure approach to the synthesis of novel chiral cryptands derived from methyl 4,6-*O*-benzylidene-α-D-manno-, -α-D-galactoand -α-D-gluco-pyranoside. J Chem Soc Chem Comm 1983, 1884.

[23] Jurczak, J., Ostaszewski, R., & Sałański, P. High-pressure approach to the synthesis of *N, N'*-dimethyl diazacoronands. J Chem Soc Chem Comm 1989, 184–185.

[24] Jurczak, J., & Ostaszewski, R. A new method for the synthesis of *N, N'*-dimethyl diazacoronands: high-pressure alkylation of α,ω-secondary diamines with α,ω-di-iodo compounds. Synth Commun 1989, 19, 2175–2179.

[25] Sałański, P., Ostaszewski, R., & Jurczak, J. The high pressure synthesis of chiral *N, N'*-dimethyldiazacoronands derived from L- and D-tartaric acid. High Press Res 1994, 13, 35–39.

[26] Okamoto, Y., & Shimizu, H. Steric hindrance under high pressure. reactions of 2,4,6-tri-*t*-butyl-*N*-methylaniline with alkyl iodides. J Am Chem Soc 1968, 90, 6145–6148.

[27] Okamoto, Y., & Shimizu, H. Sterically hindered reactions under high pressure. Reaction of 2,4,6-tri-*t*-butyl *N*-methylaniline with methyliodide. Tetrahedron Lett 1968, 9, 2751–2753.

[28] Tsukube, H., Minatogawa, H., Munakata, M., Toda, M., & Matsumoto, K. High-pressure functionalization of diaza-crown ethers: new synthesis of Ag^+ ion-specific binders. J Org Chem 1992, 57, 542–547.

[29] Matsumoto, K., & Toda, M. High pressure functionalization of diaza-crown ether and thioether. High Press Res 1992, 11, 145–152.

[30] Matsumoto, K., Hashimoto, M., Toda, M., & Tsukube, H. Ag^+ Ion-selective lariat ethers: high pressure syntheses and cation recognition properties. J Chem Soc Perkin Trans 1995, 1, 2497–2502.

[31] Matsumoto, K., Minatogawa, H., Toda, M., & Munakata, M. Synthesis of double-armed azaoligocycles based upon high pressure aromatic nucleophilic substitution reactions. Heterocycles 1990, 31, 1217–1220.

[32] Kotsuki, H., Sakai, H., & Shinohara, T. A new facile method for the synthesis of 4-dialkylaminopyridine derivatives by high-pressure-promoted amination of 4-chloropyridine . Synlett 2000, 116–118.

[33] Jenner, G. High pressure catalyzed addition of lactams to siloxydienes. High Press Res 2000, 18, 239–244.

[34] Kotsuki, H., Wakao, M., Hayakawa, H., Shimanouchi, T., & Shiro, M. Synthesis of novel chiral diazole derivative ligands for the enantioselective addition of diethylzinc to benzaldehyde. J Org Chem 1996, 61, 8915–8920.

[35] Yamada, T., Manabe, Y., Miyazawa, T., Kuwata, S., & Sera, A. Peptide synthesis under high pressure. Coupling reactions of *N*-(benzyloxycarbony1)-amino acid *N*-hydroxysuccinimide esters with *N*-(carboxymethy1)amino acid diesters. J Chem Soc Chem Comm 1984, 1500–1501.

[36] Nakahara, Y., Matsuo, I., Ito, Y., Ubagai, R., Hojo, H., & Nakahara, Y. High-pressure-promoted Fmoc-aminoacylation of *N*-ethylcysteine: preparation of key devices for the solid-phase synthesis of peptide thioesters. Tetrahedron Lett 2010, 51, 407–410.
[37] Matsumoto, K., Hashimoto, S., Uchida, T., Okamoto, T., & Otani, S. Direct aminolysis of nonactivated and thermally unstable esters at high pressure. Angew Chem Int Ed Engl 1989, 122, 1357–1363.
[38] Gante, J., Kalthof, U., Klärner, F-G., & Weber, T. Peptide synthesis under high pressure. Angew Chem Int Ed Engl 1990, 29, 1025–1026.
[39] Klärner, F-G., Kalthof, U., & Gante, J. Peptide synthesis at high pressure: activation volume of a peptide coupling, synthesis of a glutathione derivative. J Prakt Chem 1997, 339, 359–364.
[40] Jurczak, J., Kasprzyk, S., Sałański, P., & Stankiewicz, T. High-pressure approach to the synthesis of macrocyclic diamides. High Press Res 1992, 11, 139–143.
[41] Tarnowska, A., Jarosz, M., & Jurczak, J. High-pressure synthesis of cryptands via double amidation reaction of diazacoronands with active esters of α,β-dicarboxylic acids. Synthesis 2004, 369–372.
[42] Azad, S., Kumamoto, K., Uegaki, K., Ichikawa, Y., & Kotsuki, H. A new practical method for the synthesis of unsymmetrical ureas via high-pressure-promoted condensation of 2,2,2-trichloroethyl carbamates (Troc-carbamates) with amines. Tetrahedron Lett 2006, 47, 587–590.
[43] Margetić, D., Zrinski Antonac, I., Glasovac, Z., Eckert-Maksić, M., & Maksimović, Lj. Reactions of dimethyl carbonate with aliphatic amines under high pressure. Synth Commun 2011, 41, 2283–2289.
[44] Kotsuki, H., Iwasaki, M., & Nishizawa, H. A new powerful method for the transformation of lactams into ω-amino-carboxamides under high pressure conditions. Tetrahedron Lett 1992, 33, 4945–4948.
[45] Taguchi, Y., Tsuchiya, T., Oishi, A., & Shibuya, I. Synthesis of β-lactams from isocyanates and vinyl ethers under high pressure. Bull Chem Soc Jpn 1996, 69, 1667–1672.
[46] Chmielewski, M., Kałuźa, Z., Bełżecki, C., Sałański, P., & Jurczak, J. The synthesis of optically pure β-lactams derived from sugars. High-pressure [2 + 2] cycloaddition of toluene-4-sulphonyl isocyanate to glycals. Tetrahedron Lett 1984, 24, 4797–4800.
[47] Chmielewski, M., Kałuźa, Z., Bełżecki, C., Sałański, P., Jurczak, J., & Adamowicz, H. High-pressure (2 + 2) cycloaddition of toluene-4-suphonyl isocyanate to glycals. Tetrahedron 1985, 41, 2441–2449.
[48] Chmielewski, M., Kałuźa, Z., Sałański, P., & Jurczak, J. High-pressure [2 + 2]cycloaddition of acyl isocyanates to glycals. High Press Res 1992, 11, 171–176.
[49] Chmielewski, M., Kałuźa, Z., Mostowicz, D., Bełżecki, C., Baranowska, E., Jacobsen, JP., Sałański, P., & Jurczak, J. High- and atmospheric pressure cycloaddition of trichloroacetyl isocyanate to 3,4-di-*O*-acetyl-L-rhamnal. Tetrahedron 1987, 43, 4555–4563.
[50] Aben, RWM., Limburg, EP., & Scheeren, HW. A one step synthesis of novel 4-alkoxy-β-lactams by high pressure promoted [2 + 2] cycloadditions of enol ethers and isocyanates derived from amino acids. High Press Res 1992, 11, 167–170.
[51] Kumamoto, K., Misawa, Y., Tokita, S., Kubo, Y., & Kotsuki, H. High-pressure-promoted condensation of isothiocyanates with aminopyridines: efficient synthesis of pyridine - thiourea conjugates as building blocks for hydrogen-bonding receptors. Tetrahedron Lett 2002, 43, 1035–1038.
[52] le Noble, WJ., & Chang, Y-S. Kinetics of reactions in solutions under pressure. XXVI. Fragmentation of chloroacetylhydrazide. J Am Chem Soc 1972, 94, 5402–5406.
[53] Holan, G., Evans, JJ., & Linton, M. Formation of benzimidazoles at high pressure. J Chem Soc Perkin Trans 1977, 1, 1200–1203.

[54] Jenner, G. Comparative study of physical and chemical activation modes. The case of the synthesis of β-amino derivatives. Tetrahedron 1996, 52, 13557–13568.
[55] Jenner, G., & Gacem, B. High-pressure effect on organic reactions in fluorophobic media. J Phys Org Chem 2003, 16, 265–270.
[56] Jenner, G. Kinetic effect of pressure on Michael and Diels-Alder reactions in aqueous solutions. J Phys Org Chem 1999, 12, 619–625.
[57] Fedotova, A., Crousse, B., Chataigner, I., Maddaluno, J., Rulev, AYu., & Legros, J. Benefits of a dual chemical and physical activation: direct aza-Michael addition of anilines promoted by solvent effect under high pressure. J Org Chem 2015, 80, 10375–10379.
[58] Jenner, G., & Salem, RB. Kinetic effects in water and ethylene glycol. Application to high pressure organic synthesis. New J Chem 2000, 24, 203–207.
[59] Uddin, MI., Nakano, K., Ichikawa, Y., & Kotsuki, H. A novel method for the high-pressure-promoted, uncatalyzed aza-Michael reaction of nitrogen heterocycles with enones in water. Synlett 2008, 1402–1406.
[60] Jenner, G. Comparative study of physical and chemical activation modes. The case of the synthesis of β-amino derivatives. Tetrahedron 1996, 43, 13357–13368.
[61] Jenner, G. Catalytic high pressure synthesis of hindered β-aminoesters. Tetrahedron Lett 1995, 36, 233–236.
[62] Jenner, G., & Ben Salem, R. The cause of the rate acceleration by diethyl ether solutions of lithium perchlorate (LPDE) in organic reactions. application to high pressure synthesis. Tetrahedron 1997, 53, 4637–4648.
[63] Azad, S., Kobayashi, T., Nakano, K., Ichikawa, Y., & Kotsuki, H. Efficient Brønsted acid-catalyzed aza-Michael reaction of amides and ureas with α,β-unsaturated enones under high-pressure conditions. Tetrahedron Lett 2009, 50, 48–50.
[64] Rulev, AYu., Maddaluno, J., Plé, G., Plaquevent, J-C., & Duhamel, L. High pressure vs. thermal activation in the conjugate addition of amines: a new access to spirocyclamines. J Chem Soc Perkin Trans 1998, 1, 1397–1401.
[65] Rulev, A., & Maddaluno, J. Spiroaziridination of cycloalkylidene esters under high pressure. Eur J Org Chem 2001, 2569–2576.
[66] Rulev, AY., & Maddaluno, J. Thermal and hyperbaric addition of *N, N-* and *N, O-*binucleophiles on cycloalkylidenic bromo esters. J Phys Org Chem 2002, 15, 590–598.
[67] Rulev, AYu., Yenil, N., Pesquet, A., Oulyadi, H., & Maddaluno, J. Tandem aza-Michael additions under high pressure: a shortcut to the azanorbornyl skeleton. Tetrahedron 2006, 62, 5411–5416.
[68] Larsson, R., Scheeren, HW., Aben, RWM., Johansson, M., & Sterner, O. Total synthesis of transtaganolide E and F: insight in the biosynthesis of the transtaganolides. Eur J Org Chem 2013, 6955–6960.
[69] Tsypysheva, IP., Lobov, AN., Kovalskaya, AV., Petrova, PR., Ivanov, SP., Rameev, SA., Borisevich, SS., Safiullin, RL., & Yunusov, MS. Aza-Michael reaction of 12-*N*-carboxamide of (–)-cytisine under high pressure conditions. Nat Prod Res 2015, 29, 141–148.
[70] Rulev, AYu., Kotsuki, H., & Maddaluno, J. High pressure promoted aza-Michael addition of primary and secondary amines to α-substituted acrylates. Green Chem 2012, 14, 503–508.
[71] d'Angelo, J., & Maddaluno, J. Enantioselective synthesis of β-amino esters through high-pressure-induced addition of amines to α,β-ethylenic esters. J Am Chem Soc 1986, 108, 8112–8114.
[72] Dumas, F., Mezrhab, B., & d'Angelo, J. Investigating the π-facial discrimination phenomenon in the conjugate addition of amines to chiral crotonates: a convenient basis for the rational design of chiral auxiliaries. J Org Chem 1996, 61, 2293–2304.

[73] Dumas, F., Fressigne, C., Langlet, J., & Giessner-Prettre, C. Theoretical investigations of the influence of pressure on the selectivity of the Michael addition of diphenylmethaneamine to stereogenic crotonates. J Org Chem 1999, 64, 4725–4732.

[74] Moura, S., Thomassigny, C., Ligeour, C., Greck, C., Joseph, D., Drège, E., & Dumas, F. Asymmetric aza-Michael addition under ultra-high pressure: short bias to polyhydroxylated piperidines. Green Chem 2011, 13, 1812–1818.

[75] Matsumoto, K., & Ikemi-Kono, Y. High-pressure reaction of 2 -halogenopyridines with dimethyl acetylenedicarboxylate. J Chem Soc Chem Comm 1978, 543–544.

[76] Matsumoto, K., Uchida, T., Ikemi, Y., & Acheson, RM. High-pressure cycloaddition of aza-aromatic compounds. Heterocycles 1981, 16, 138.

[77] Jenner, G. The pressure dependence of sterically hindered organic reactions of ketones. High Press Res 2002, 22, 511–514.

[78] Jones, WH., Tristram, EW., & Benning, WF. High pressure studies: oximes of hindered ketones. J Am Chem Soc 1959, 81, 2151–2154.

[79] Andersen, B., & Grønlund, F. High pressure kinetics of glucose mutatoration studied by optical rotation. Acta Chem Scand A 1979, 33, 275–280.

[80] Isaacs, NS., & Coulson, M. Effect of pressure on processes modelling the Maillard reaction. J Phys Org Chem 1996, 9, 639–644.

[81] Jeżewski, A., Jurczak, J., Lidert, Z., & Tice, CM. Synthesis of herbicidal 3-substituted-4(3*H*)-pyrimidinones under high pressure. J Het Chem 2001, 38, 645–648.

[82] Jenner, G. High pressure and lanthanide catalyzed synthesis of enamino compounds. Tetrahedron Lett 1996, 37, 3691–3694.

[83] Jenner, G. Effect of high pressure on Biginelli reactions. Steric hindrance and mechanistic considerations. Tetrahedron Lett 2004, 45, 6195–6198.

[84] Gaynor, SP., Gunter, MJ., Johnston, MR., & Warrener, RN. Probing the dimensions of semi-rigid inner functionalised U-shaped bis-porphyrin cavities. Org Biomol Chem 2006, 4, 2253–2266.

[85] Bengelsdorf, IS. High pressure-high temperature reactions. I. The trimerization of aromatic nitriles. J Am Chem Soc 1958, 80, 1442–1444.

[86] Cairns, TL., Larchar, AW., & McKusick, BC. The trimerization of nitriles at high pressures. J Am Chem Soc 1952, 74, 5633–5636.

[87] Jarre, W., Bieniek, D., & Korte, F. Synthese und Reaktionen von Symmetrisch Substituierten *s*-Triazinen unter Hohem Druck. Tetrahedron 2975, 31, 619–623.

[88] Shibuya, I., Oishi, A., & Yasumoto, M. Reaction of disubstituted cyanamides with formamides under high pressure. Heterocycles 1998, 48, 1659–1662.

[89] Edmondson, SJ., Hill, JS., Isaacs, NS., & Mitchell, PCH. High-pressure conditions for improved synthesis of phthalocyanines. J Chem Soc Dalton Trans 1990, 1115–1118.

[90] Brower, KR. Kinetic study of aromatic nucleophilic substitution under high pressure. II. Variation of solvent and reactant charge type. J Am Chem Soc 1959, 81, 3504–3507.

[91] Barret, IC., & Kerr, MA. The high-pressure S_NAr reaction of *N*-*p*-fluorobenzyl-2-chlorobenzimidazole with amines; an approach to norastemizole and analogues. Tetrahedron Lett 1999, 40, 2439–2442.

[92] Hashimoto, S., Otani, S., Okamoto, T., & Matsumoto, K. Aminolysis of halogenopyridines at high pressures. Heterocycles 1988, 27, 319–322.

[93] Poulsen, S-A., Young, DJ., & Quinn, RJ. High-pressure synthesis of enantiomerically pure C-6 substituted pyrazolo[3,4-*d*]pyrimidines. Bioorg Med Chem Lett 2001, 11, 191–193.

[94] Kotsuki, H., Kobayashi, S., Matsumoto, K., Suenaga, H., & Nishizawa, H. High pressure organic chemistry; XII. A convenient synthesis of aromatic amines from activated aromatic fluorides. Synthesis 1990, 1147–1148.

[95] Kotsuki, H., Kobayashi, S., Suenaga, H., & Nishizawa, H. High pressure organic chemistry; XI. A new convenient synthesis of aromatic amines from activated phenols. Synthesis 1990, 1145–1146.
[96] Gomes, ATPC., Freire, PC., Domingos, CRM., Neves, MGPMS., Cavaleiro, JAS., Almeida Paz, FA., Saraiva, JA., & Tomé, AC. Synthesis under high hydrostatic pressure – a new method to prepare 5,10,15,20-tetrakis[4-(substituted amino)-2,3,5,6-tetrafluorophenyl] porphyrins. J Porph Phthaloc 2016, 20, 1–13.
[97] Meguro, M., Asao, N., & Yamamoto, Y. Ytterbium triflate and high pressure-mediated ring opening of epoxides with amines. J Chem Soc Perkin Trans 1994, 1, 2597–2601.
[98] Kotsuki, H., Hayashida, K., Shimanouchi, T., & Nishizawa, H. High-pressure organic chemistry. 19. High-pressure-promoted, silica gel-catalyzed reaction of epoxides with nitrogen heterocycles. J Org Chem 1996, 61, 984–990.
[99] Kotsuki, H., Hayakawa, H., Wakao, M., Shimanouchi, T., & Ochi, M. Synthesis of novel chiral diazole ligands for enantioseleetive addition of diethylzinc to benzaldehyde. Tetrahedron Asymm 1995, 6, 2665–2668.
[100] Pirkle, WH., Hauske, JR., Eckert, CA., & Scott, BA. High pressure assisted synthesis. evidence for nucleophilic displacement on 2,2,2-trifluoro-1-phenylethyl tosylate. J Org Chem 1977, 42, 3101–3103.
[101] Chan, HSO., Neuendorf, AJ., Ng, S-C., Wong, PML., & Young, DJ. Synthesis of fully sulfonated polyaniline: a novel approach using oxidative polymerisation under high pressure in the liquid phase. J Chem Soc Chem Comm 1998, 1327–1328.
[102] Brown, CL., Muderawan, IW., & Young, DJ. High-pressure synthesis of oligoanilines. Synthesis 2003, 16, 2511–2517.
[103] Trošelj, P., Jušinski, I., Margetić, D., & Eckert-Maksić, M. Synthesis of highly basic hexasubstituted biguanides by environmentally friendly methods. Synlett 2013, 24, 2540–2544.
[104] Oishi, A., Taguchi, Y., & Shibuya, I. Synthesis of azetidines from imines with 2,3-dihydrofuran under high pressure. Rev High Press Sci Technol 1998, 7, 1253–1255.
[105] Aben, RWM., Smit, R., & Scheeren, JW. High-pressure-promoted (2 + 2) cycloadditions of imines with electron-rich alkenes. A simple route to azetidines and β-amino carbonyl compounds. J Org Chem 1987, 52, 365–370.
[106] le Noble, WJ., White, EH., & Dzadzic, PM. Effect of pressure on the stereochemistry of ion pair collapse in the *N*-nitrosamide deamination. J Am Chem Soc 1976, 98, 4020–4021.
[107] Giel-Pietraszuk, M., Gdaniec, Z., Brukwicki, T., & Barciszewski, J. Molecular mechanism of high pressure action on lupanine. J Mol Struct 2007, 826, 120–125.
[108] Laila, A., & Isaacs, NS. Effect of pressure on the reaction between 3-methyl-1-*p*-tolyl-triazene and benzoic acid; a kinetic study. J Prakt Chem 1996, 338, 691–694.
[109] Diab, S., Noël-Duchesneau, L., Sanselme, M., Kondo, Y., De Paolis, M., & Chataigner, I. High pressure elicits unexpected transformations of plain nitroaromatics with 4-(cyclohex-1-en-1-yl)morpholine. Eur J Org Chem 2018, 18, 2048–2052.

6 C–O, C–S and other bonds

Introduction

Among the high-pressure preparations and reactions of heteroatom bonds covered in this chapter, the largest number of examples deals with the C–O bonds (synthesis of ethers and esters) and C–S bonds (thioethers), whereas other types of carbon–heteroatom or bisheteroatom bonds are less studied. Many other examples of C–O bond-forming reactions are given in cycloaddition chapters (hetero Diels–Alder and dipolar cycloadditions).

6.1 C–O bonds

6.1.1 Ethers

An early account of the formation of ether C–O bonds with the aid of high pressure is given by Fawcett [1]. For the reaction between sodium ethoxide and ethyl iodide in ethanol at 3 kbar and 15–30 °C, authors have demonstrated the beneficial increase in the reaction rates, going from ambient pressure to 3 kbar. This synthetic work on diethyl ether has been further extended to 12 kbar in the follow-up publication by Gibson [2]. Furthermore, esterification of ethyl alcohol with acetic anhydride at 3 kbar and 15–45 °C in ethanol was reported in the same paper. Ethyl acetate reaction was additionally carried out at 8.5 kbar and 80 °C in toluene. Acetone, hexane and amyl ether were also used as solvents at 3 kbar and temperatures of 40–80 °C.

An example of double ethereal C–O bond-forming reaction is ketalization described by Dauben (Scheme 6.1) [3]. Even sterically congested cycloalkanones such as

https://doi.org/10.1515/9783110556841-006

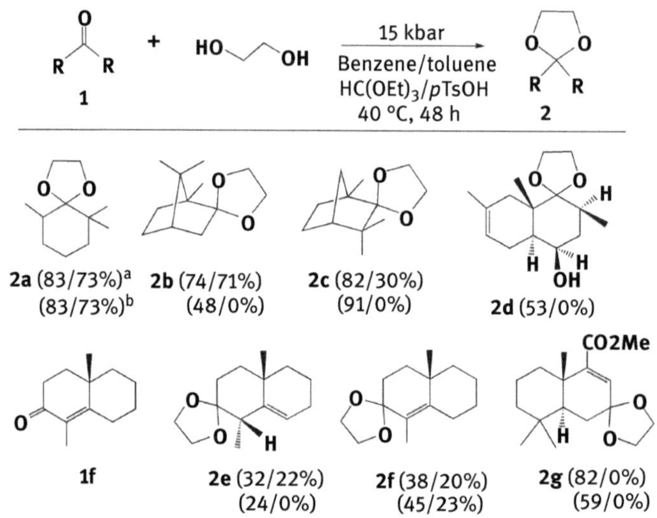

[a] Yields at 15 kbar/1 bar;[b] Method B: $(TMSOCH_2)_2$/TMSOTf, CH_2Cl_2, 20 °C, 48 h, yields at 15 kbar/1 bar

Scheme 6.1: Acetalization of cyclic ketones.

2,2,6-trimethylcyclohexanone, camphor and fenchone at 15 kbar were converted to ketals in protic conditions in the presence of ethylene glycol and *p*-toluenesulfonic acid (Method A). Triethyl orthoformate was added to the reaction mixture as the in situ water trap. Interestingly, bicyclic enone **1f** gave the mixture of ketal **2f** (38%) and the isomerized ketal **2e** (32%). At ambient pressure, in refluxing benzene conditions ketals **2a–g** were prepared in much lower yields or not at all. When the temperature for ketalization of fenchone and **1f** was decreased to 20 °C, yields of products **2c** and **2f**(**2e**) at 15 kbar were drastically reduced to 22% and 16(5)%, respectively, which could be raised to 62% and 34(27)% by elongation of the pressurizing time to 148 h. It could be noticed that the decrease in the temperature does not prevent isomerization of ketals **2f** to **2e.** A significant increase in reaction rates was noticed in the follow-up paper for the acetalization of fenchone with ethylene glycol to **2c** when the pressure was elevated from 1 kbar (28%) to 15 kbar (82%) [4]. Aprotic conditions (Method B) featuring the reaction of ketones with 1,2-bis[(trimethylsilyl)-oxylethane and the catalyst trimethylsilyl trifluoromethane sulfonate in dichloromethane under 15 kbar pressure afforded ketals in high yields. Under ambient pressure, experiments at –78 °C for 48 h products were formed in lower yields or not detected.

High-pressure ketalization was further studied by Kotsuki (Scheme 6.2) [5]. A number of ketones **3** was subjected to noncatalytic reaction with trimethyl orthoformate in anhydrous methanol at 8 kbar at elevated temperatures (40–100 °C). This method is advantageous over the existing methods as being simple, completely neutral and with applicability to a variety of ketones. Chemoselectivity is demonstrated by

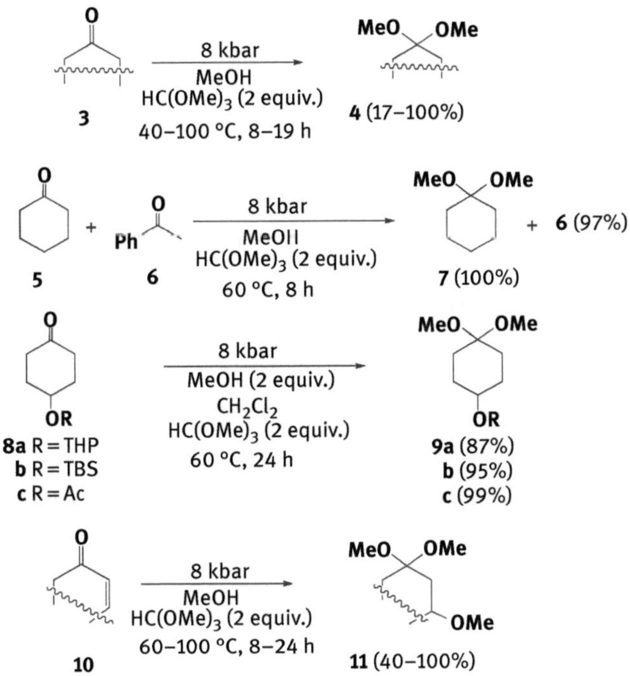

Scheme 6.2: Ketalization reaction.

exclusive ketalization of aliphatic ketone cyclohexanone **5** in the presence of aromatic acetophenone. This method is compatible with acid-sensitive functionalities such as tetrahydropyranyl, *tert*-butyldimethylsilyl and acetyl groups. Further experiments with trimethyl orthoformate as water scavenger showed that the oxy-Michael ketalization of enone conjugated systems **10** is equally well promoted by the pressure.

Besides the formation of ether bonds, high pressure was applied by Kotsuki to the ether cleavage reactions and formation of esters (Scheme 6.3) [6]. Transformation was accomplished with acyl halides at high pressure without using Lewis acid catalysts under almost neutral conditions. High yields in many of example reactions are the consequence of the acceleration of ionization (formation of the oxonium ion complex **13**) and bond-forming processes (the nucleophilic attack of halide) at elevated pressures. In addition, the pressure enhances the solvolysis reaction of acyl halides, which has a negative activation volume [7]. Cyclic and acyclic ethers reacting with acyl chlorides or bromides at 8 kbar and elevated temperatures furnish the ω-chloro- or ω-bromoesters **14**. In these conditions, regioselectivity known from the classical conditions does not change in the reaction of 2-methyltetrahydrofuran **15** with benzoyl or pivaloyl chloride. Using both reagents, the primary chloride **17** was obtained as the major isomer, suggesting that an S_N2 cleavage by chloride occurs predominantly at the less hindered side of C–O bond (Path a). The increase in the bulkiness

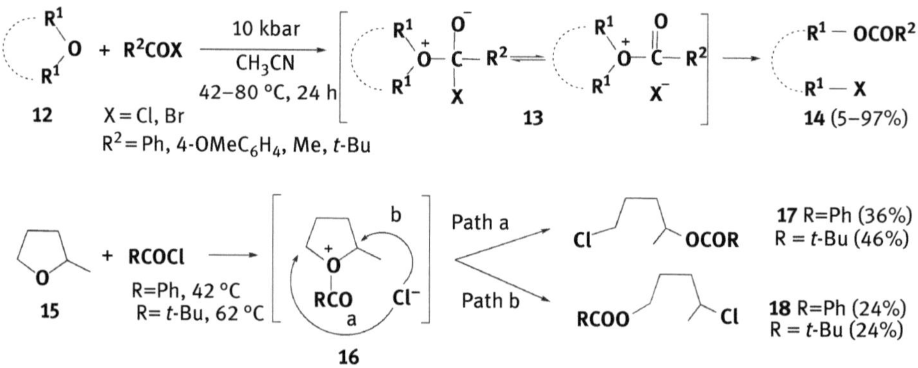

Scheme 6.3: Ether cleavage reaction.

of acyl chloride when pivaloyl chloride was applied raises the amount of the primary chloride **17**. Similarly, stereochemistry of the reactions of epoxides follows the same trends seen at ambient pressure reactions. The epoxide opening reaction of cyclohexene oxide and epichlorohydrin gave the products arising from the *trans*-diaxial opening of the epoxide ring. Regioselectivity was observed in the C–O cleavage reaction of epichlorohydrin, where the halide ion attack is exclusive at the primary position.

Kotsuki has reported another epoxide hydrolysis method under neutral conditions employing water in conjunction with high pressure (Scheme 6.4) [8]. This transformation includes the breaking of C–O bond and concomitant formation of new C–O bond. Since this type of reaction proceeds through ionic intermediates, it is facilitated by high pressure. Selected examples include cyclohexane **21** oxide, which was effectively hydrolyzed to *trans*-1,2-cyclohexanediol **22** with 40-fold excess of water. Although acetone and THF are solvents of choice, the reaction also proceeds in methanol and acetonitrile. In the case of methanol, an ether side-product **23** was obtained. When indene oxide **24** was subjected to hydrolysis, the *trans*-diol product **25** was accompanied by formation of acetonide **26** in 1:3 ratio, indicating that the condensation of epoxide with acetone is supported by high pressure. Ether linkage cleavage was also obtained in the case of hydrolysis of 1,2-epoxy-3-cyclohexanone **27**. This reaction produced mixture of the expected diol **28** and **29**. It was established that **28** and **29** are in the equilibrium, and compression of **28** in water provided **29** in 37% yield. Under high pressure reaction conditions, norbornane epoxide afforded rearranged diols **31** and **32**, and this behavior is in accordance to the literature. Scheeren also reported that the ether cleavage of the C–O bond of the 7-oxabicyclo [2.2.1]heptane moiety is promoted at 15 kbar employing acetic trifluoroacetic anhydride as the acetylating agent (Chapter 2, Scheme 6.2) [9].

Previously described reactions of ether linkage hydrolysis are based on the work by Le Noble who established a beneficial effect of high pressure on the rates of reactions of hydrolysis of epoxides. A detailed kinetic study of acid-catalyzed hydrolysis of

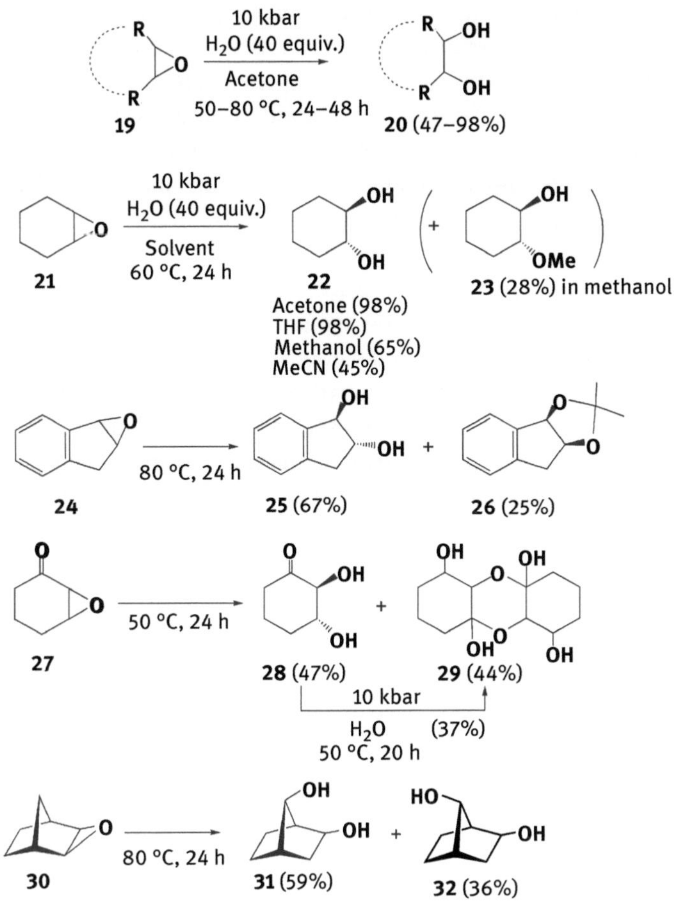

Scheme 6.4: Epoxide hydrolysis.

trimethylene oxide and epichlorohydrin was carried out in water over a pressure range of 7 kbar (Scheme 6.5) [10]. Authors have found that activation volumes for both reactions are negative (volume of activation at zero pressure, $\Delta V_0^* = -5.5$ and -8.3 cm^3 mol^{-1}, respectively). These results imply an A2 mechanism, characterized by

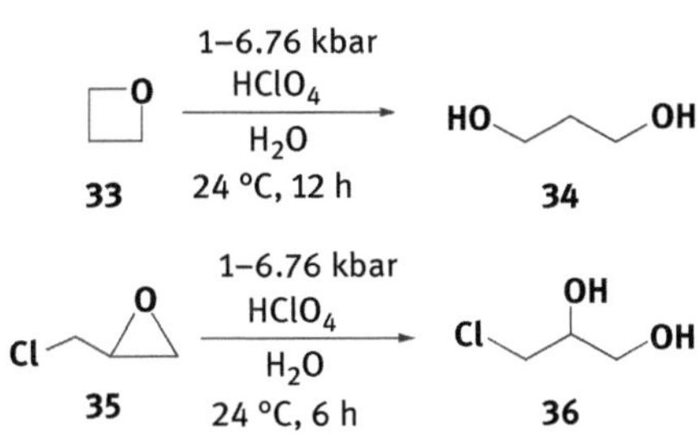

Scheme 6.5: Cyclic ether hydrolysis.

the bimolecular nucleophilic displacement of the protonated oxygen atom by a water molecule (bond making appears to progress faster than the bond breaking and accordingly ΔV_0^* is negative), thus correcting a widely accepted A1 mechanism.

The O-alkylation of sodium phenoxide with allyl chloride well demonstrates the capability of high-pressure conditions to influence the reaction outcomes. In this system, three possible products **39–41** could be formed, the first one arising from O-alkylation and the others from C-alkylation (Scheme 6.6) [11]. As shown by Le Noble, the compositions of reaction mixtures change by applying high pressure and using various solvents. The results are rationalized by the solvation of the oxygen atom, which hinders the approach of the alkylating agent to that position, but not to the *ortho-* and *para-*carbon atoms. Exclusive O-alkylation could be diverted by solvents that are highly effective as a proton donor in hydrogen bonding. The application of pressure enhances solvation, causing a decrease in the total volume as the solvent near the ion becomes oriented and compressed. This was shown by increased C-alkylation in water, an effective H-bonding solvent. On the other hand, pressure causes moderate C-alkylation in methanol (an ineffectively H-bonding solvent), but in 1,2-dimethoxyethane (a nonprotic solvent) C-alkylation at 5.5 kbar was not observed.

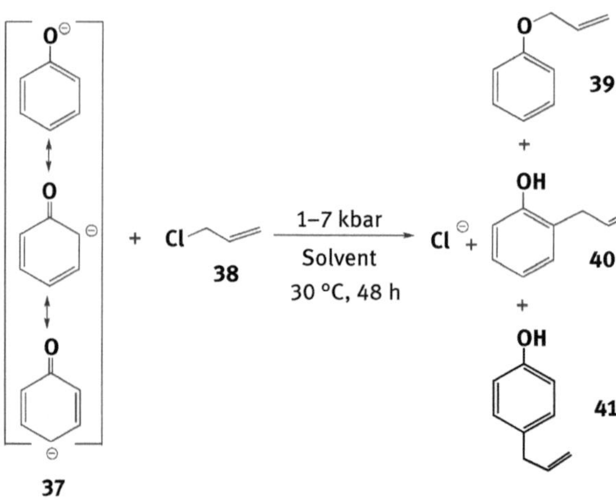

	H_2O	DME	MeOH
39	63/42.5[a]	100/100	92/100
40	18/16	0/0	5.1/0.4
41	39.5/21	0/0	2.9/0.1

[a]Yields (%), at 5.5 kbar/1 bar

Scheme 6.6: O-alkylation of sodium phenoxide with allyl chloride.

Similar pressure effects on O/C-alkylation in the reaction of 1-methyl-2-naphthoxide with benzyl bromide were observed by Le Noble in the follow-up study (Scheme 6.7) [12]. More intense solvation in methanol diminishes the extent of O-alkylation with an increase in pressure, whereas C/O-alkylation ratio remained unchanged in DME.

The reaction of 2,3,5-tri-*O*-benzoyl-D-ribofuranosyl chloride **46** with hydroxyuracil in conventional conditions provided C–O–N analogues of nucleosides **48** and **49** in

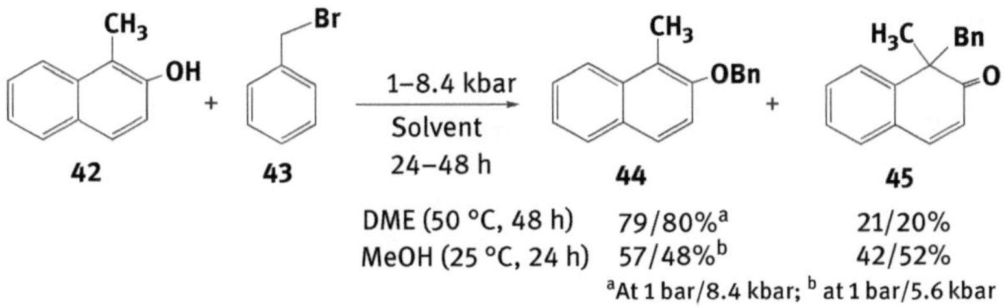

Scheme 6.7: O-alkylation of 1-methyl-2-naphthoxide with benzyl bromide.

9.2 and 1% yield, respectively. When this synthesis was carried at 10 kbar, yield was significantly improved, although similar 1:9 α, β mixture was obtained (Scheme 6.8) [13]. Further elevation of pressure to 15 kbar afforded marginally better yield (51%).

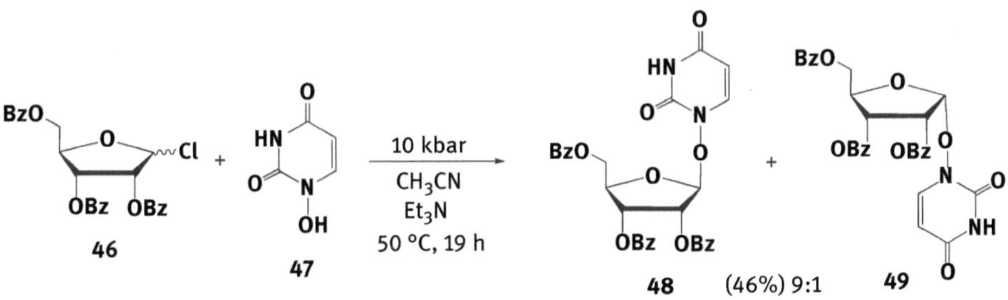

Scheme 6.8: O-alkylation of nucleosides.

Glycosyl C–O bond-forming reaction of sterically hindered alcohols (triethylcarbinol and 1-tert-butylcyclohexanol) with the halogen reactants, unreactive *O*-benzoyl glucosyl bromide monosaccharide **50** was studied by Dauben (Scheme 6.9) [14]. To achieve glycosylation, 2,4,6-collidine was added as the base and tetraethylammonium bromide and/or silver trifluoromethanesulfonate as activators. Under high-pressure conditions, glycosylation took place and products were obtained, with AgOTf being by far superior activator. The AgOTf-activated reaction at ambient pressure afforded the β-glycoside **51a**, whereas at 15 kbar the yield of **51a** was diminished, along with the significant amount of orthoester **52a.** The ability of high pressure in overcoming steric hindrance is demonstrated in the glycosylation of the extremely hindered alcohol **51b** and production of the β-glycoside **51b** in 49% yield, without the formation of orthoester **52b.** Glycosylation reaction of substrate **50** with the 4-hydroxy group in 2-acetamido sugar **53** at ambient pressure provided the orthoester **54** and *N*-glucosyl compound **55,** whereas at 15 kbar only **54** was obtained. Similarly, glycosylation of galactosyl bromide **56** with **53** gave only the orthoester product **57.** The suppression of the formation of *N*-glycosylide was rationalized by the existence

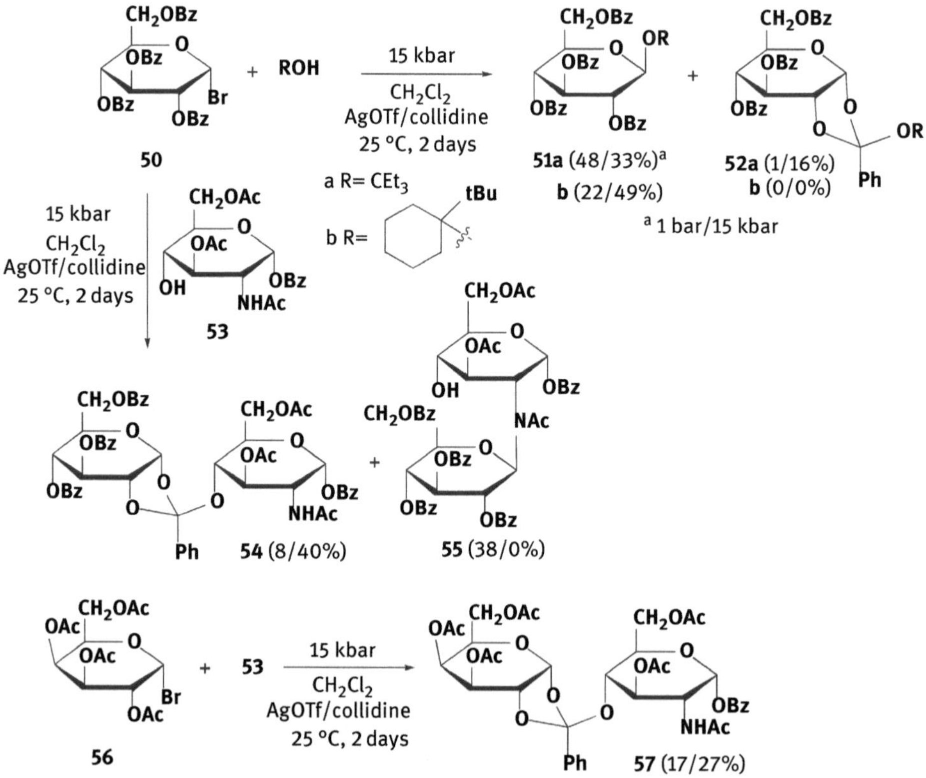

Scheme 6.9: Glycosylations of unreactive alcohols.

of the bicycloacyloxonium intermediate under high pressure. The formation of this intermediate is enhanced by pressure as a result of a negative $\Delta V^{\neq}$ for the isomerization of a glycosyl cation to an acyloxonium ion (–11 $cm^3\ mol^{-1}$) [15]. The stability of the bicycloacyloxonium intermediate minimizes the attack at the anomeric center, leading to a glycosidic or an *N*-glycosyl product [16].

Pressure exerts the positive directive toward products in the nucleophilic addition of linear primary alcohols to acrylic compounds **59** (nitriles, amides, ketones and esters) (Scheme 6.10) [17]. In the presence of tri-*n*-butylphosphine as the base, this oxa-Michael-type reaction provides conjugate addition, which is especially notable for the sterically congested substrates that poorly react at ambient pressure. This reaction is multistep process in which the mechanistic steps involving the formation of alkoxy carbanion and bimolecular S_N2 reaction undergo rate acceleration by pressure. The advantages of HP are well demonstrated for the addition of primary alcohols to cinnamonitrile **62**, with a yield increase as bulkiness of alcohol grows. Another oxa-Michael reaction of alcohols on *α*,*β*-unsaturated ketones was carried out by Hayashi at 2 kbar to obtain the corresponding alcohols in the presence of catalytic amounts of DMAP and $LiClO_4$ [18].

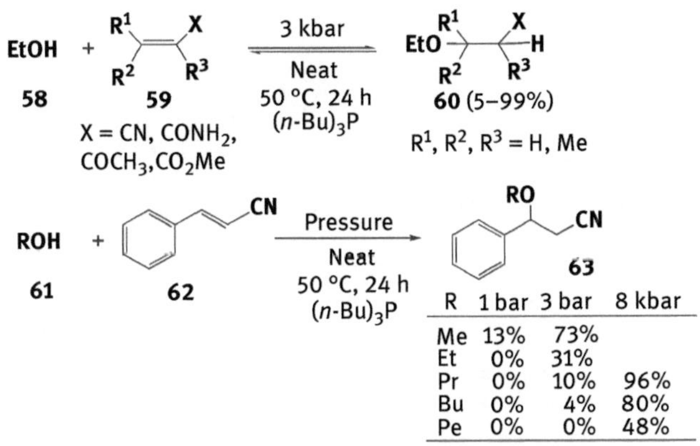

Scheme 6.10: Oxa-Michael reaction.

Matsumoto has realized that high-pressure addition of alcohols to nitriles forms new C–O bond producing the methylbenzimino ether **65** (Scheme 6.11) [19]. This product in given reaction conditions trimerizes to triphenyl-1,3,5-triazine **66** and mixtures of two products were always obtained. The ratio **65/66** depends on the benzonitrile/methanol ratio – the largest amount of triazine was formed with 7.26 excess of methanol, while a further increase of methanol again favors the dominant formation of **65**. Kinetic measurements allowed the estimation of volumes of activation for the formation of methylbenzimino ether **65** and triazine **66**. Both values are negative and large, $\Delta V^{\neq} = -17.9$ and -31 $cm^3\ mol^{-1}$, respectively, thus indicating that the formation of products is facilitated by pressure.

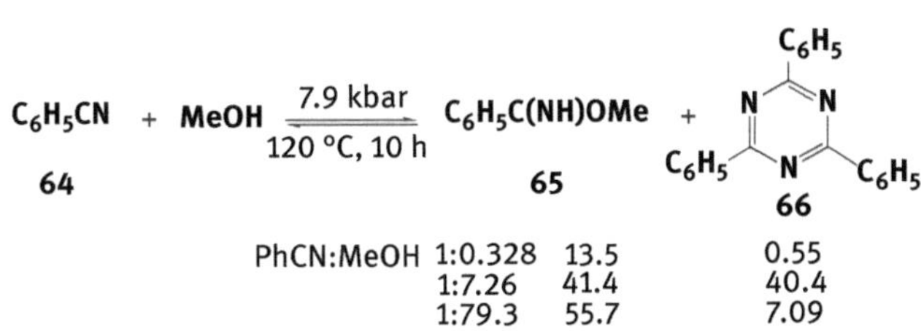

Scheme 6.11: Addition of alcohols to nitriles.

The C–O bond could effectively be formed by cycloaddition reactions, as exemplified in chapters on (hetero)cycloadditions. An additional example that features the [2 + 2] cycloaddition of ketene acetals was published by Scheeren (Scheme 6.12) [20]. Here, high-pressure-promoted cycloadditions of ketene acetals **67a,b** with α,β-unsaturated aldehydes and ketones led to the formation of three types of products: oxetanes, dihydropyrans and cyclobutane products. It was established that high pressure has a strong

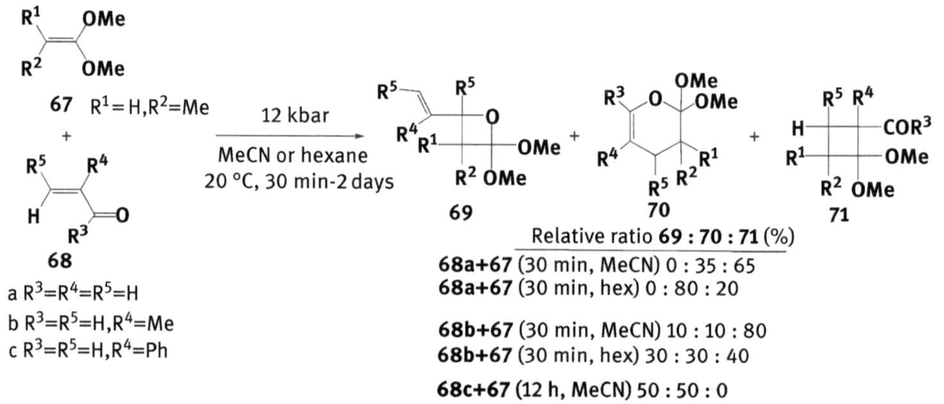

Scheme 6.12: The formation of the C–O bond by ketene acetals cycloadditions.

accelerating effect, and the nature of solvent influences the distribution of products. The selection of results indicates that cyclobutanes **71** are the dominant product when *α*,*β*-unsaturated aldehydes without large *β*-substituents are used and the reaction is performed in a polar solvent (acetonitrile). In the nonpolar hexane formation of oxetanes **69** and dihydropyrans **70** become more important. This solvent influence was rationalized by the reaction mechanism. Dihydropyrans **70** are formed in a concerted [4 + 2] cycloaddition via less polar transition state, while oxetanes **69** and cyclobutanes **71** are formed via the dipolar intermediates, which are favored in polar solvent.

6.1.2 Esters

Several papers dealing with transformations of the ester C–O bond were published. In an early account, Newitt has studied the influence of pressure on the esterification of ethanol with acetic acid and on the hydrolysis of various esters (Scheme 6.13) [21]. It was found that in an uncatalyzed reaction, the application of high pressure has a beneficial effect on the reaction equilibrium. Furthermore, it was established that by applying pressure there is a strong increase in the tendency of hydrolysis of studied ethyl esters by water alone or by dilute acids.

The aminoacylation of the phenylalanine-specific t-RNA with phenylalanine and methionine could be carried out at 6 kbar [22]. Jurczak has demonstrated that this esterification reaction takes place at the 3′ end of the t-RNA molecule. However, the yields of Phe-t-RNA Phe or Met-t-RNA Met are about 10 times higher by the enzymatic aminoacylation reaction than that in the high-pressure conditions.

High-pressure hydrolysis method developed by Yamamoto features nearly neutral reaction conditions and high efficiency (Scheme 6.14) [23, 24]. *N*-methylmorpholine

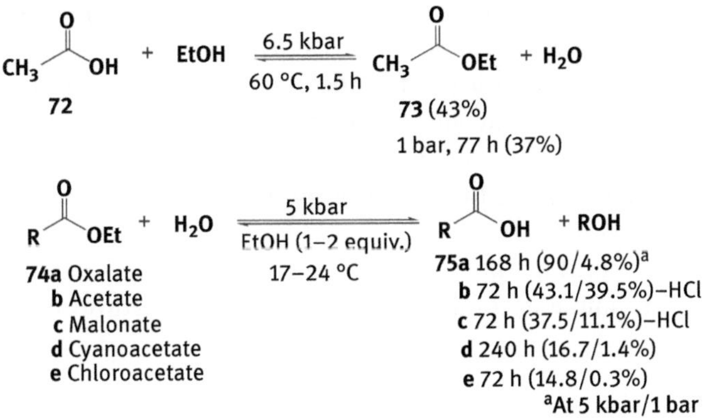

Scheme 6.13: Esterification and ester hydrolysis.

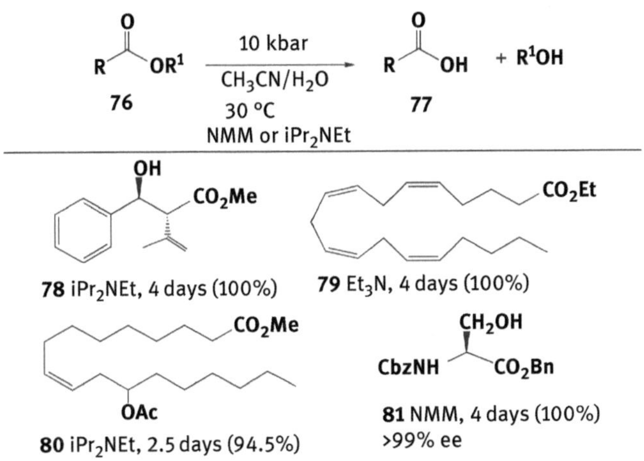

Scheme 6.14: Esters hydrolysis.

or iPr_2NEt was applied as catalysts and hydrolysis was carried out at room temperature. In these conditions, a high chemoselectivity and regioselectivity was achieved. *β*,γ-Unsaturated ester **64** gave single product, instead of a number of products obtained in the ordinary aqueous conditions caused by the isomerization of the double bond. Ethyl arachidonate **65** was quantitatively hydrolyzed, without side reactions. Depending on the solvent, substrate **66** could be regioselectively hydrolyzed at methyl ester side (CH_3CN/H_2O) or at acetate functionality (in methanol). Difficulties in the hydrolysis of amino acid **67** are due to its water solubility, which are overcome at high pressure and application of only 3–4 equivalents of water. This hydrolysis in identical conditions at ambient pressure did not occur.

Transesterification of prednisolone 21-acetate **82**, 21-butyrate and 21-pivalate with methanol in the presence of triethylamine as the catalyst at 11 kbar proceeds smoothly to prednisolone **83** in 78–100% yield (Scheme 6.15) [25]. In identical high-pressure conditions, esters with large steric hindrance such as prednisolone 11-acetate **84** and 21-tetrahydrophthalate were hydrolyzed at slower rate (76% and 81%, respectively) with the concomitant formation of by-products. Hydrolysis of the budesonide esters such as 21-acetate **85** was achieved in acetonitrile in 40–84% yield. Here described mild method allows hydrolysis of hindered steroidal ester compounds that in ambient conditions do not undergo hydrolysis or poorly hydrolyze.

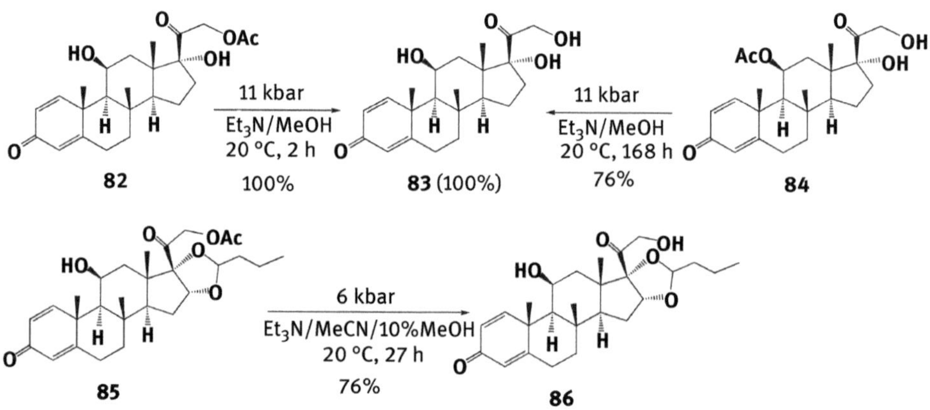

Scheme 6.15: Transesterifications of prednisolone and budesonide esters.

Novel transterification protocol was developed by Jurczak based on kinetic data (volume of activation for transterification reaction $\Delta V^{\neq}$ = ca. 12 cm^3), which suggests that this process is facilitated by high pressure. A series of benzoate esters readily undergo base catalyzed transesterification to methyl benzoate (Scheme 6.16) [26]. Jurczak has found that the optimal base for this transformation carried out at 10 kbar is DBU. Menthyl benzoate also readily undergoes transesterifications with ethanol, while propan-1-ol was less reactive (15%) and no reaction occurred with trifluoroethanol, propan-2-ol and *tert*-butanol. Finally, L-phenylalanine benzyl ester **91** transesterification with methanol proceeded in quantitative yield, with only a slight racemization (95% ee).

Pressure may alter the mechanistic path of photochemical decarboxylation of aryl esters. Inoue has found that the irradiation of chiral aryl ester mesityl (*S*)-(-)-2-methylbutanoate **93** affords alkylmesitylene **94** in good yields, as well as a smaller amount of 2,4,6-trimethylphenol **95** (Scheme 6.17) [27]. Applying pressure increases the **94**/**95** ratio from 5:1 to >50, likely by the mediation of the equilibrium between

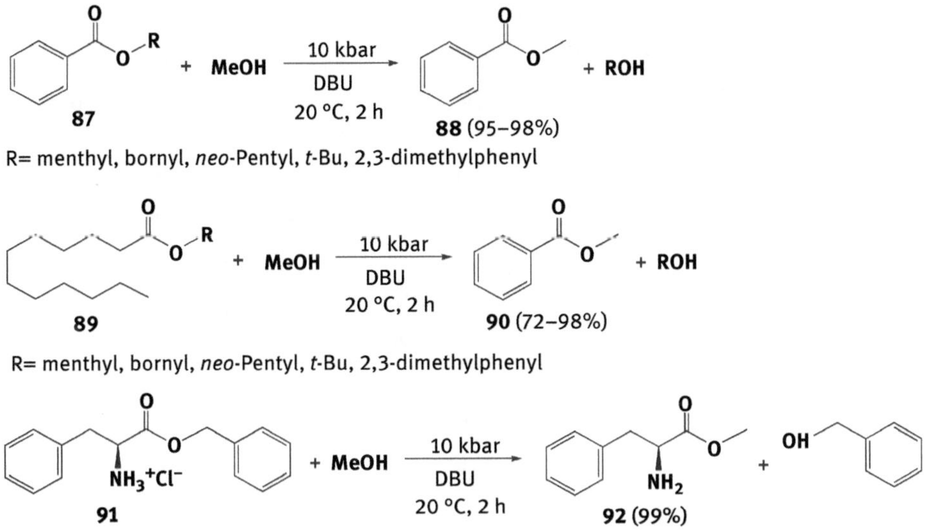

Scheme 6.16: Transesterification reactions.

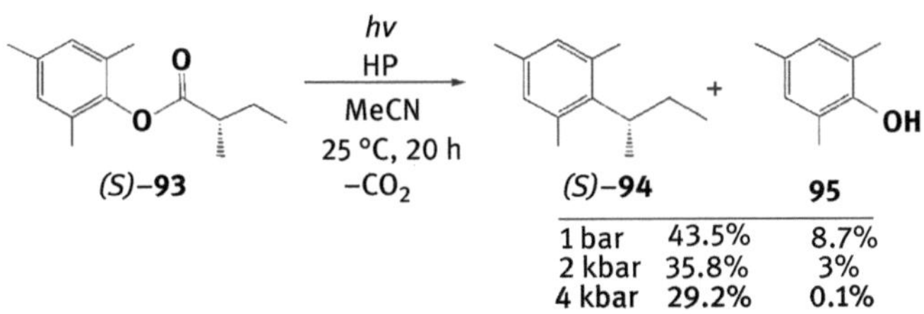

Scheme 6.17: Photodecarboxylation reactions.

proposed radical precursors of **94** and **95**. At the same time, the ee of **93** of this asymmetric photochemical transformation remained unaffected (>99%).

6.1.3 Alcohols

New C–O bonds were formed in solvolysis reaction of heptachlor **96**. Korte has found that this base-catalyzed high-pressure solvolysis produced several products possessing C–O bonds (Scheme 6.18) [28]. The dominant product in the reaction mixture was alcohol **97**. Highly stable epoxide ring of Dieldrin **101** was hydrolyzed and transformed in the mixture of ether and alcohol products **102–106** by high-pressure acid catalytic treatment. Some of these products could not be prepared in classical reaction conditions.

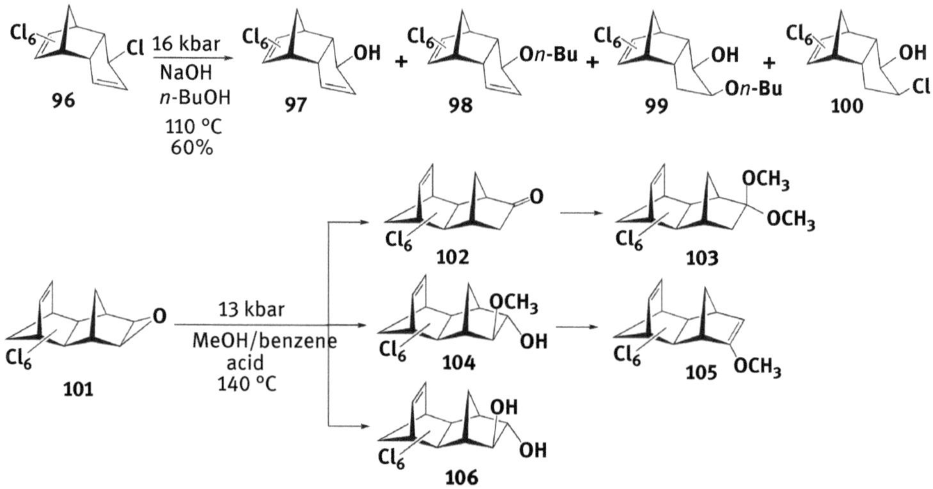

Scheme 6.18: Solvolysis of heptachlor and epoxide opening of Dieldrin.

6.2 C–S bond

Several chemical reactions for the C–S bond formation under high-pressure conditions were reported in the literature. Interesting is the reaction of carbon disulfide with norbornene at 10 kbar and 100 °C that affords tetrathiafulvalene derivative **108** (*syn-* or *anti-*), as described by Plieninger (Scheme 6.19) [29]. Similar reaction with norbornadiene afforded the corresponding product **110**, although in much smaller yield, due to the polymerization. Dicyclopentadiene **111** and dihydrocyclopentadiene **113**, as well as norbornene **115** reacted in the similar manner. Cyclocondensation with 2,3-dimethylbicyclo[2.2.2]octa-2,5-dienedicarboxylate **117** provided tetrathiafulvalene **118** in very small yield via retro-Diels–Alder fragmentation of the initially formed adduct. Reaction mechanism for the formation of tetrathiafulvalenes likely involves 1,3-dipolar cycloaddition of CS_2 onto the C = C bond, with the formation of the carbene intermediate that dimerizes in the following reaction step.

The related cyclocondensations of CS_2 reported by Okamoto involves reactions of triple bond of propiolates **119a–c** in neat CS_2-producing tetrathiafulvalenes **120** in high yields (Scheme 6.20) [30].

Different cyclocondensation product was reported in the reaction of CS_2 published by Jenner (Scheme 6.21) [31]. In this synthesis, tetrathiafulvalene **123** was obtained by CS_2 reaction with methyl propionate. The obtained small yield was not improved by the addition of Lewis acid catalyst ($LiClO_4$). At 1 bar the reaction does not take place.

Another cyclocondensation of CS_2 with cyano bond was published by Yasumoto (Scheme 6.22) [32]. High-pressure addition of CS_2 to dialkylcyanamide afforded 2,6- bis

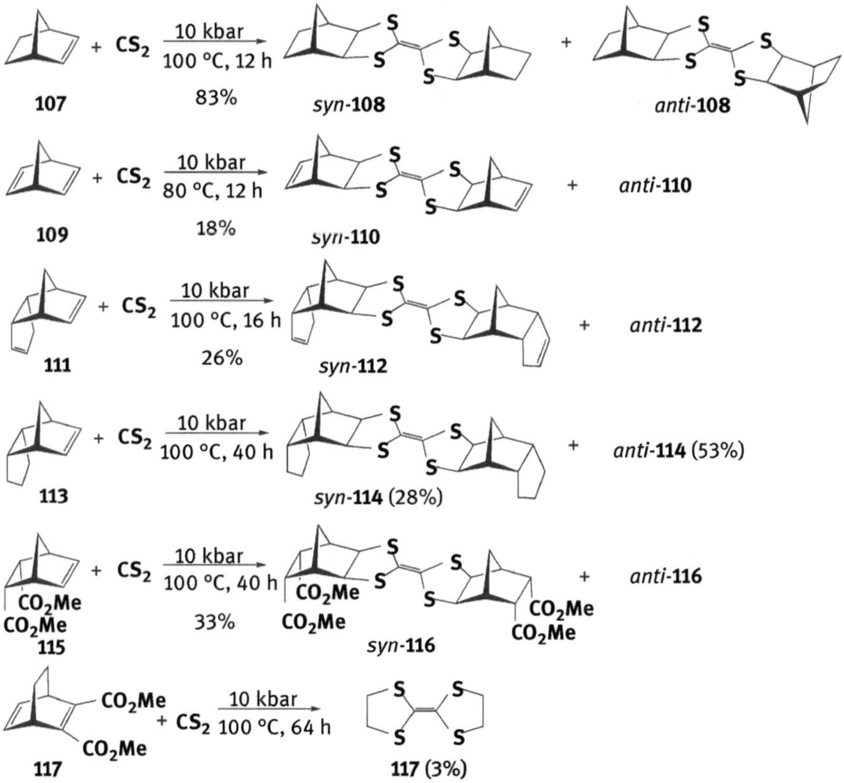

Scheme 6.19: Reactions of CS_2 with norbornenes.

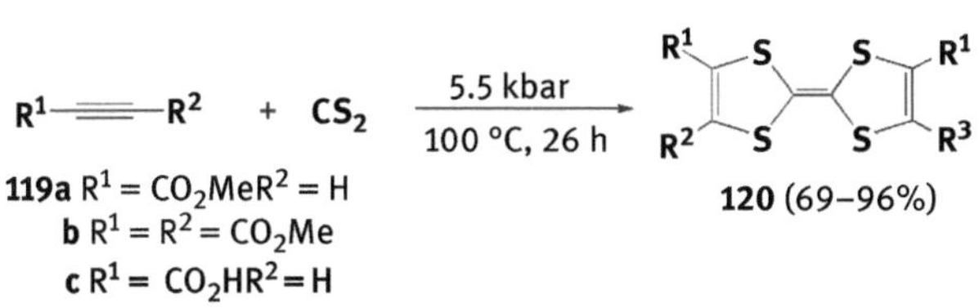

Scheme 6.20: Cyclocondensation of CS_2 with methyl propionate.

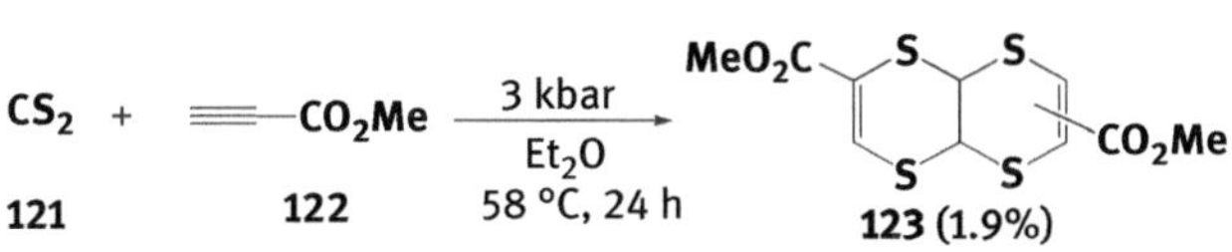

Scheme 6.21: Cyclocondensation of CS_2 with methyl propionate.

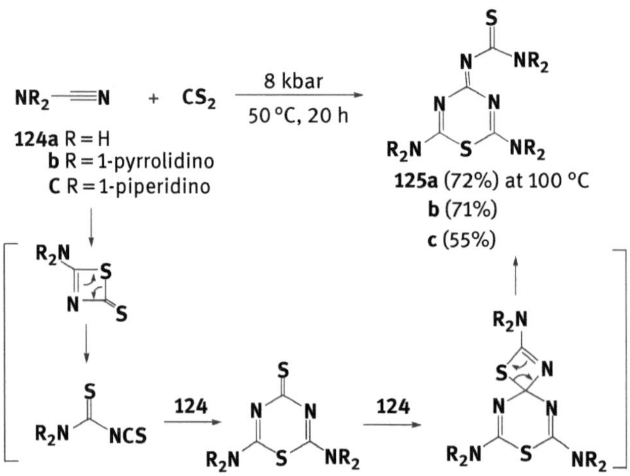

Scheme 6.22: Cycloaddition of CS_2 with dialkylcyanamides.

(dialklamino)-4-dialkylthiocarbamoylimino-1,3,5-thiadiazine **125** in high yields. In the case of dimethylcyanamide, two side-products were also obtained: 2,6-bis(dimethylamino)-[1,2,4]dithiazolo[1,5-*b*][1,2,4]dithiazole-6a-S^{IV} (4.6%) and 5-dimethylamino-1,2,4-dithiazole-3-thione (5.5%). The reaction mechanism that involves three equivalents of cyanamide starts with [2 + 2] cycloaddition, followed by rearrangement [4 + 2] and [2 + 2] cycloadditions.

High pressure was also applied in derivatization of the platinum complexes. Young has reported that the S-alkylation of binuclear platinum complex $[Pt_2(\mu\text{-}S)(\mu\text{-}SC_3H_7)(dppp)_2]$ **126** was accelerated by employing high pressure (Scheme 6.23) [33]. At 10 kbar, by the reaction with ethyl 6-bromohexanoate, ethyl 3-bromopropionate and allyl bromide heterodi-μ-thiolato binuclear platinum complexes **127** and **128** were obtained in high yield. When $[Pt_2(\mu\text{-}S)_2(dppp)_2]$ complex **129** was subjected to the S-

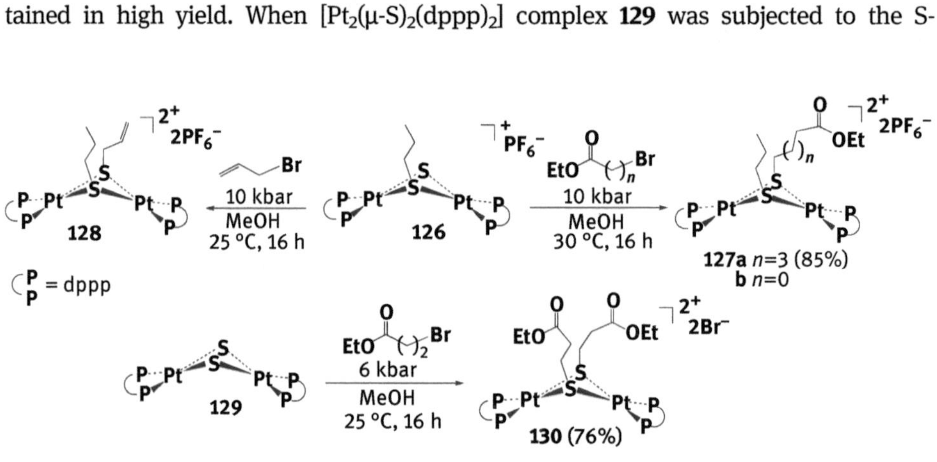

Scheme 6.23: The formation of the S-alkylated Pt complexes.

alkylation with ethyl 3-bromopropionate at 6 kbar resulted in complete dialkylation to form the doubly bridging ester thiolate complex **130**, whereas incomplete reaction occurs and the mixture of mono- and bis-S-alkylation product was obtained at ambient pressure.

The ability to form C–S bonds on the platinum complexes in high-pressure conditions exploited Young in the synthesis of dithiacyclophanes (Scheme 6.24) [34]. The reaction of $[Pt_2(\mu\text{-}S)_2(L)_2]$complexes **120** with and excess of α,α'-dichloro-*o*-xylene at 15 kbar provided mixtures of products, where 3,8-dibenzo-1,6-dithiacyclodecane product **133** was isolated. The involvement of platinum complexes in the catalytic cycle was supported by the isolation of the *ortho*-xylyl-bridged dithiolated intermediates **132**. Evident is the importance of high pressure to perform the reaction, since ambient pressure reaction of **131a** produces only **133a**.

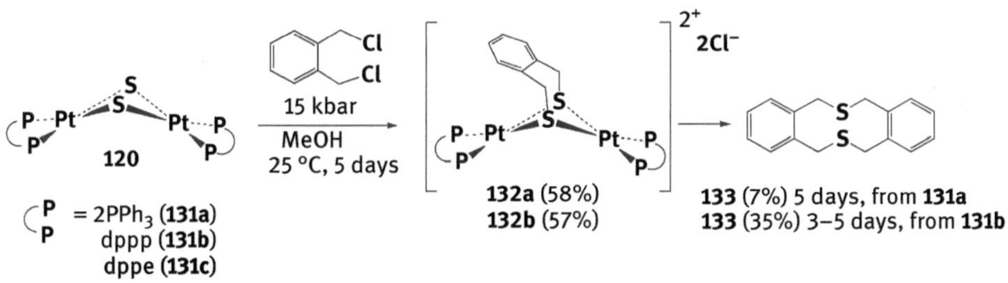

Scheme 6.24: Synthesis of 3,8-dibenzo-1,6-dithiacyclodecane.

An efficient way for the transformation of alcohols to sulfides was developed by Kotsuki et al. (Scheme 6.25) [35]. Sterically hindered alcohols were converted to the corresponding thiols **135** by employing PhSSPh and Bu_3P at 10 kbar. All reported reactions proceed sluggishly at the ambient pressure, even after the much longer reaction times.

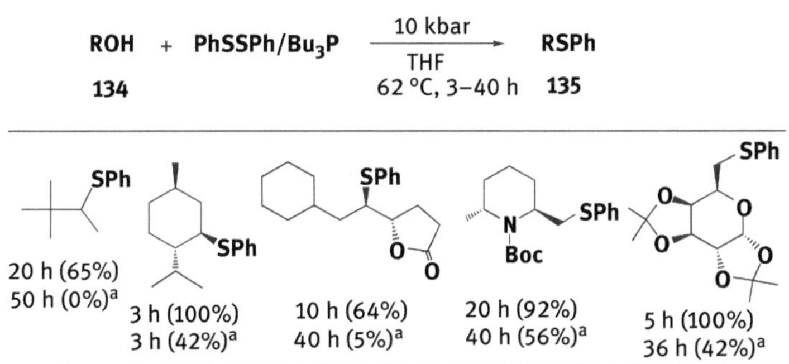

[a]Ambient pressure

Scheme 6.25: Synthesis of thiols.

The volume of reaction for the addition of phenylmethylthiol to acetyl cyanide was determined by Isaacs (Scheme 6.26) [36]. Kinetic measurements from 1 bar to 1.25 kbar in $CDCl_3$ at 25 °C afforded volume of reaction $\Delta V = -13$ cm^3 mol^{-1}. This value is associated with a transition state in which association and partial bonding between base and cyanide occurs. On the other hand, the volume of reaction for an intramolecular hydroxyketone-hemiacetal interconversion **138**⟷**139** is determined to be $\Delta V = -3.2$ cm^3 mol^{-1} in DMSO/$CDCl_3$ at 25 °C. This number is much smaller than that reported for intermolecular ester hydrolysis process of ethyl acetate in aqueous acetone: −9, −13 and −17 cm^3 mol^{-1} in 80, 70 and 57% water, respectively [37].

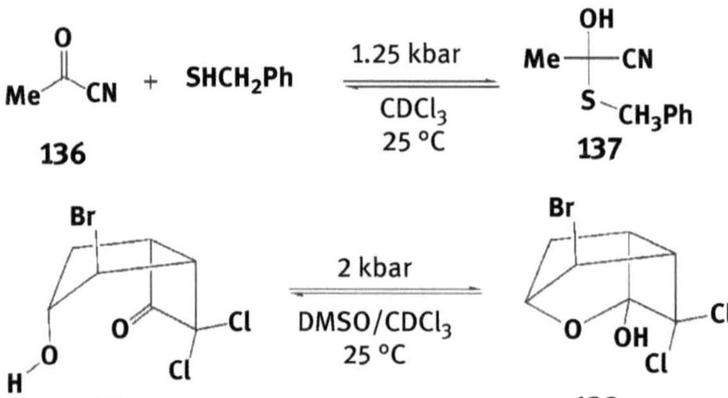

136 137 138 139

Scheme 6.26: The addition of phenylmethylthiol to acetyl cyanide.

Nucleophilic polymerization of thietanes to polythietanes initiated by the neutral nucleophiles was conveniently performed at 15 kbar. Young has shown that quantitative yield of polythietane **141** was obtained with the aid of triphenylphosphine at moderate temperature (Scheme 6.27) [38]. At ambient conditions, the reaction does provide only 2% of the product at much higher temperature. Similarly, thietane **142a** was polymerized. For monomer thietanes **142b, 143** and **144**, more harsh conditions were required, using the stronger nucleophile hexamethylphosphorous triamide and longer reaction time. Even under these conditions, conversions were moderate, and thietane **144** failed to react.

Cheletropic reaction of dimethylbutadiene and sulfur dioxide involving C–S bond-forming/breaking processes was studied by Isaacs (Scheme 6.28) [39]. Spectrometric kinetic measurements of synthesis of sulfolene **146** at pressures within the range 1 bar –1 kbar in the high-pressure apparatus with optical window cell have established that the volumes of activation and reaction are $\Delta V^{\neq} = -35$ cm^3 mol^{-1} and $\Delta V = -33$ cm^3 mol^{-1}. These values indicate that the transition state volume for this reaction is similar to that of the product **146**. Both values are similar to those for Diels–Alder reactions. The observed difference between $\Delta V^{\neq}$ and ΔV, where $\Delta V^{\neq}$ is slightly more negative than ΔV, was explained by attractive secondary orbital

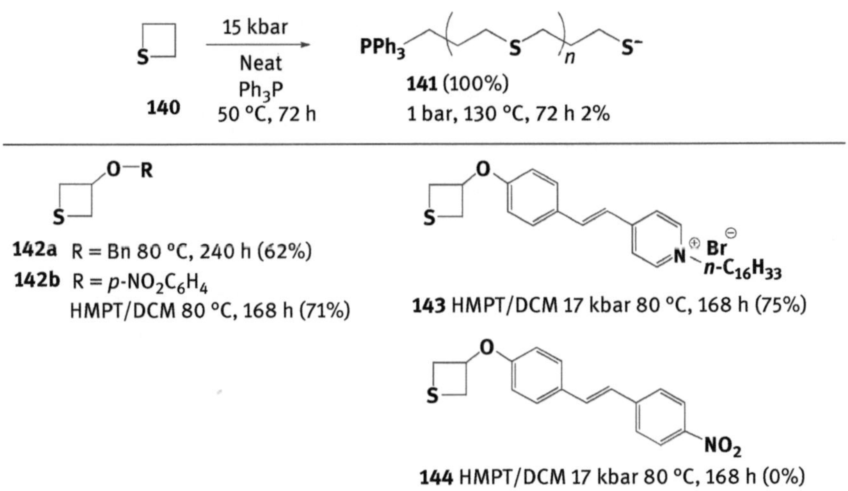

Scheme 6.27: Polymerization of thietanes.

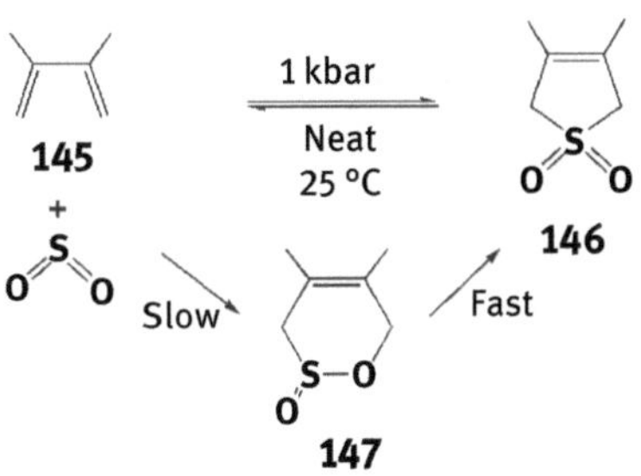

Scheme 6.28: Cheletropic reaction of sulfur dioxide.

interactions in the transition state. This might also be a two-step process, involving initial slow [4 + 2] cycloaddition across the S=O bond, which is followed by fast rearrangement of sulphenolactone **147**.

6.3 C–Sn bond

The reduction of ketones with stoichiometric amount of trialkyltin hydride under high pressure is summarized in Chapter 7. In addition to these reactions, several examples of hydrostannation of C–C double bond was described by Rahm et al. (Scheme 6.29) [40]. High-pressure reactions were carried out without a radical initiator. A high-pressure reaction of phenylcyclohexene with tin deuteride provided *anti*-hydrostannation product **151** in high yield. This result was a great improvement in terms of yields in comparison with thermal conditions. Heating at 180 °C for extended time provides **151** in only 30% yield, whereas the addition of a radical initiator was

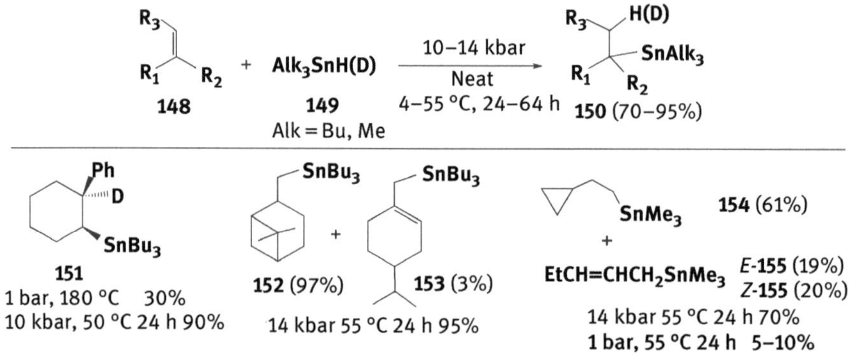

Scheme 6.29: Hydrostannation of alkenes.

even less yielding. In the case of *β*-pinene, the reaction proceeded in 95% yield giving two products. The ring-opened product **153** was a minor component, whereas the bicyclic product **152** that could not be prepared in previous synthetic studies constituted 97% of reaction mixture. The reaction of vinylcyclopropane with trimethyltin hydride could be steered toward the retention of the cyclic structure **154** by using high pressure in combination with low temperature. These conditions are advantageous to the atmospheric pressure employing radical initiator, where open-ring products are exclusively obtained. Without the radical initiator, hydrostannation reactions at atmospheric pressure proceed in very low yields.

High pressure was used by Matsumoto to achieve very mild reaction conditions for the allylation of aldehydes with buten-2-enylstannanes **157** (Scheme 6.30) [41]. Reactions take place at room temperature under neutral conditions, giving adducts **158** that could be hydrolyzed to homoallyl alcohols. The *threo*-isomer is predominantly formed from the *trans*-buten-2-enylstannanes, indicating that the mechanism goes via a six-membered cyclic transition state involving coordination of Sn atom to the carbonyl oxygen. This preference for *threo*-isomer is in contrast to ambient pressure reactions catalyzed by Lewis acids that provide *erythro*-alcohols by preventing the coordination of the Sn atom.

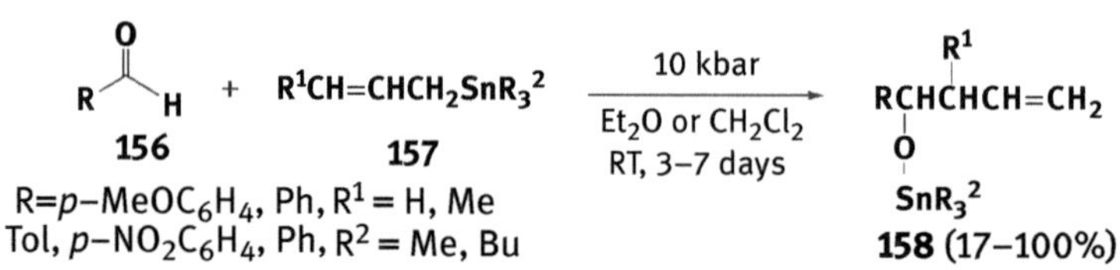

Scheme 6.30: Allylation of aldehydes with but-2-enylstannanes.

Stereospecific *syn* addition of aldehyde with the stereoselective preference for *erythro* product **160** was observed in high-pressure-facilitated benzaldehyde addition

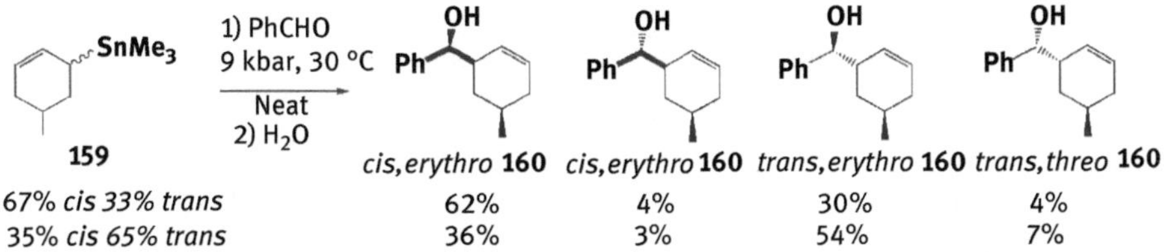

Scheme 6.31: The addition of benzaldehyde to 5-methylcyclohex-2-enyltrimethylstannane.

to 5-methylcyclohex-2-enyltrimethylstannane **159** (Scheme 6.31) [42]. Reactions conducted at various temperatures and starting with different isomeric mixtures indicate a cyclic transition state (S_Ei') mechanism. The *erythro* preference could be explained by unfordable steric interactions between phenyl and methyl groups in the transition state leading to the *threo* product. For comparison, additions conducted in classical conditions catalyzed by BF_3–OEt_2 are *cis* selective with the slight *threo* diastereoselectivity.

The evidence for a six-membered cyclic transition state for allylation of aldehydes with but-2-enylstannanes **162** was provided by Isaacs [43]. Kinetic study of the addition of chloral to aryltributylstannane at pressures varying from 1 to 800 bar (Scheme 6.32) provided an activation volume of -33.4 ± 0.6 cm^3 mol^{-1}, which is the characteristic value for the Diels–Alder reactions and indicates a single-step, concerted process.

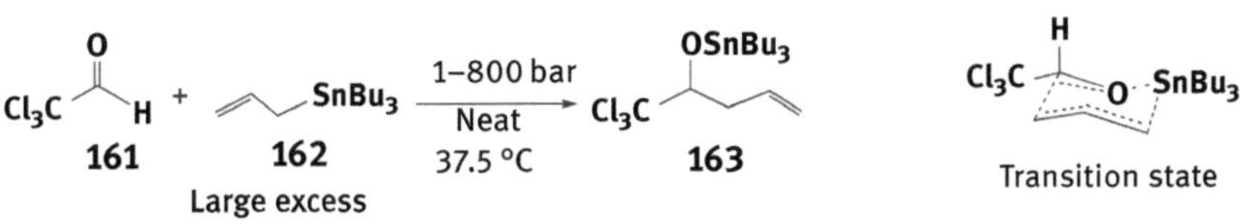

Scheme 6.32: Allylation of chloral with aryltributylstannane.

6.4 C–Si bond

Several sterically demanding alcohols are protected by Dauben at 15 kbar (Scheme 6.33) [44]. High-pressure reactions of alcohols **164a–d** with TBDMSCl with pyridine base afforded TBDMS-protected alcohols **165a–d** in high yields. Experiments showed that these ethers could not be achieved at atmospheric conditions, but only at room temperature and elevated pressure. A significant increase in yields was found by Dauben for the etherification of linalool **164a** when the pressure was elevated from 1 kbar (0%) to 15 kbar (97%).[4] Control experiments carried out at 1 bar (45 °C, 18–44 h, using DMAP) provided the corresponding silyl ether **165a** in <5%, whereas **165b–d** could not be achieved. In identical high-pressure conditions, alcohol **164a** also reacted

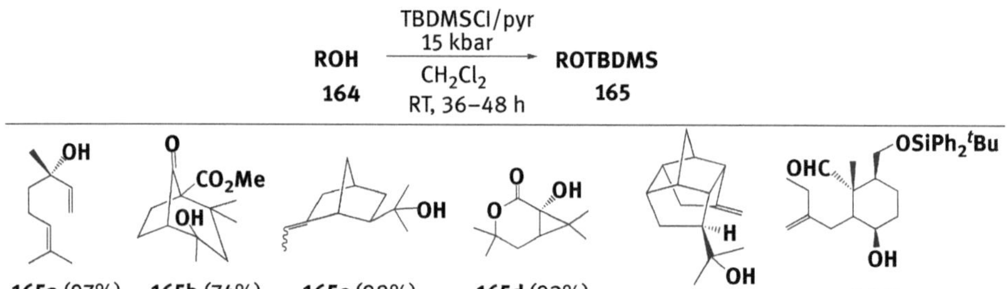

165a (97%) **165b** (74%) **165c** (98%) **165d** (92%) **165e** **165f**

Scheme 6.33: Protection of alcohols.

with *t*-Bu(Ph)$_2$SiCl, affording after 24 h of pressurization the TBDPSCl ether in 91% yield. Acetate and bezoyl esters of alcohols **164a–d** including **164e** and **164f** were also prepared at 15 kbar employing Ac$_2$O/pyr for 24–48 h obtaining acetates in high yield (84–98%) which are markedly better than reactions which were carried out at 1 bar (28–81%). Similarly, the Bz$_2$O/pyr reactions conducted for 24–36 h afforded benzoates in 83–98% yield, with atmospheric experiments less successful (control 49–86%). Alcohols **164a** and **164b** could also be protected as ethers by employing methoxyethoxymethyl chloride (MEM/*i*-Pr$_2$NEt, 36 h) to obtain the corresponding ethers in 62 and 98% yields, respectively. Described protections of sterically congested and thermally labile alcohol substrates demonstrate the synthetic utility of high pressure for reactions that could not be achieved by other conditions.

High-pressure conditions were also used for the cleavage of the O–Si bond located at the sterically hindered positions. Deprotection of *tert*-butyldiphenylsilyl (TBDPS) ethers was carried out by Ito at 10 kbar using 10% HF/pyridine in DMF (Scheme 6.34) [45]. Desilylation of ether **166a** has shown to be promoted when high pressure is applied: conversion of 96% was obtained after applying pressure for 12 h, whereas atmospheric pressure reaction provided no product. The rate acceleration by high pressure is likely the consequence of the negative volume of activation ($\Delta V^{\neq}$) for the presumed mechanism initially involving a nucleophilic attack of fluoride to silicon. Reactions are shown to be chemoselective in the presence of other alcohol protecting groups (benzyl, *p*-methoxybenzyl, isopropylidene, cyclohexylidene, allyl, trityl and phthaloyl). In the conditions applied, *O*- and *S*-glycosidic bonds remained unaffected. Clean conversion to alcohol was also obtained in the case of the difficult desilylation of trisaccharide **166f**, which at 1 bar gave no product, whereas the use of TBAF/AcOH reagent caused a significant migration of the acetyl group and formation of the mixture of products.

Effective Si-migration under high pressure was observed by Yamamoto and Matsumoto (Scheme 6.35) [46]. O-silylated ketene acetals **168** undergo silicon migration from oxygen to carbon at α-position, forming α-silylated ketones and esters **169** in high yields. At atmospheric pressure, the reaction does not take place. This

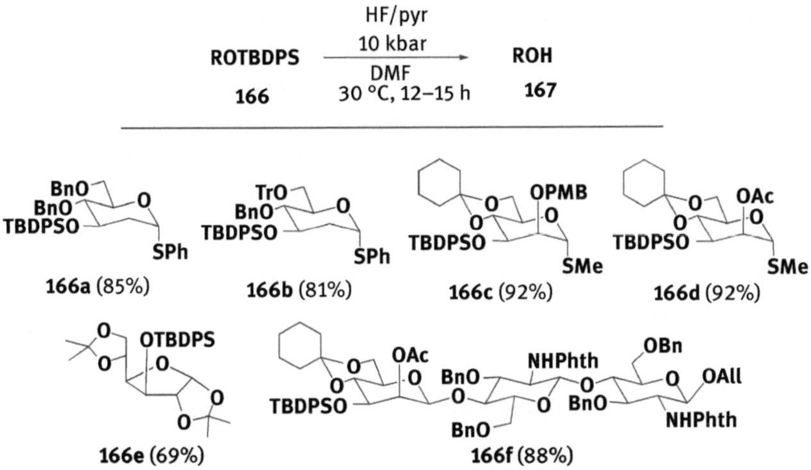

Scheme 6.34: Deprotection of TBDPS ethers.

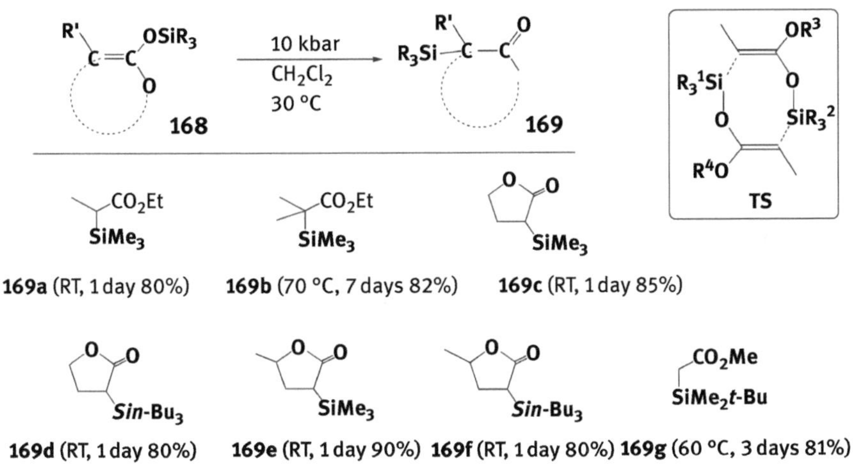

Scheme 6.35: Si-migration under high pressure.

process undertaken at room temperature involves O–Si bond cleavage and the formation of new C–Si bond. Evidence for the intermolecular reaction mechanism was obtained in pressurization of mixture of **168c** and **168e**, with the complete scrambling of the silyl groups in four possible products **169c–169f**. The cyclic transition state for this process would involve two molecules with the coincident double Si-migration to the neighboring molecule, which is, due to the negative volume of activation, supported by high pressure.

Desilylation of TBDMS ethers occurs in another example described by Kotsuki, which proceeds with the simultaneous C–O cleavage (Scheme 6.36) [47]. When dinucleotide pdCpA **170** was subjected to pressure of 8 kbar with an excess of ammonium hydroxide, deprotection of all TBDMS and cyanoethyl groups takes place. Superiority of the conditions is further exemplified by the employment of tetra-*n*-butyl ammonium salt leading to quantitative yield of product **171.**

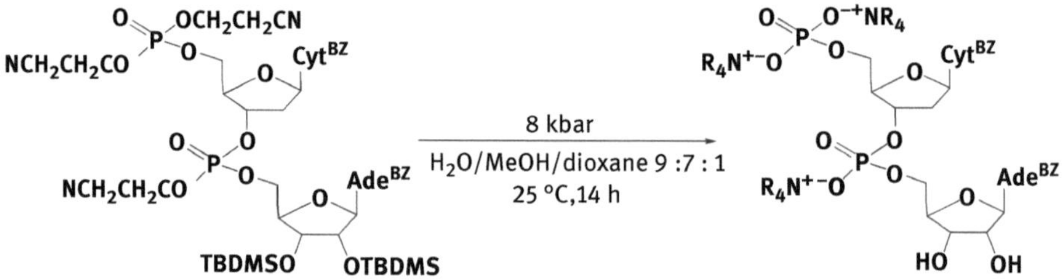

170

171 (aq. NH_4OH, 60%)
(aq. *n*-Bu_4OH, 97%)

Scheme 6.36: Deprotection of dinucleotide.

6.5 C–Se bond

High-pressure conditions were applied to hetero Diels–Alder cycloaddition reaction of selenobenzophenone **172** to ascertain concreteness of the process (Scheme 6.37) [48]. Klärner studied reaction where **172** acts as a dienophilic component in [4 + 2] hetero Diels–Alder cycloaddition with the simultaneous formation of new C–Se and C–C bonds across the C=Se bond. When *trans,trans*-hexa-2,4-diene reacted at ambient pressure, *cis*-**173** was obtained selectively. In the cycloaddition reaction with *cis, trans*-hexa-2,4-diene, the same stereochemistry was observed, due to the loss of stereochemistry of the conjugated diene. However, cycloaddition carried out at 11–12 kbar afforded the stereoisomeric mixture of *cis*-**173** and *trans*-**173** in 4:6 ratio. It was rationalized that two possible cycloaddition reaction mechanisms (concerted and multistep) are well separated in energy. High pressure was required to kinetically accelerate the concerted HDA reaction to compete with the multistep addition route.

6.6 C–B bond

Hydroboration of olefins could also be facilitated by employing high-pressure conditions. Okamoto has demonstrated the beneficial rate acceleration in hydroboration of sterically congested olefins and formation of products that could not be

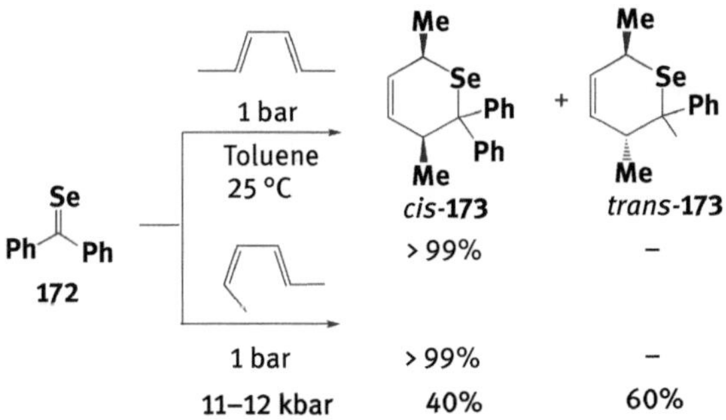

Scheme 6.37: Hetero Diels–Alder cycloadditions of selenobenzophenone.

achieved at the standard pressure (Scheme 6.38) [49]. The reaction of tetramethylethylene with borane at ambient pressure provides thexylborane **175**, which was converted to trithexylborane **176** with excess of **175** at 5.5 kbar pressure. Trialkylborane **179** was synthetized analogously over two reaction steps, using sterically more demanding 1,2-dimethyldi-*tert*-butylethylene as a substrate (mixture of *E*- and *Z*-isomers in 63:37 ratio). Finally, hydroboration of the steroidal system of 5*β*-chol-9(11)-ene **180** at 1 bar showed the formation of some monoalkylborane product **181**, where signals in IR spectra became stronger after the reaction was performed at 5.5 kbar.

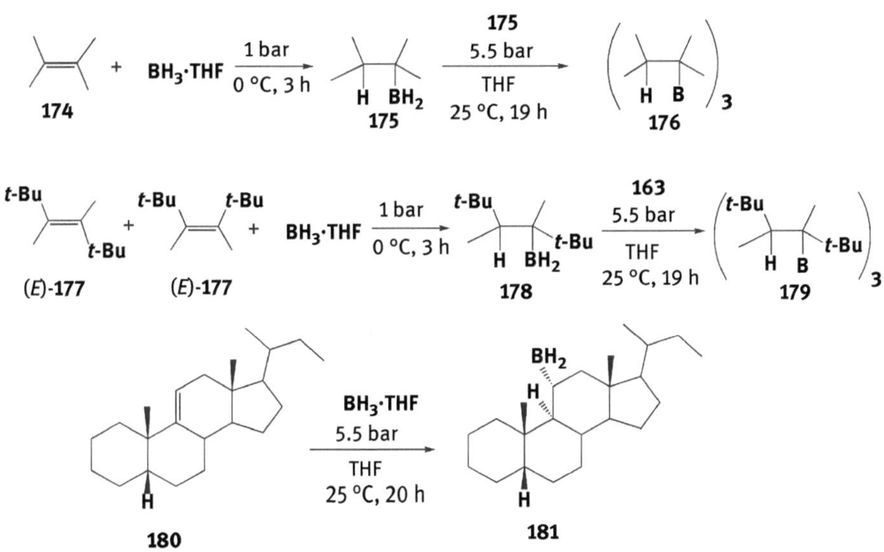

Scheme 6.38: Hydroboration of olefins.

Three different types of olefins were investigated by Maddaluno in hydroboration reactions (Scheme 6.39) [50]. Hydroboration of 1-bromopropene **182** by pinacol borane

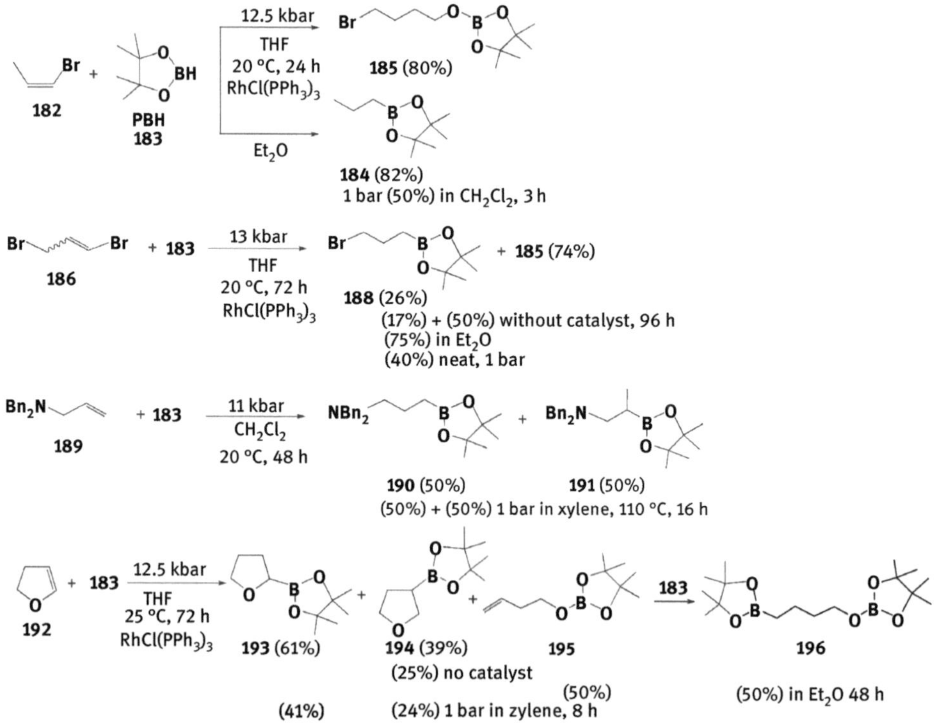

Scheme 6.39: Hydroboration of olefins.

with the addition of 0.5% of Wilkinson catalyst $RhCl(PPh_3)_3$ at 12.5 kbar provided depending on the solvent 4-bromobutoxyborane **185** (THF) or 1-propylboronate **184** (ether). At ambient pressure, product **184** was obtained in 50% yield. When 1,3-dibromopropene **186** used as a mixture of (*E*)- and (*Z*)-isomers ($E{:}Z \approx 6{:}4$) was subjected to hydroboration conditions, mixtures of **188** and **185** were obtained in THF, whereas reactions carried out in diethyl ether or neat produced only **188.** Hydroboration of *N,N*-dibenzylallylamine by pinacol led to the formation of mixture of regioisomers **190** and **191**, while the reaction in THF at 9 kbar gave only unreacted **189.** In ambient conditions, similar results were obtained as for reaction at 11 kbar. Several hydroboration products **193–195** were obtained in the reaction of 3,4-dihydrofuran and the reaction outcome could be controlled by the modification of reaction conditions.

6.7 C–P bonds

A selection of the results of the alkylation of triphenylphosphine with halides under high pressure is given in Scheme 6.40. Dauben has developed a synthetic method

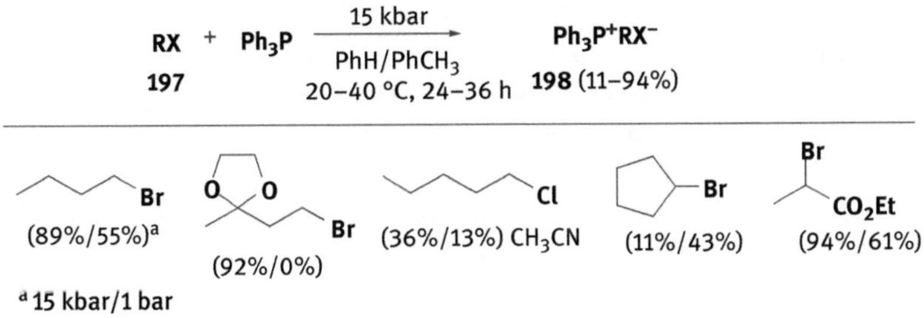

Scheme 6.40: Alkylation of triphenylphosphine.

for the preparation of Wittig phosphonium salts at ambient temperature where 15 kbar conditions supersede thermal reactions [51]. Besides benzene/toluene mixtures, acetonitrile was also found to be a suitable solvent. These results are in accord with the kinetic study in which a significant increase in yields was found for the alkylation of bromide with triphenylphosphine when the pressure was elevated from 1 kbar (1%) to 15 kbar (92%) [4].

Synthesis of phospholes was improved at high-pressure conditions. Cheletropic reaction of butadienes **199a–c** with phenyldibromophosphine at 7 kbar provided the formation of two new C–P bonds of the corresponding phospholenium salts **201**. These could be conveniently transformed to phospholes **202** in high yields (Scheme 6.41) [52]. For this process, Isaacs has determined an exceedingly large negative volume of activation $\Delta V^{\neq} = -60\ \mathrm{cm^3\ mol^{-1}}$, which is responsible for the significant reaction time reduction from 12 days at room pressure to 2 h at 7 kbar. This value is greater than that either for a Menschutkin reaction (–30 to $-50\ \mathrm{cm^3\ mol^{-1}}$) or Diels–Alder reactions ($-35\ \mathrm{cm^3\ mol^{-1}}$). It suggests a concerted cycloaddition reaction mechanism taking place simultaneously with the buildout of ionic charges. This electrostriction process affords an extra volume reduction. The volume of reaction is smaller ($-38\ \mathrm{cm^3\ mol^{-1}}$).

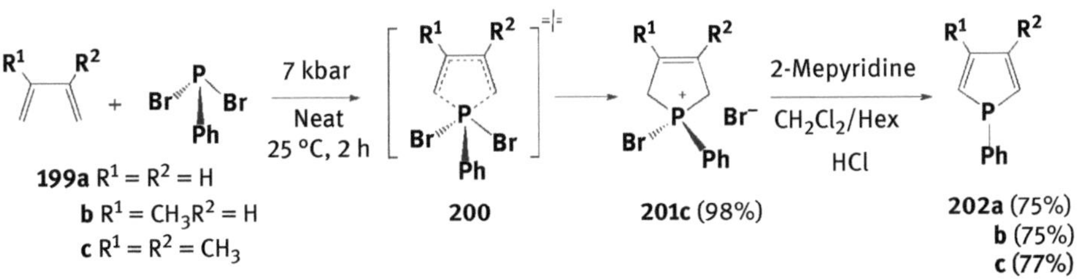

Scheme 6.41: Synthesis of phospholes.

Phospholes, which were prepared by the above-described procedure, were used in Diels–Alder cycloaddition reactions. Isaacs has shown that cycloadditions of 3,4-

dimethyl-1-phenylphosphole, 3-methyl-1-phenylphosphole and 1-phenylphosphole **203** with fumaronitrile at 9 kbar gave 7-phosphabicyclo[2.2.1]hept-2-ene structures **189** (Scheme 6.42) [53]. In the case of derivatives **203a** and **203b**, mixtures of **204** and oxidized product **205** were obtained. The reaction of **203c** with diphenylketene afforded spiro product **206**. The identical compound was obtained in the reaction of **203b**, demonstrating that all carbon atoms of phosphole are lost during the process. The proposed reaction mechanism involves four equivalents of diphenylketene and starts with a nucleophilic attack of phosphorus atom on keten, followed by further additions and rearrangements, where [4 + 2] cycloreversions serve to eliminate of phosphole carbons [54]. Reactions of phospholes with diazomethane provided cycloadducts in the form of oxides **207**. Interestingly, when benzaldehyde was used in the high-pressure reaction, phosphole oxide dimer **208** and *trans*-stilbene were isolated.

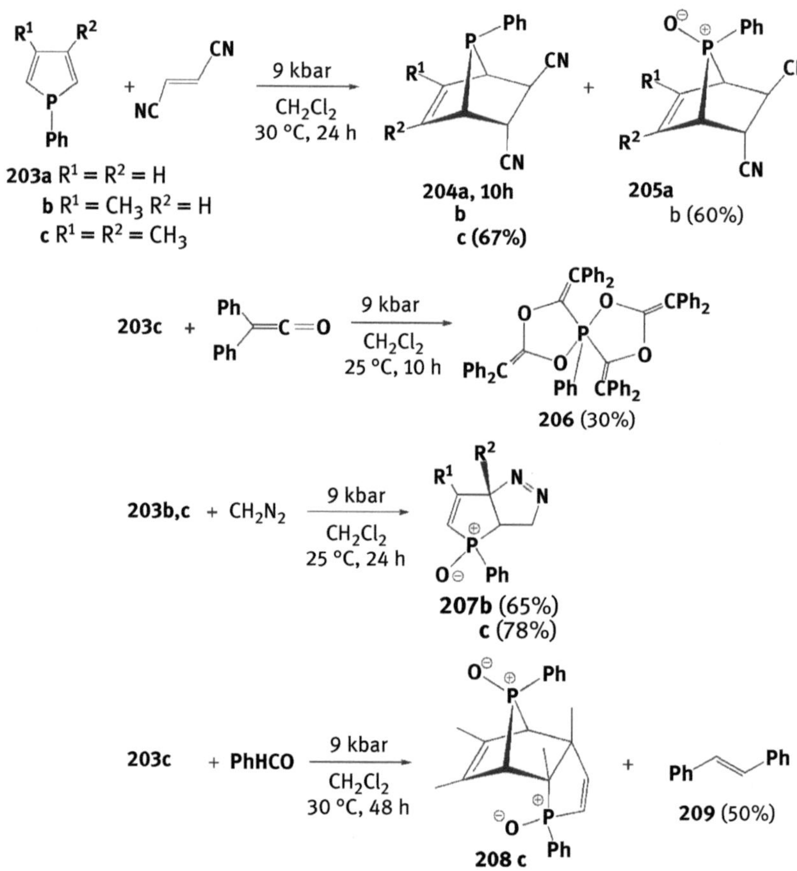

Scheme 6.42: Reaction of phopholes.

The mechanism of hydrolysis of phosphonium salts by alkali producing the phosphine oxide and toluene was examined by Isaacs (Scheme 6.43) [55]. Kinetic measurements in the pressure range of 1 bar to 1 kbar were used to obtain volumes of activation for reactions: alkaline decomposition of triphenylbenzylphosphonium bromide ($\Delta V^{\neq}$ = + 31 cm^3 mol^{-1}) and alkaline decomposition of triphenyl(*p*-nitro) benzylphosphonium bromide ($\Delta V^{\neq}$ = + 32 cm^3 mol^{-1}). Reaction rates are strongly retarded by pressure and activation volumes are some of the largest positive values known in the literature, whereas the volume or reaction is + 16 cm^3 mol^{-1}. These results suggest a two- or three-step mechanism that involves desolvation and fragmentation. Each of these processes makes a contribution to an increase in volume. For the related reaction, a hydrolysis of diethyl *p*-nitrophenyl phosphate negative $\Delta V^{\neq} = -16$ cm^3 mol^{-1} was found and a $B_{AC}2$ mechanism was established.

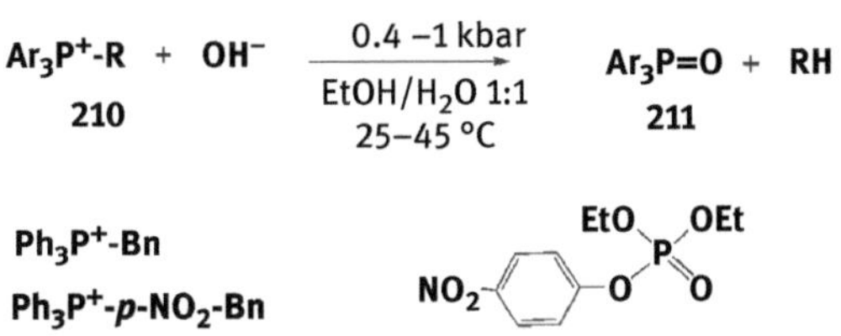

Scheme 6.43: Hydrolysis of phosphonium salts.

6.8 C–Halogen bonds

Formation of C–Br bond is involved in brominolysis of stannanes. Volumes of activation for Et, 1-Bu and 2-Bu substituents were determined ($\Delta V^{\neq}$ = −56, −65 and −47 cm^3 mol^{-1}, respectively) and also the corresponding volumes of reactions (ΔV = + 20, + 23 and + 26 cm^3 mol^{-1}, respectively) (Scheme 6.44) [56]. Obtained values of $\Delta V^{\neq}$ indicate a similar transition state for all substituents. Very high dipole moment was indicated by electrostrictive volume changes (about −70 cm^3 mol^{-1}) and corresponds to fully developed charge in transition state of the S_E2 reaction and retention mode **213**. For the related reaction, iododestannylation of tetramethyltin the volume of activation of −50 cm^3 mol^{-1} (in dibutyl ether, 25 °C, 1–600 bar) was established [57]. For this reaction, the transition state with the inversion of configuration was suggested.

The effects of high pressure on the competition between fluoride-anion substitution (S_N2) and elimination (E2) reaction processes was studied by Gerdes (Scheme 6.45) [58]. Both processes have negative volumes of activation, thus being accelerated by pressure; however, some selective rate enhancement of one reaction over another was achieved. It was found that by the conduction of reactions with tetrabutylammonium fluoride in THF at higher pressures,

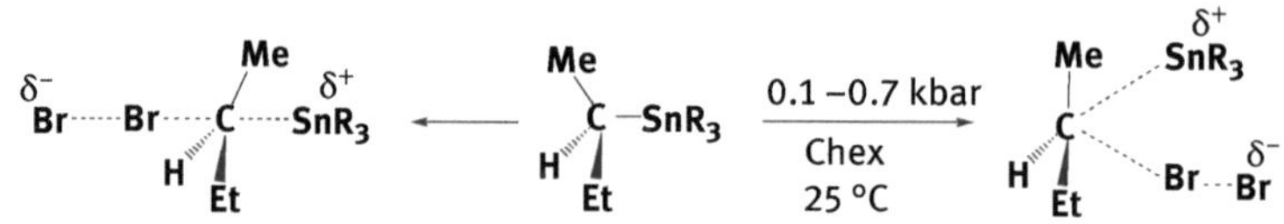

214 212 213

R = Et, 1-Bu, 2-Bu

Scheme 6.44: Brominolysis of stannanes.

215 216 217

	216	217
218 $ArCH_2CH_2OTs$	69 /38%[a]	0 /26%
219 $ArCH_2CH(OTs)CH_3$	34 /38%	9 /34%
220 $ArCH_2CH_2CH_2OTs$	89 /84%	0 / 0%
221 $ArCH_2CH_2CH(OTs)CH_3$	63 /32%	25 / 0%

[a] 15 kbar / 1 bar, 128 h

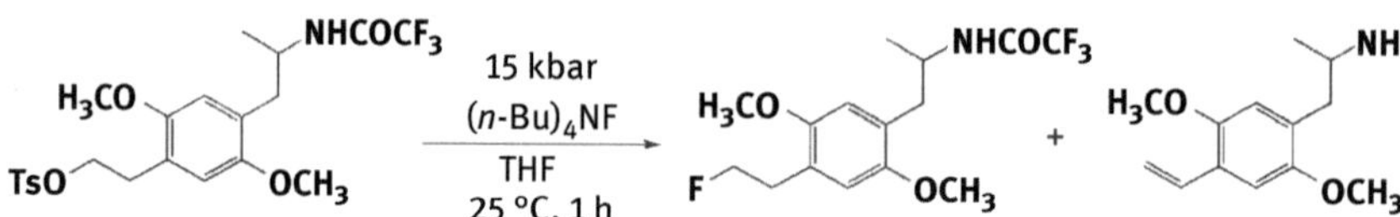

222

223 (75%)
1 bar, 168 h (43%)

224 (0%)
(23%)

Scheme 6.45: Fluoride-anion substitution of tosylates.

the ratio of substitution/elimination products is shifted in favor of substitution products **216**. The optimized conditions were applied to the production of the amphetamine precursor **223** in high pressure conditions and provided substitution product in high yield (75%), whereas the formation of styrene product **224** was not detected. At ambient pressure, both fluoride **223** and styrene **224** were produced in the overall lower yield.

6.9 C–Metal bond

High-pressure techniques were also applied to coordination chemistry [59, 60]. The formation of metal–carbon bond in organometallic complexes is also facilitated by high pressure. Terminal acetylenes with $FeH_2(dmpe)_2$ **225** form an acetylide hydride metal complex, $FeH(C{\equiv}CPh)(dmpe)_2$ **226**, and react further with the metal bis (acetylide) complex Fe(C≡CPh)(dmpe)$_2$ **227**. However, the formation of complex is synthetically difficult to stop at the stage of the acetylide hydride metal complex **226**. Applying of high pressure to the synthesis of organometallic complexes, Field has achieved selective formation of **226** (Scheme 6.46) [61]. At 7.5 kbar in short time, quantitative conversion to **226** takes place, and if the pressurization is continued, **227** started to form. For a comparison, at ambient pressure, this reaction proceeds very slowly and with low yield.

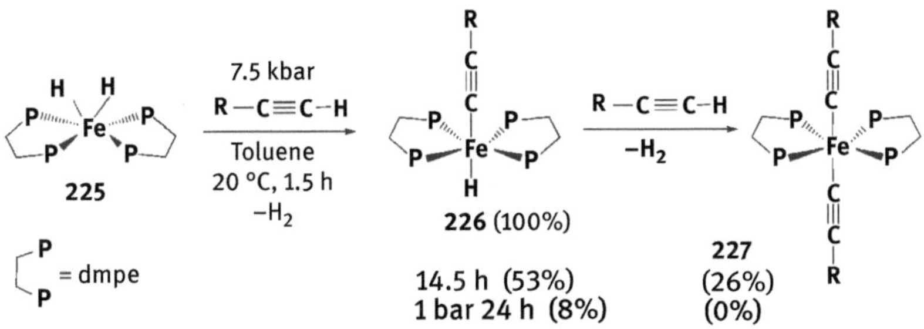

Scheme 6.46: Synthesis of iron complex.

In the follow-up publication, Field and coworkers have prepared analogous type of iron acetylide complex (Scheme 6.47) [62]. The reaction of excess diphenylbutadiyne with FeH$_2$(dmpe)$_2$ **225** in producing iron metallocycle **229** is slow at ambient conditions – after two days at room temperature and pressure product **229** was obtained in approximately 41% yield. When high pressure was applied, within 4 h at 8 kbar complex **229** was obtained in 92% yield, while after 1.5 h, a mixture of **228** and **229** is present.

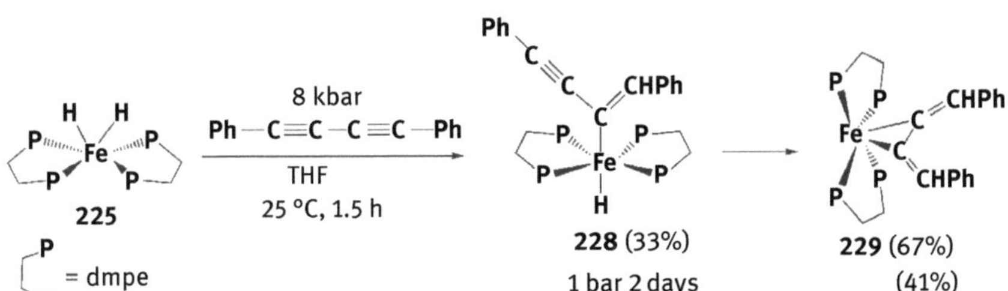

Scheme 6.47: Synthesis of iron complex.

6.10 O–O bond

Jurczak has found that energy needed to accomplish deoxygenation of methyl ether *N*-oxides **230–235** was significantly reduced by the employment of high pressure

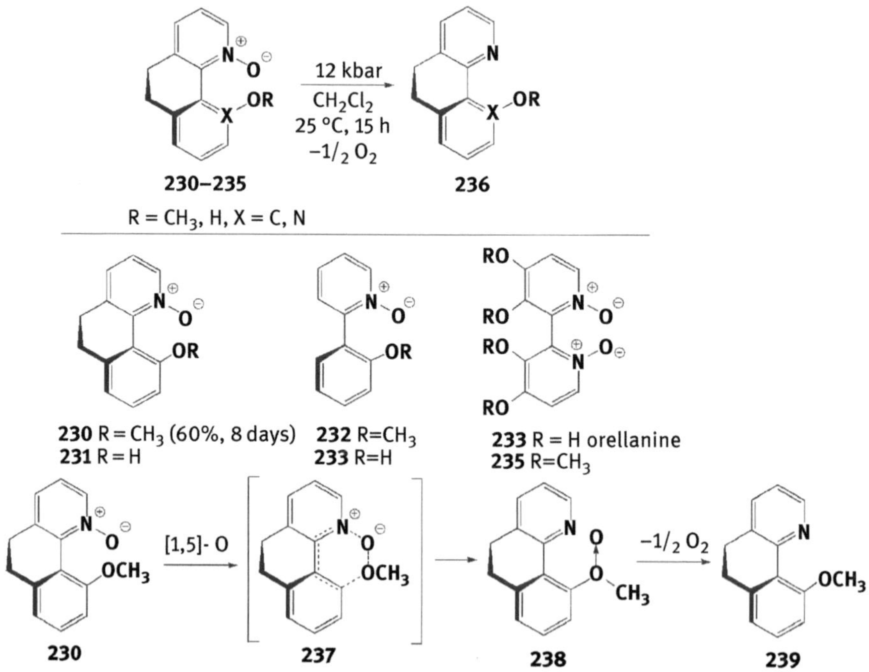

Scheme 6.48: Deoxygenation of methyl ether *N*-oxides.

(Scheme 6.48) [63]. At 1 bar and higher temperatures, alkaloid orellanine was found to decompose easily to the corresponding tertiary base orelline. Orellanine **234** and its methyl ether derivative **235** as well as orellanine model methyl ethers of *N*-oxides **230** and **232** when pressurized at 12 kbar at 50 °C undergo deoxygenation, but with various efficacy. In these conditions, reaction of 5-methoxy-9,10-dihydro-4-azaphenanthrene-*N*-oxide **230** was completed, but also at room temperature within 8 days. At ambient temperature, pressurization of **230** yielded deoxygenated product **239** in the form of hypochlorite salt. The existence of chlorine in the product indicates that under high-pressure conditions the solvent oxidation takes place, which was not observed at ambient pressure. In this reaction, side product 5-hydroxy-9,10-dihydro -4-azaphenanthrene-*N*-oxide **231** was also obtained (40% yield), which is less susceptible to deoxygenation than **230**. The demethylation of **215** also takes place at 1 bar, however, with the simultaneous deoxygenation. The ability to lower reaction temperature to room conditions by high pressure (required temperature for these substrates is well above 100 °C at atmospheric pressure) is related to negative activation volume $\Delta V^{\neq}$ of this process. These rate acceleration pressure effects are well in line with presumed mechanism with initial sigmatropic [1,5]-O shift and contraction of the activation volume. By this process, unstable peroxide intermediate is formed, which then expels oxygen.

References

[1] Gibson, RO., Fawcett, EW., & Perrin, MW. The effect of pressure on reactions in solution I-sodium ethoxide and ethyl iodide to 3,000 kg^{-1} cm^{-2}. II-Pyridine and ethyl iodide solution to 8,000 kg^{-1} cm^{-2}. Proc R Soc Lond A 1935, 150, 223–240.

[2] Williams, EG., Perrin, MW., & Gibson, RO. The effect of pressure up to 12,000 kg^{-1} cm^{-2} on reactions in solution. Proc R Soc Lond A 1936, 154, 684–703.

[3] Dauben, WG., Gerdes, JM., & Look, GC. Organic reactions at high pressure. Conversion of cyclic alkanones and enones to 1,3-dioxolanes. J Org Chem 1986, 51, 4964–4970.

[4] Dauben, WG., Gerdes, JM., & Look, GC. Organic reactions at high pressure: the effect of pressure change upon reaction rates of bimolecular processes. Synthesis 1986, 532–535.

[5] Kumamoto, K., Ichikawa, Y., & Kotsuki, H. High-pressure-promoted uncatalyzed ketalization of ketones and oxy-Michael/ketalization of conjugated enones. Synlett 2005, 14, 2254–2256.

[6] Kotsuki, H., Ichikawa, Y., Iwasaki, M., Kataoka, M., & Nishizawa, H. High pressure organic chemistry; XV. Noncatalyzed cleavage reaction of ethers with acyl halides. High Pressure Res 1992, 11, 107–118.

[7] Jee, J-G., & Ree, T-Y. The solvolysis of benzoyl chloride in water-acetone mixtures under high pressure. Bull Korean Chem Soc 1987, 8, 31–39.

[8] Kotsuki, H., Kataoka, M., & Nishizawa, H. High pressure-promoted uncatalyzed hydrolysis of epoxides. Tetrahedron Lett 1993, 34, 4031–4034.

[9] Beusker, PH., Aben, RWM., Seerden, J-PG., Smits, JMM., & Scheeren, HW. Exploration of high-pressure cycloadducts of furans and citraconic anhydride as precursors for CD-Ring fragments of paclitaxel and its analogues. Eur J Org Chem 1998, 2483–2492.

[10] Le Noble, WJ., & Duffy, M. The effect of pressure on the rates of acid-catalyzed hydrolysis of trimethylene oxide and of epichlorohydrin. J Phys Chem 1964, 68, 619–622.

[11] Le Noble, WJ. The effect of pressure on the homogeneous alkylation of phenoxide ion. J Am Chem Soc 1963, 85, 1470–1472.

[12] Le Noble, WJ. The effect of pressure on the benzylation of 1-Methyl-2-naphthoxide ion. J Phys Chem 1964, 68, 2361–2362.

[13] Grochowski, E., Stepowska, H., Salański, P., & Jurczak, J. High-pressure approach to the synthesis of "C-O-N" analogues of nucleosides. Carbohydr Res 1988, 177, 244–246.

[14] Dauben, WG., & Köhler, P. High-pressure glycosylations of unreactive alcohols and the formation of *N*-glycosyl collidinium salts. Carbohydr Res 1990, 203, 47–56.

[15] Kochetkov, NK., Zhulin, VM., Malysheva, NM., Klimov, EM., & Makarova, ZG. The effect of high pressure on glycosylation under Helferich conditions. Carbohydr Res 1987, 167, C8–C10.

[16] Kochetkov, NK., Zhulin, VM., Klimov, EM., Malysheva, NM., Makarova, ZG., & Ott, AY. The effect of high pressure on the stereospecificity of the glycosylation reaction. Carbohydr Res 1987, 164, 241–254.

[17] Jenner, G. Synthesis of hindered functionalized ethers via high-pressure addition of alcohols to acrylic compounds. Tetrahedron Lett 2001, 42, 4807–4810.

[18] Hayashi, Y., & Nishimura, K. Application of high pressure induced by water-freezing to the Michael reaction of alcohols with *β,β*-Enones. Chem Lett 2002, 31, 296–297.

[19] Kurabayashi, M., Yanagiya, K., & Yasumoto, M. Reaction of nitriles under high pressure. I. The formation of triphenyl-1,3,5-triazine *via* the equilibrium among benzonitrile, alcohols and benziminoethers. Bull Chem Soc Jpn 1971, 44, 3413–3418.

[20] Aben, RWM., & Scheeren, HW. High pressure-promoted cycloadditions of ketene acetals and *α,β*-unsaturated aldehydes and ketones. Tetrahedron Lett 1985, 26, 1889–1892.

[21] Newitt, DM., Linstead, RP., Sapiro, RH., & Boorman, EJ. Liquid-phase reactions at high pressures. Part I. Hydrolysis of esters, and the Knoevenagel reaction. J Chem Soc 1937, 876–883.

[22] Krzyżaniak, A., Barciszewski, J., Sałański, P., & Jurczak, J. The non-enzymatic specific aminoacylation of transfer RNA at high-pressure. Int J Biol Macromol 1994, 16, 153–158.
[23] Yamamoto, Y., Furuta, T., Matsuo, J., & Kurata, T. Cleavage of esters under nearly neutral conditions at high pressure. Chemo- and regioselective hydrolysis in organic solvents. J Org Chem 1990, 56, 5737–5738.
[24] Furuta, T., Matsuo, J., Kurata, T., & Yamamoto, Y. Cleavage of esters under nearly neutral conditions at high pressure. High Press Res 1993, 11, 93–106.
[25] Kroszczyński, W., Olszewska, E., Salański, P., & Jurczak, J. Effective high-pressure cleavage of sterically hindered steroid esters. Helv Chim Acta 2004, 87, 1488–1492.
[26] Romanski, J., Nowak, P., Kosinski, K., & Jurczak, J. High-pressure transesterification of sterically hindered esters. Tetrahedron Lett 2012, 53, 5287–5289.
[27] Mori, T., Weiss, RG., & Inoue, Y. Mediation of conformationally controlled photodecarboxylations of chiral and cyclic aryl esters by substrate structure, temperature, pressure, and medium constraints. J Am Chem Soc 2004, 126, 8961–8975.
[28] Roemer-Mähler, J., Bieniek, D., & Korte, F. Beitrage zur ökologischen chemie LIV. Solvolyse von heptachlor und dieldrin unter hohen drucken. Chemosphere 1973, 2, 31–33.
[29] Plieninger, H., Heuck, CC., & Bühler, R. Über die umsetzung von bicycloheptenderivaten mit schwefelkohlenstoff unter hohem druck – III. Über hochdruckversuche. Tetrahedron 1972, 28, 73–81.
[30] Rice, JE., & Okamoto, Y. One-step synthesis of substituted tetrathiofulvalenes under high pressure. J Org Chem 1981, 46, 446–447.
[31] Jenner, G. Lithium perchlorate-diethyl ether as a medium in organic addition reactions. An ersatz to high pressure methodology? High Pressure Res 1993, 11, 257–262.
[32] Tsuchiya, T., Yasumoto, M., Shibuya, I., & Goto, M. Novel preparation of 1,3,5-thiadiazine derivatives by cyclocondensation of carbon disulphide with dialkylcyanamides under high pressure. J Chem Soc Perkin Trans 1990, 1, 1218–1219.
[33] Huay Chong, S., Young, DJ., & Andy Hor, TS., Pressure-assisted hetero- and homodialkylation of sulfide in $[Pt_2(\mu\text{-}S)_2dppp)_2]$: one-pot conversion of $\{Pt_2(\mu\text{-}S)_2\}$ into $\{Pt_2(SR)_2\}$ and $\{Pt_2(SR)(SR')\}$. Chem Asian J 2007, 2, 1356–1362.
[34] Huay Chong, S., Young, DJ., & Andy Hor, TS. Combinative use of high-pressure, metal-templating and sulfur-nucleophilicity towards dithiacyclophane synthesis and its complex intermediates. J Organomet Chem 2006, 691, 349–355.
[35] Kotsuki, H., Matsumoto, K., & Nishizawa, H. High pressure-promoted transformation of alcohols into the corresponding phenylsulfides with Bu_3P-PhSSPh. Tetrahedron Lett 1991, 32, 4155–4158.
[36] Isaacs, NS., Rzepa, HS., Sheppard, RN., Lobo, AM., Prabhakar, S., & Merbach, AE. Volumes of reaction for the formation of some analogues of tetrahedral intermediates. J Chem Soc Perkin Trans 1987, 2, 1477–1482.
[37] Tonnet, ML., & Whalley, E. Effect of pressure on the alkaline hydrolysis of ethyl acetate in acetone-water solutions. Parameters of activation at constant volume. Can J Chem 1975, 53, 3414–3418.
[38] Puspawati, MN., & Young, DJ. High-pressure synthesis of novel thietane monomers and polymers. High Pressure Res 2008, 28, 675–679.
[39] Isaacs, NS., & Laila, A. The volume profile for a cheletropic reaction. J Phys Org Chem 1994, 7, 178–180.
[40] Rahm, A., Degueil-Castaing, M., & Pereyre, M. Organometallic chemistry under high pressure: trialkyl-tin hydride addition to unsaturated systems. J Organomet Chem 1982, 23, C29–C32.
[41] Yamamoto, Y., Maruyama, K., & Matsumoto, K. Organometallic high pressure reactions. Allylation of aldehydes with allylic stannanes under mild conditions. J Chem Soc Chem Commun 1983, 489–490.

[42] Isaacs, NS., Marshall, RL., & Young, DY. The stereochemistry of high pressure aldehyde addition to allylstannanes. Tetrahedron Lett 1992, 33, 3023–3024.

[43] Isaacs, NS., Maksimovic, Lj., Rintoul, GB., & Young, DY. The high pressure addition of aldehydes to allylstannanes: a 'Diels-Alder Like' transition state. J Chem Soc Chem Commun 1992, 1749–1750.

[44] Dauben, WG., Bunce, RA., Gerdes, JM., Henegar, KE., Cunningham, Jr AF., & Ottoboni, TB. Organic reactions at high pressure. A mild method for the placement of protecting groups on hindered and sensitive alcohols. Tetrahedron Lett 1983, 24, 5709–5712.

[45] Matsuo, I., Wada, M., & Ito, Y. Desilylation under high pressure. Tetrahedron Lett 2002, 43, 3273–3275.

[46] Yamamoto, Y., Maruyama, K., & Matsumoto, K. A new procedure for the synthesis of β-silylated esters and lactones. Migration of silicon from oxygen to carbon at high pressure. Organomet 1984, 3, 1583–1585.

[47] Kataoka, M., Fukui, C., Mimoto, A., Kuge, H., Honke, K., & Kotsuki, H. A new strategy for synthesis of the dinucleotide pdCpA: a convenient method for the deprotection of cyanoethyl, TBDMS, and benzoyl groups in one step at high pressure. Heterocycles 2015, 91, 1164–1169.

[48] Erker, G., Hock, R., Krüger, C., Werner, S., Klärner, F-G., & Artschwager-Perl, U. Synthesis and cycloadditions of monomeric selenobenzophenone. Angew Chem Int Ed Engl 1990, 29, 1067–1068.

[49] Rice, JE., & Okamoto, Y. Hydroboration of sterically hindered olefins under high pressure. J Org Chem 1982, 47, 4189–4194.

[50] Colin, S., Vaysse-Ludot, L., Lecouvé, J-P., & Maddaluno, J. Combining high pressure and catalysis: pinacol- or catecholborane hydroboration of functionalized olefins. J Chem Soc Perkin Trans 2000, 1, 4505–4511.

[51] Dauben, WG., Gerdes, JM., & Bunce, RA. Organic reactions at high pressure. Preparation of wittig phosphonium salts at ambient temperature. J Org Chem 1984, 49, 4293–4295.

[52] Isaacs, NS., & El-Din, GN. An improved synthesis of phospholes. Synthesis 1989, 967.

[53] Isaacs, NS., & El-Din, GN. Cycloadditions and other reactions of phospholes under high pressures. Tetrahedron 1989, 45, 7083–7092.

[54] Isaacs, NS., El-Din, GN., & Drew, MGB. The reaction between a phosphole and diphenylketene at high pressure; an unexpected outcome. Recl Trav Chim Pays-Bas 1989, 108, 120–121.

[55] Isaacs, NS., & Abed, OH. Profiles and mechanisms for the alkaline decomposition of phosphonium salts and phosphate esters. Tetrahedron Lett 1986, 27, 1209–1210.

[56] Isaacs, NS., & Laiala, AH. Reactions at high pressure 13; volume profile for the brominolyses of tetra-alkylstannanes. Tetrahedron Lett 1984, 25, 2407–2408.

[57] Isaacs, NS., & Javaid, K. Reactions at high pressure, IV: iododestannylation of tetramethyltin. Tetrahedron Lett 1977, 18, 3073–3076.

[58] Gerdes, JM., Keil, RN., Shulgin, AT., & Mathis, CA. High pressure nucleophilic fluoride-ion substitution reactions: formation of fluoroalkylbenzenes. J Fluor Chem 1996, 78, 121–129.

[59] Kotowski, M., & van Eldik, R. Applications of high pressure techniques to mechanistic studies in coordination chemistry. Coord Chem Rev 1989, 93, 19–57.

[60] Hubbard, CD. Feeling pressure: a chemical kinetics response. High Press Res 2003, 23, 217–224.

[61] Field, LD., & George, AV. The use of high reaction pressures in the synthesis of acetylide complexes of iron. J Organomet Chem 1993, 454, 217–219.

[62] Field, LD., George, AV., Hambley, TW., Turner, P., & Wallace, BJ. Metallocycle synthesis accelerated by high pressure. J Organomet Chem 1997, 540, 95–99.

[63] Antkowiak, R., Antkowiak, WZ., Bańczyk, I., Czerwiński, G., Jurczak, J., Raczko, J., & Salański, P. Steric requirements and high pressure influence on the thermal deoxygenation of *N*-oxides of oreallanine-like structure. Heterocycles 1994, 39, 485–490.

7 Oxidation and reduction reactions

Oxidation and reduction reactions in high-pressure conditions were the topic of study of several research groups as these represent important chemical transformations.

7.1 Oxidation reactions

A study of the reaction mechanism of Baeyer–Villiger oxidation of ketones by *m*-chloroperbenzoic acid (mCPBA) was carried out by Jenner (Scheme 7.1) [1]. A small positive effect of elevated pressure on the reaction yields was established, which was uniformly increased by a factor of 2–3 (in comparison with atmospheric pressure). However, the correlation between pressure steric acceleration and steric demands of substituents on ketone was not established, which is in variance with other high-pressure reactions. Kinetic measurements were used to elucidate mechanistic aspects of the reaction. Slightly negative values of activation volumes ($\Delta V^{\neq}$, from −2 to −8 $cm^3\ mol^{-1}$) determined at 1 kbar and reaction volumes indicate the mechanism associated with a fully concerted rate-determining step (involving simultaneous electron flow with concomitant migration and O–O cleavage in transient intermediate mCPBA/ketone adduct).

The 3-aryl-2-*tert*-butyloxaziridines were successfully employed as oxidants in the oxidation reaction of sulfones under high pressure (Scheme 7.2) [2]. All reactions were carried out neatly as homogeneous mixtures without any solvent added. Oxidation produced mixtures of sulfoxides **7**, *N*-*tert*-butylarylaldimine **8**, *N*-*tert*-butyl -α-arylnitrone **9** and aryl aldehyde **10**. The product distribution varies depending on the substituents and it is dramatically changed between 4 and 8 kbar (Table 7.1). Whereas atmospheric oxidation produced targeted sulfone in minor quantities, with nitrones as major product, synthesis under high pressure revert the product distribution in favor of sulfones. Higher product yields as compared to atmospheric reactions were explained in steric terms as high pressure overcomes the steric hindrance imposed by the bulkiness of the oxaziridine substituents and brings reactants sufficiently close to react.

Significantly larger activation volumes were determined by Isaacs for Swern–Moffatt oxidation of cyclohexanol by dimethyl sulfoxide (DMSO)/acetic anhydride and DCC/acetic anhydride methods ($\Delta V^{\neq} = -25.2$ and $-34\ cm^3\ mol^{-1}$, respectively, at 1 kbar) (Scheme 7.3) [3]. Strongly negative values of volumes of activation were explained by the multistep mechanism, in which the slow process is the alcohol displacement of the intermediate alkoxysulfonium ion in the second reaction step. The volume of activation

https://doi.org/10.1515/9783110556841-007

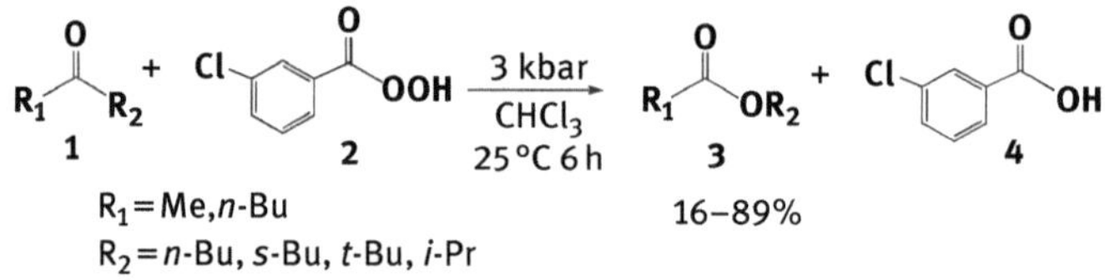

Scheme 7.1: Baeyer–Villiger oxidation of ketones by mCPBA.

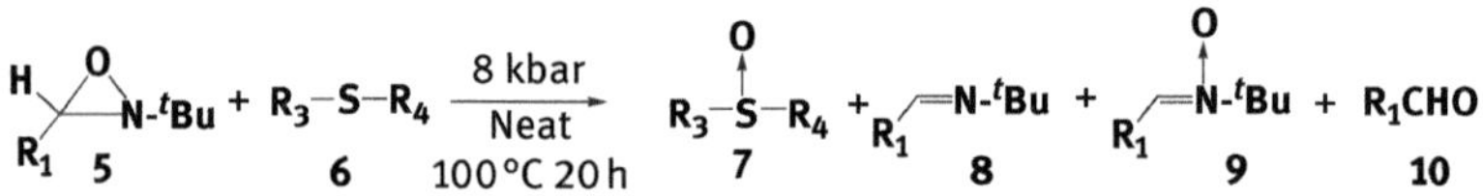

Scheme 7.2: Thioether oxidation by 3-aryl-2-*tert*-butyloxaziridines.

Table 7.1: Oxidation of sulfides with oxaziridines (8 kbar, 100 °C, 20 h).

Oxaziridine 5	Sulfide 6				Yields (%)		
R_1	R_2	R_3	7	8	9	10	Conditions
Ph	Ph	CH_3	2	8	54	18	1 atm
			3	6	81	12	4 kbar
			7	22	6	22	40 °C
			64	62	18	6	
	CH_3	CH_3	25	53	36	8	
	Ph	Ph	3	5	38	20	
4-$NO_2C_6H_4$	Ph	CH_3	42	62	18		
4-$CH_3C_6H_4$			6	6	58		
4-$CH_3OC_6H_4$			8	7	63	15	
2-Pyridyl			37	66	8		
3-Pyridyl			18	36	23	8	
4-Pyridyl			39	51	13		

is the sum of all reaction steps, first one being the formation of the sulfonium ion. It was demonstrated that the application of pressure to these oxidation reactions has practical value in promoting difficult oxidations. When oxidation reaction of cyclohexanol by DMSO/acetic anhydride was carried out at ambient pressure, even after

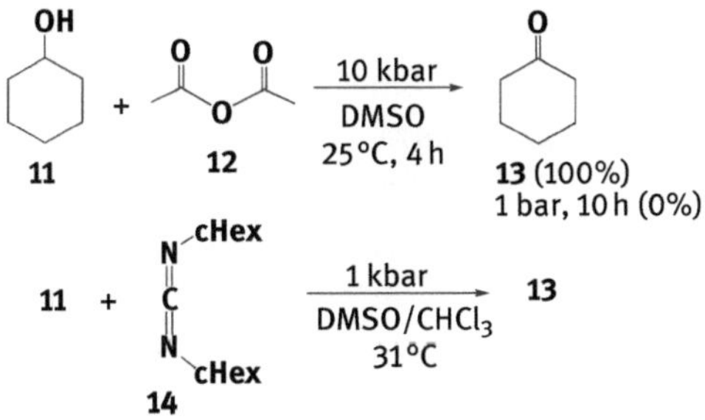

Scheme 7.3: Swern–Moffatt oxidation of cyclohexanol.

10 h product was not detected. On the other hand, at 10 kbar, cyclohexyl alcohol was fully converted to cyclohexanone.

7.2 Reduction reactions

Reduction of ketones with stoichiometric amount of tributyltin hydride under high pressure (10 kbar) provided the corresponding secondary alcohols (Scheme 7.4) [4]. Under thermal conditions and ambient pressure reaction, yields were significantly lower. Whereas the reduction of PhCH(Me)C(=O)-*i*-Pr ketone did not cause the change in diastereomeric ratio in high-pressure conditions, the *cis/trans* ratio of alcohols **19** has changed from 9/91 to 20/80 when pressure was applied to reduction of ketone **18**. These results are the effect of pressure on sterically hindered reactions.

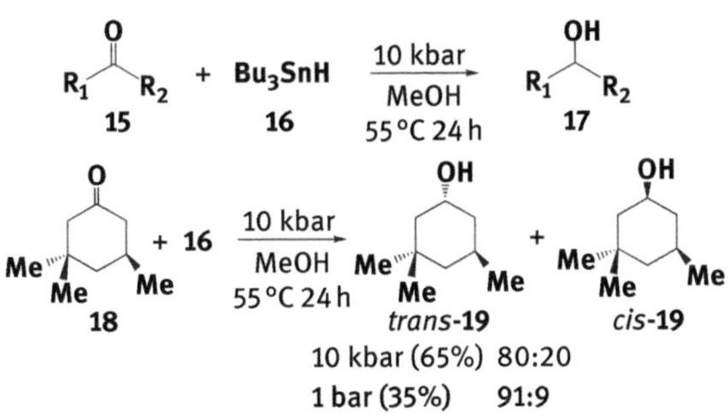

Scheme 7.4: Reduction of ketones with tributyltin hydride.

In the following publication, Rahm has shown that the ratios between the yields obtained for ketone reductions with Bu_3SnH at 1 kbar and at atmospheric pressure are from 1.8 to 31.5, showing the rate acceleration caused by pressure (Scheme 7.5)

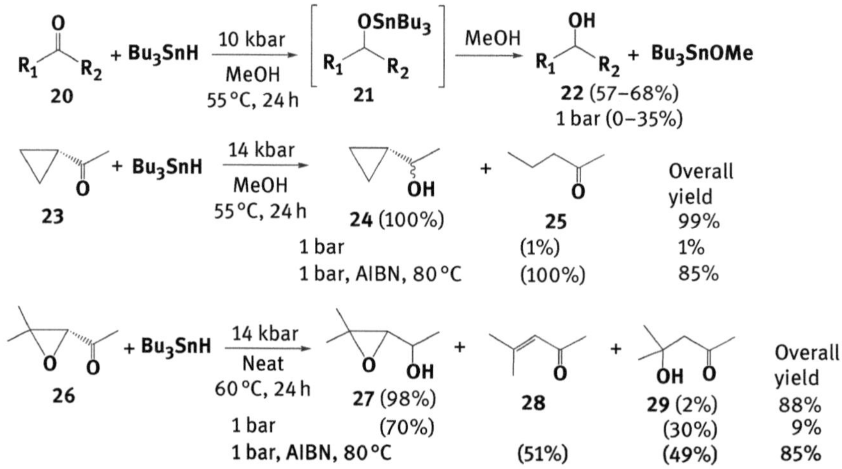

Scheme 7.5: Reduction of ketones with tributyltin hydride.

[5]. Reduction of sterically the most congested ketone **20** (R_1 = *t*-Bu, R_2 = $CH(CH_3)Ph$) did not proceed at atmospheric pressure, but at 1 kbar reduction to alcohol was obtained in 57% yield. The most likely mechanism in a polar solvent such as methanol is the ionic one. Reductions of cyclopropyl and α,β-epoxy ketones **23** and **26** strongly depend on the reaction conditions applied. While in methanol the polar mechanism is the most likely, in neat conditions free radical mechanism could take place. In methanol, by the reduction of **23** at 14 kbar cyclopropylcarbinol was obtained as a sole product, whereas in neat reactions mixture of alcohol and acyclic ketone was obtained. Acyclic ketone was formed as a sole product with neat reagents and with the radical initiator. Similar product outcomes were found for **26.**

Several *p*-quinones have been reduced under high-pressure conditions. Chloranil **31** was reduced by the hydride transfer between leuco-Crystal Violet **30** and chloranil (Scheme 7.6) [6]. The process was found by Isaacs to be isotope dependent ($\Delta V^{\neq}(H) = -25$ and $\Delta V^{\neq}(D) = -35$ cm^3 mol^{-1}) and the k_H/k_D isotope effect diminishes from 11.5 at 1 bar to 8 at 2 kbar. These differences are explained by the increased solvation of the proton under pressure diminishes the extent of the tunneling.

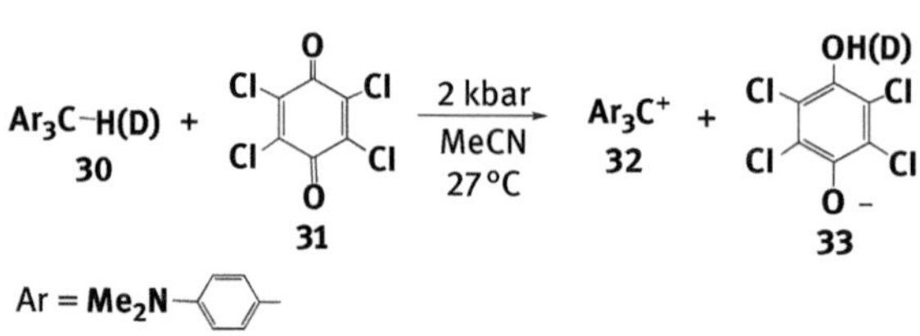

Scheme 7.6: Reduction of *p*-quinones.

Some evidence in favor of an ionic reaction mechanism involving bimolecular transfer of a hydride ion from tetralin to the quinone was obtained by high-pressure studies carried out by Brower (Scheme 7.7) [7]. Activation volume for the reaction of thymoquinone with tetralin was determined to be $\Delta V^{\neq} = -28$ cm^3 mol^{-1}. Similarly, large negative value was determined for the reaction of thymoquinone with 1,4-cyclohexadiene ($\Delta V^{\neq} = -33$ cm^3 mol^{-1}) [8] suggesting a transition state which is bimolecular and ionic. It is known that bimolecular hydride transfer reactions are rate controlling and have negative activation volumes.

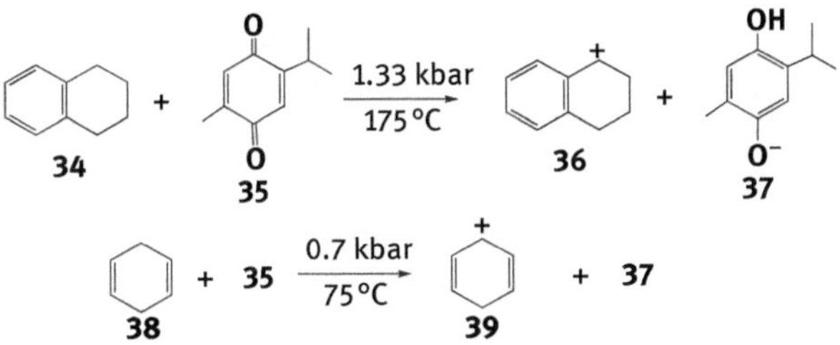

Scheme 7.7: Reduction reactions of thymoquinone.

Reduction of quinones by ascorbic acid forming dehydroascorbic acid and hydroquinones was studied by Isaacs (Scheme 7.8) [9]. Determined volumes of activation are very negative ($\Delta V^{\neq} = -20$, -16 and -14 cm^3 mol^{-1}) for reactions with *p*-benzoquinone, 1,2-dichloro-4,5-dicyano-1,4-benzoquinone (DDQ) and 1,2-naphthoquinone, respectively. The $\Delta V^{\neq}$ values are pH dependent and these are determined in acidic conditions, which changed at higher pH to -4 cm^3 mol^{-1} for reaction involving *p*-benzoquinone. Volume of reaction for *p*-benzoquinone is -12.2 cm^3 mol^{-1}, and these data indicate the formation of charges in transition state.

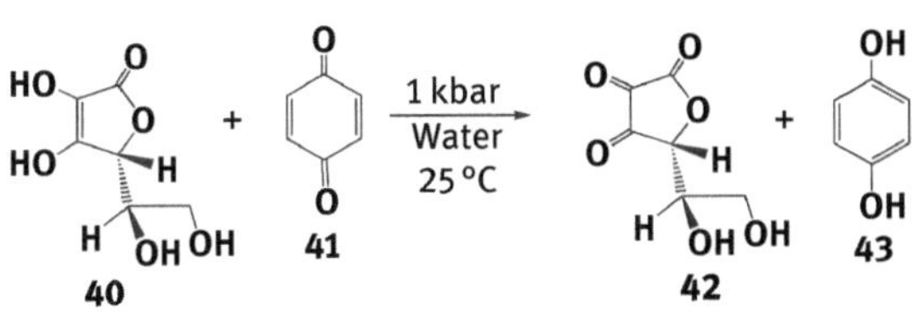

Scheme 7.8: Reduction of benzoquinone by ascorbic acid.

Reduction of a series of aryl ketones **44** by hydrogen generated with the aid of Ni–Al alloy and high pressure proceeded with high yields and with the improved selectivity in comparison with other reaction conditions investigated (Scheme 7.9) [10]. At atmospheric pressure, obtained conversions were significantly lower and

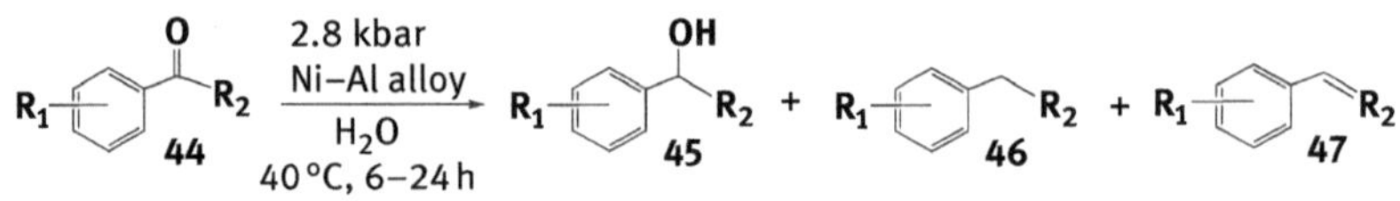

R_1 = 4-F, 4-Cl, 3-Me, 4-Me, 2,4-diMe, 3-NO_2
R_2 = Me, Ph

Conversion (23–100%) selectivity (67–100%)
1 bar (0–100%) sel 76–97%
MW 2 h 80 °C (0–100%) sel 0–58%
Sonication 2 h RT (0–100) sel 0–100%

Scheme 7.9: Metal-catalyzed reduction of aryl ketones by hydrogen.

4-fluorobenzophenone does not react at all, whereas at 2.8 kbar conversion to the corresponding alcohol **45** was 64%. Sonication and microwave irradiation were also employed to initiate reduction, but conversions or selectivities were not as good as in high-pressure conditions. In the optimal conditions, reaction was selective and nitro substituent was not significantly affected. This method was found to be general to reduce other types of ketones (acetophenones, aliphatic and cyclic ketones), with the diminished common problem of nonselective overhydrogenation and conversions ranging from 23% to 100% and selectivities 48–100%.

References

[1] Jenner, G. High pressure mechanistic diagnosis in Baeyer-Villiger oxidation of aliphatic ketones. Tetrahedron Lett 2001, 42, 8969–8971.

[2] Shimizu, M., Shibuya, I., Taguchi, Y., Hamakawa, S., Suzuki, K., & Hayakawa, T. Oxidation of sulfides to sulfoxides with 3-aryl-2-*tert*butyloxaziridines under high pressure. J Chem Soc Perkin Trans 1 1997, 3491–3492.

[3] Isaacs, NS., & Laila, LH. The effect of pressure on Swern-Moffatt oxidations. J Phys Org Chem 1991, 4, 639–642.

[4] Rahm, A., Degueil-Castaing, M., & Pereyre, M. Organometallic chemistry under high pressure: trialkyl-tin hydride addition to unsaturated systems. J Organomet Chem 1982, 23, C29-C32.

[5] Degueil-Castaing, M., Rahm, A., & Dahan, N. Reduction of ketones by tributyltin hydride: the effect of high pressure on steric hindrance and rearrangement processes. J Org Chem 1986, 51, 1672–1676.

[6] Isaacs, NS., Javaid, K., & Rannala, E. Reactions at high pressure. Part 5. The effect of pressure on some primary kinetic isotope effects. J Chem Soc Perkin Trans 2 1978, 709–711.

[7] Pajak, R., & Brower, KR. Effects of pressure and isotopic substitution on the rate of reaction of coal with tetralin. J Org Chem 1982, 47, 1889–1893.

[8] Pajak, R., & Brower, KR. Kinetic isotope effects and pressure effects in several hydrogen-transfer reactions of tetralin and related compounds. J Org Chem 1985, 50, 2210–2216.

[9] Isaacs, NS., & van Eldik, R. A mechanistic study of the reduction of quinones by ascorbic acid. J Chem Soc Perkin Trans 2 1997, 1465–1467.

[10] Tomin, A., Lazarev, A., Bere, MP., Redjeba, H., & Török, B. Selective reduction of ketones using water as a hydrogen source under high hydrostatic pressure. Org Biomol Chem 2012, 10, 7321–7326.

8 Multicomponent reactions

Multicomponent reactions were carried out in high-pressure conditions and many of them are covered in a review by Scheeren and coworkers [1]. These include Mannich, Passerini, Ugi, Strecker and Horner–Wadsworth–Emmons reactions, as well as tandem cycloadditions.

The multicomponent Mannich reaction of carbonyl compounds with CH_2Cl_2 as a C_1 synthon and secondary amines [pyrrolidine, piperidine, DIPEA (diisopropylethyl amine), Et_2NH] was carried out at pressures of 6–9 kbar by Matsumoto (Scheme 8.1) [2]. This synthetic procedure allows the synthesis of Mannich bases that are highly sensitive to heat, such as from pinacolone. In these conditions, even the sterically demanding phenyl isopropyl ketone reacts with a base such as diisopropylamine.

In similar reaction conditions, the multicomponent Mannich reaction with indole was carried out (Scheme 8.2) [3]. Reaction presumably involves in the first reaction step the addition of CH_2Cl_2 on secondary amine, which is followed by the formation of an imine reactive intermediate. Reactivity of 2-methylindole was greatly diminished in comparison to indole (50–98%), and Mannich products $R^1 = R^2 = Et$ and $R^1 = R^2 = (CH_2)_5$ were obtained in only 18% and 7% yields, respectively. Interestingly, the course of the reaction of 2-phenyl indole with diethylamine differs; instead ofproducing Mannich base, bis(2-phenylindolyl)methane **9** and 2-phenyl-3-hydroxymethylindole **10** were formed through different reaction paths (Scheme 8.3).

The Passerini reaction, that is, a three-component reaction of isocyanide with carboxylic acids and carbonyl compounds to yield compounds **14** in a single step is carried out by Jenner with the aid of high pressure (Scheme 8.4) [4]. There is a clear influence of pressure on reactions involving ketone and isocyanide reactants with increased steric bulk. The increase of pressure from ambient to 3 kbar has a beneficial effect on the reaction yields, which are at least doubled, as in the case of moderately bulky ketones and isocyanides. Reactions involving more bulky *t*-butyl isocyanide were clearly more pressure sensitive, especially with the increasing structural complexity of substituent R_1. For instance, the Passerini reaction of *t*-butyl isocyanide with pinacolone **12d** proceeds in 13% yield at 3 kbar and 73% at 6 kbar, and does not proceed at all at ambient pressure due to steric inhibition. The large effect of pressure is also evident in the reaction of *t*-butyl carboxylic acid with ketone **12b** and *t*-butyl isocyanide, whose yield increases from 7% at atmospheric pressure to 52% at 3 kbar.

A four-component reaction, a condensation of amino acid, ketone, amine and isocyano ester (Ugi Reaction) under high pressure was carried out by Matsumoto to obtain tripeptide products possessing sterically very hindered α,α-diisopropylglycine and α,α-diphenylglycine units (Scheme 8.5) [5]. Two synthetic approaches were used (methods A and B). In the reaction of 1,2-diisopropyl ketone **33a** and *t*-butyl

https://doi.org/10.1515/9783110556841-008

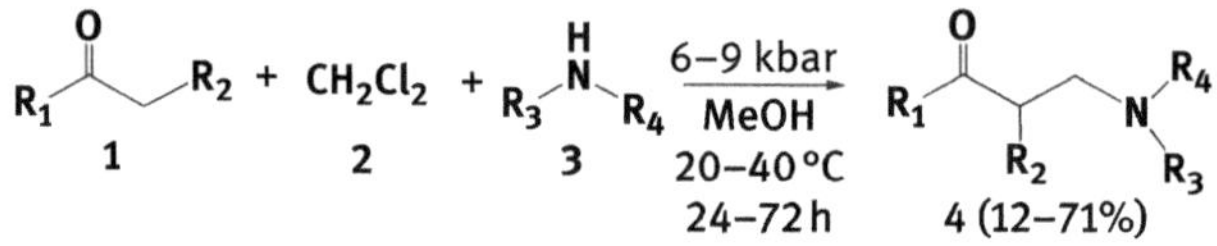

Scheme 8.1: Mannich reaction of carbonyl compounds with CH_2Cl_2 as a C_1 synthon and secondary amines.

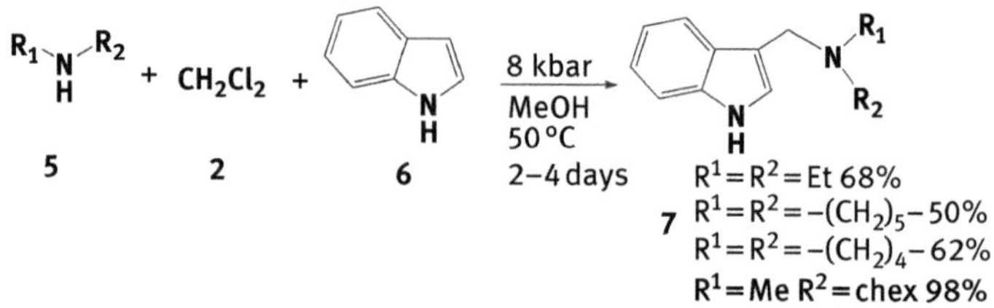

Scheme 8.2: Multicomponent Mannich reaction of indole.

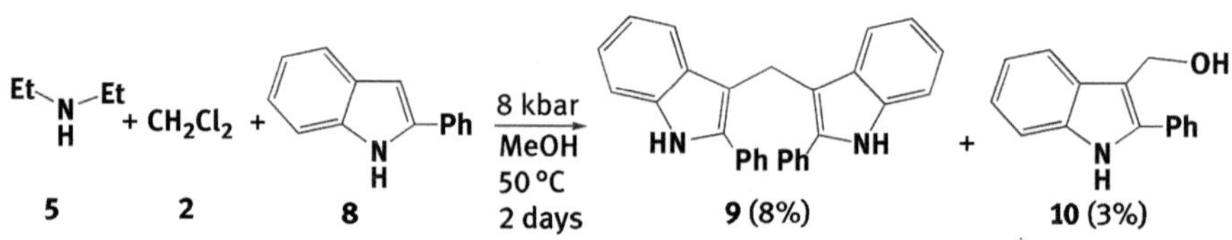

Scheme 8.3: Multicomponent Mannich reaction of 2-phenylindole.

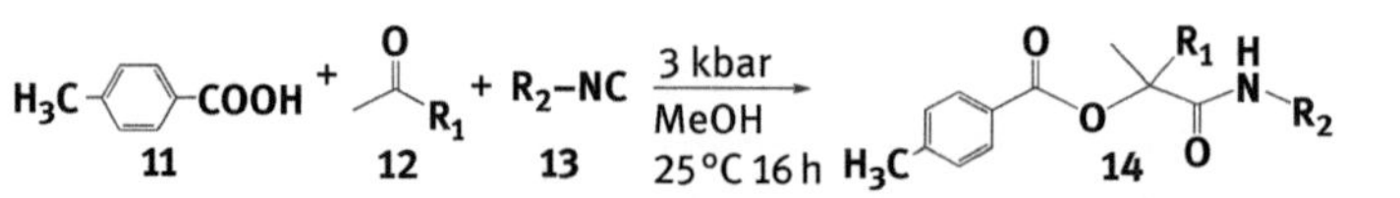

	1 bar	3 kbar
14a R_1=Ph, R_2=chex	12%	27%
b R_1=Et, R_2=*t*-Bu	7%	52%
c R_1=*i*-Bu, R_2=*t*-Bu	1%	12%
d R_1=R_2=*t*-Bu	0%	13% (73%, 6 kbar)

Scheme 8.4: Effects of pressure on the Passerini reaction in relation to steric bulk.

isocyanoacetate **44a** (Method A), three main products were obtained: Ugi product 35a (4%) and byproducts from the Passerini reaction **36b** (15%) and **36a** (12%) in which benzylamine did not participated. Product **36b** is the result of the transesterification of *t*-butyl ester **34a** or **36a** with methanol, although *t*-butyl esters are scarcely transesterified at atmospheric pressure. The Passerini reaction is suppressed by Method B, where Schiff base was prepared first and then subjected to pressurization with **31** and **34b**. In this reaction, yield of **36b** has improved from 6% (method A) to

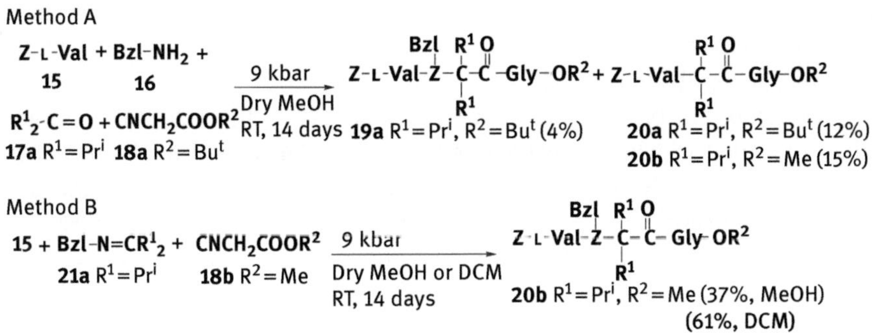

Scheme 8.5: Tripeptide synthesis by the Ugi reaction.

37% in methanol and further to 61% in dichloromethane. This effect is likely due to a solvent effect related to the low polarity of dichloromethane, suggesting that charge separation may occur in the transition state of the Ugi reaction.

Domino multicomponent [4 + 2]/[3 + 2] cycloaddition reactions of 2-methoxybuta-1,3-diene and β-nitrostyrene **25** studied by Scheeren were proved to be an efficient access to heteropolycyclic systems (Scheme 8.6) [6]. In this multicomponent system at 15 kbar, 2-methoxybuta-1,3-diene **24** reacted as the diene, β-nitrostyrene **25** as the dienophile as well as the heterodipolarophile to afford tricyclic products **22, 6a** and **6b.** The same products could be obtained by sequential high-pressure cycloadditions of **24** and **25** in the first step (to obtain cyclic enol ether **26**). Further reaction of **26** with **25** in domino [4 + 2]/[3 + 2] fashion (via cyclic dipole intermediate **27**) affords **22** and **23a** and **23b** with the same regioselectivity. A good example of the

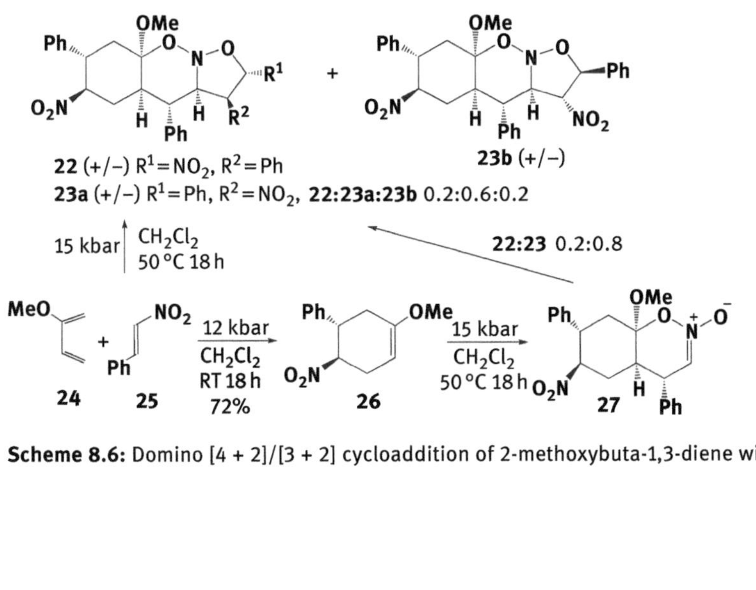

Scheme 8.6: Domino [4 + 2]/[3 + 2] cycloaddition of 2-methoxybuta-1,3-diene with β-nitrostyrene.

application of tandem multicomponent cycloadditions in hyperbaric conditions to natural product derivatization is tandem Diels–Alder (DA) reaction – dipolar multicomponent reaction with steroids [7].

The extension of this procedure is a four-component [4 + 2]/[4 + 2]/[3 + 2] domino cycloaddition reaction in which 2-methoxybuta-1,3-diene **24** reacted as the diene, *N*-phenylmaleimide **28** as the dienophile, β-nitrostyrene **25** as the heterodiene and styrene **29** as the dipolarophile were carried out at 15 kbar. A mixture of two diastereomeric tetracyclic nitroso acetals **30a** and **30b** was obtained in 83% yield (Scheme 8.7).

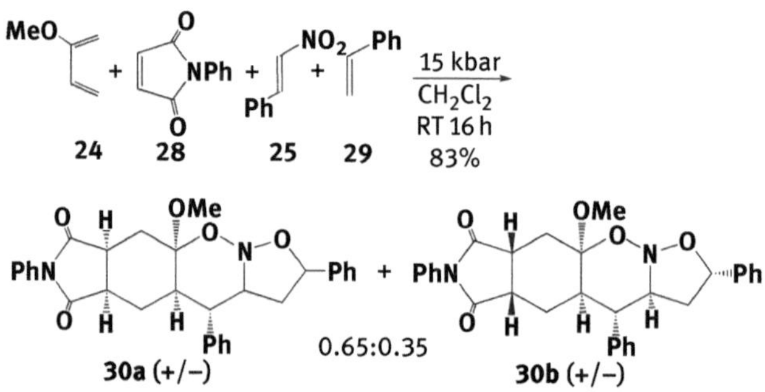

Scheme 8.7: A four-component [4 + 2]/[4 + 2]/[3 + 2] domino cycloaddition.

Tetracyclic nitroso acetal products could be also obtained by the three-component [4 + 2]/[3 + 2] domino cycloaddition reaction in which two reactants employed are cyclic (Scheme 8.19). Here, methoxycyclohexene enol ether **31** was used as the dienophile, trans-β-methyl-β-nitromethyl styrene **32** as the heterodiene and cyclohexenone **33** reacted as the dipolarophile in an one-pot synthesis. Pressurization was carried out at 15 kbar to obtain products **34a,b** in 53% yield [1].

In addition to the example given in Scheme 8.8, the reaction of tricomponent mixture of **35, 32** and 3-methyl-2-cyclohexenone **36** illustrates how high pressure can overcome steric hindrance. By the application of 15 kbar pressure, tricyclic product **37** was obtained in 69% yield as the mixture of diastereoisomers (Scheme 8.9) [1]. Similar type of products were obtained in tandem [4 + 2]/[3 + 2] three-component reactions of 1-nitro-2-heteroarylethenes, *p*-methoxybenzyl vinyl ether and methyl acrylate at 15 kbar [8].

Completely regioselective one-pot three-component tandem [4 + 2]/[3 + 2] cycloadditions of enol ethers with nitrostyrenes and resin-bound acrylate was also carried out (Scheme 8.10). This is the first reported high-pressure cycloaddition reaction on a solid support [9]. Cycloaddition readily takes place under high-pressure conditions with the dipolarophile (acrylate) attached to the Wang resin. A series of nitroso acetals **41** was prepared as mixtures of diastereomers in reasonable yields (33–52%) after the cleavage from the resin and transesterification.

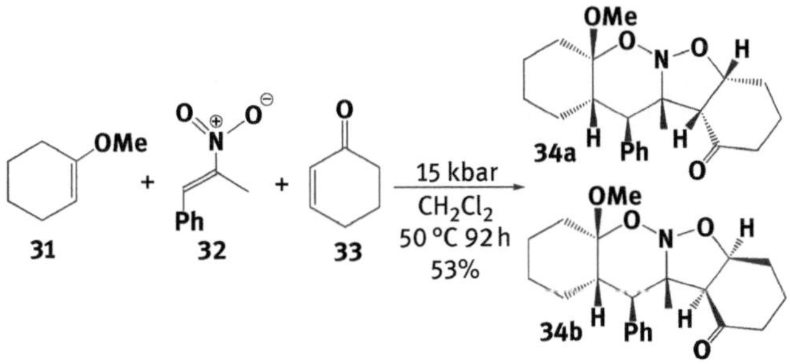

Scheme 8.8: Synthesis of tetracyclic nitroso acetals.

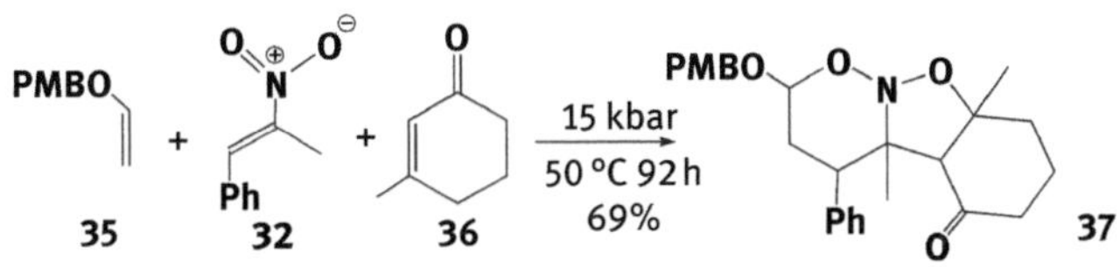

Scheme 8.9: Tandem [4 + 2]/[3 + 2] three-component reaction of sterically hindered substrate.

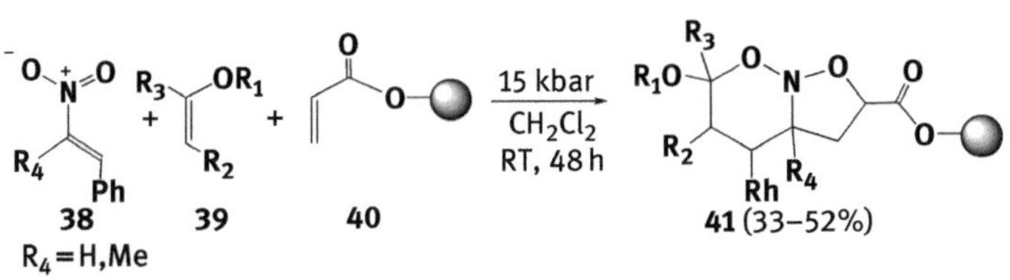

Scheme 8.10: A three-component reaction of enol ethers with nitrostyrenes and resin-bound acrylate.

Bicyclic nitroso acetals **44** were also prepared by solid phase tandem [4 + 2]/[3 + 2] cycloadditions of resin-bound nitroalkenes **42** with ethyl vinyl ether and styrene (Scheme 8.11) [10]. High-pressure reactions of **44** attached on Wang resin solid support provided cycloaddition products **44** in 29–56% yield, after cleavage from the resin and reduction to alcohol. The resin-bound nitroalkene **42** (R = Ph) also reacted with 2,3-dimethylbutadiene under 15 kbar, providing the resin-bound DA cycloadduct **46**. The use of high pressure was preferred over toluene reflux at ambient pressure in view of shorter reaction times and higher yields. 4-Methoxybenzyl (*E*/*Z*)-2-nitro-3-phenyl-2-propenoate reacted with 2,3-dimethylbutadiene **45** in refluxing toluene for 48 h yielding 73% of cycloadduct, whereas at 15 kbar and room temperature (18 h) the yield was 91%.

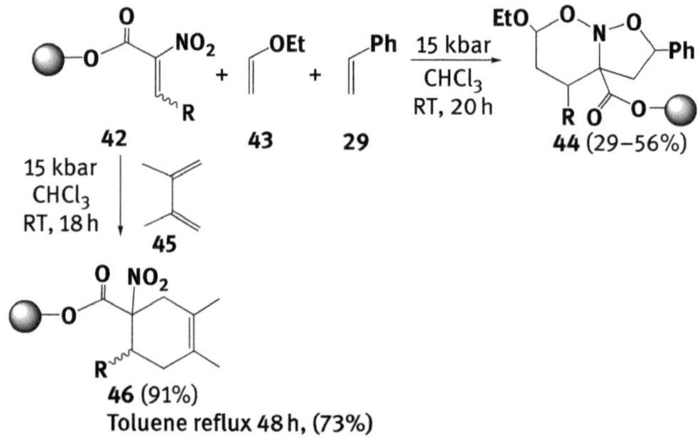

Scheme 8.11: Solid phase tandem [4 + 2]/[3 + 2] cycloadditions.

The cycloadditions reactions of nitroaromatic compounds could be extended to multicomponent domino [4 + 2]/[3 + 2] reactions of nitroheteroaromatics. Chataigner et al. have shown that the activation by high pressure allows 3-nitroindole and 3-nitropyrrole heterocycles to behave as electron-poor heterodienes in cycloaddition processes (Scheme 8.12) [11]. The benefits of the applied pressure on the cycloaddition between indole **47**, ethyl vinyl ether and methylacrylate could be illustrated in Table 8.1. The first reaction in this domino sequence promoted by high pressure is [4 + 2] inverse electron demand cycloaddition whose *endo*/*exo*-selectivity depends on the nature of the heterocycle, being *endo*-selective for indole. The subsequent [3 + 2] cycoaddition process is facially selective. Whereas the reactions of 3-nitroindole proceed at 12 kbar at room temperature, the low reactivity of 3-nitropyrrole requires more drastic conditions (16 kbar and 50 °C, 60 h) to obtain product **51** in 81% yield with *endo*/*exo*-selectivity 25:75.

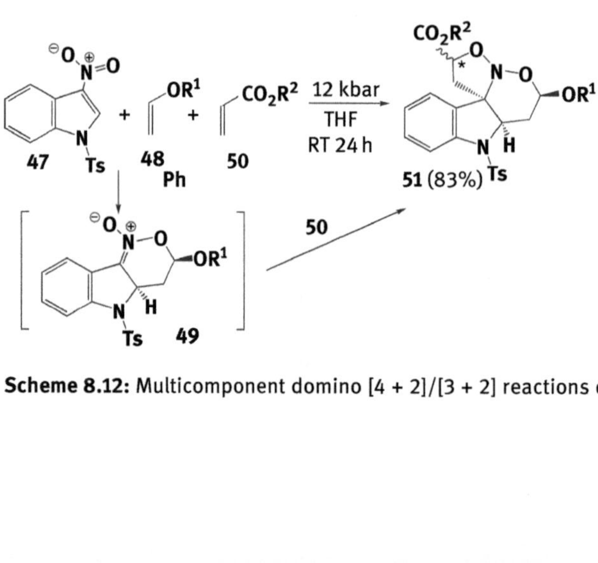

Scheme 8.12: Multicomponent domino [4 + 2]/[3 + 2] reactions of nitroheteroaromatics.

Table 8.1: Reaction of indole **47**, ethyl vinyl ether (R^1 = Et) and methylacrylate (R^2 = Me).

Solvent	Conditions	Conversion (%)	Yield (%)	dr
THF	rt, 168 h	42	20	50/50/0/0
DCM	42 °C, 168 h	53	29	50/50/0/0
Toluene	110 °C, 60 h	100	83	45/45/5/5
Toluene	MW 120 W, 130 °C, 1 h	100	75	35/35/15/15
THF	12 kbar, rt, 24 h	100	83	55/45/0/0

High pressure was also successfully applied to the Strecker three-component synthesis of α-aminonitriles (Scheme 8.13) [12]. Reactions of aniline, ketones and trimethylsilyl cyanide (TMSCN) were facilitated at 6 kbar to provide α-aminonitriles **55** in 68–99% yield. For comparison, the Strecker reaction of acetone at 6 kbar produced **55a** in 99%, whereas at atmospheric conditions, the yield was only 22%. Acetophenone reaction at high pressure, on the other hand, afforded 88% of product **55b**, while no product was obtained at normal pressure. High-pressure conditions are

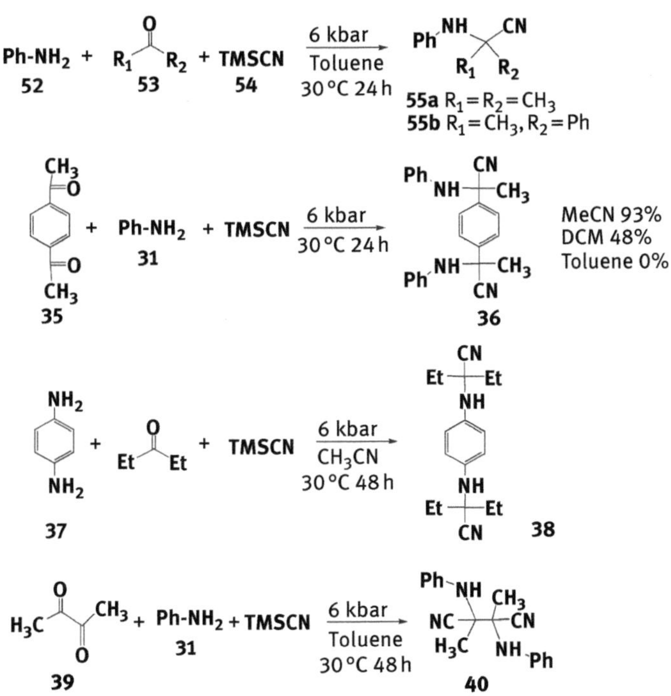

Scheme 8.13: The Strecker three-component synthesis of α-aminonitriles.

amenable to the double Strecker reaction of 1,4-diacetylbenzene **56** to obtain the corresponding double product **57** in 93% yield (Scheme 8.14). In this reaction, strong dependence of solvent on yield was found: acetonitrile was the best, whereas dichloromethane gave low yield and reaction in toluene did not proceed at all, which may be associated with the solubility of substrate at given pressure. Conversely, the reaction of 1,4-diamino benzene **58** in double Strecker reaction affords product **59** in 77% yield. Furthermore, the double Strecker reaction with butane-2,3-dione provides **61** in 25% yield [13]. A triple Strecker reaction of triamino diphenyl ether with acetone was unsuccessful at 6 kbar.

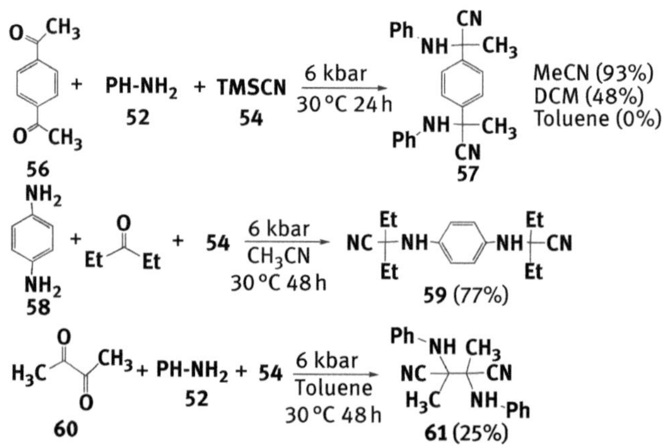

Scheme 8.14: The double Strecker reaction of 1,4-diacetylbenzene.

Advantageous effects of pressure on the Strecker synthesis of α-aminonitriles from ketones and aromatic amines were in particular evident in the case of sterically more hindered reactants [14]. For the series of reactions carried out at atmospheric pressure and 3 kbar in toluene or when ketone was at the same time used as a solvent, ratio of yields increased in some cases from 1 to 9.

Diketones also react with diamines in double Strecker reactions obtaining cyclic products (Scheme 8.15) [15]. Although in low yield, 1,2-diacetylbenzene, 1,2-diaminobenzene and TMSCN under 6 kbar produced cyclic product **63.** Similar reaction of cyclohexane-1,3-dione at 6 kbar produced 1,2,3,4,5,10-hexahydrophenazine-4a,10a-dicarbonitrile **65** and a small amount of side-product 1,2,3,4-tetrahydrophenazine **66**, which could be formed by bisdehydrocyanation reaction of **65** or by the dehydrative condensation of **64** with **63.**

The formation of aromatized products was the only reaction path proceeding in attempted double Strecker reactions of 2,3-butanedione and 9,10-phenanthrenequinone. In these reactions, at 8 kbar 2,3-dimethylquinoxaline **68** and dibenzo[a,c]phenazine **70** were produced in high yields (Scheme 8.16). It was

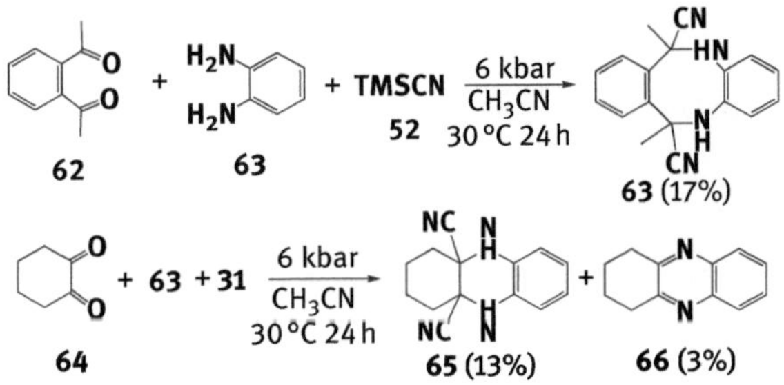

Scheme 8.15: The double Strecker reaction of diketones.

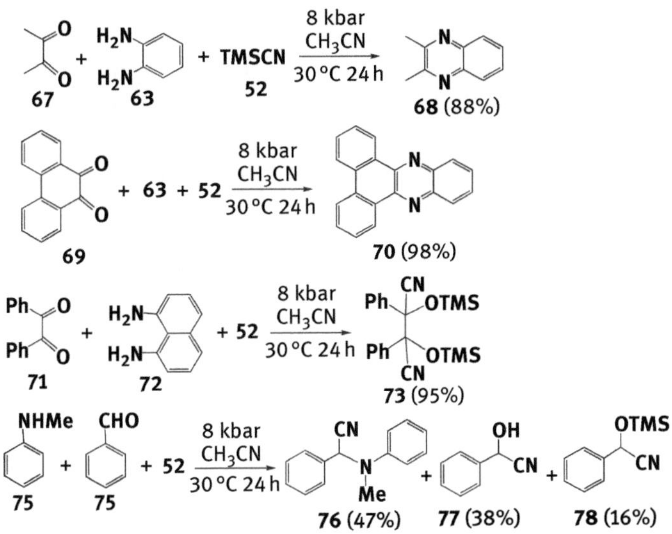

Scheme 8.16: The double Strecker reactions of 1,2-diketones.

found that these condensations do not require the presence of TMSCN. Cyanotrimethylsilylation is another side-reaction of double Strecker reactions in the case of 1,2-diphenyletanedione **71** and 1,8-diaminonaphthalene **72**. Side-product **78** of cyanotrimethylsilylation was also obtained in the reaction of sterically hindered *N*-methylaniline with benzaldehyde. The major product is of the Strecker reaction, followed by significant amount of cyanation product **77**.

The alkenylation of carbonyl compounds with phosphonates could be effectively carried out by the aid of high pressure. Reiser and coworkers have reported that aldehydes react with phosphonates **80** (Horner–Wadsworth–Emmons reaction) even at room temperature in the presence of triethylamine, without the need for the further Lewis acid catalyst activation (Scheme 8.17) [16]. In the studied set of reactants, yields varied from 5% to 47% at atmospheric pressure, and reactivity was

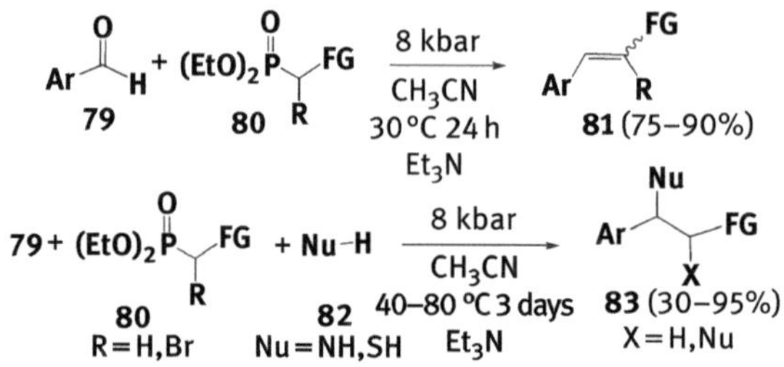

Scheme 8.17: The Horner–Wadsworth–Emmons reaction of aldehydes.

significantly enhanced at 8 kbar (to 75–90% yield, even for reactions that did not proceed at all in standard conditions). This process was further coupled with the Michael reaction by the addition of nucleophiles (amines or thiols) to the three-component system. It was found that the Michael reaction was also promoted by the application of high pressure and that all components are compatible to the development of novel domino Horner–Wadsworth–Emmons/Michael reaction. This three-component process at 8 kbar and with the moderate heating (40–80 °C) led to the formation of β-aminoesters, β-thioesters or β-thionitriles **146** (yields within the range from 30% to 95% were obtained).

Multicomponent reaction of a series of arylbutenones **84** with isopropenyl acetate **85** and methyl propiolate under 9 kbar provides **87** in high yields (Scheme 8.18) [17]. This reaction is a two-step process comprising acetylation of enones, which generates the 1,3-diene system. In the subsequent reaction step, the in situ formed diene participates in the [4 + 2] addition with methyl propiolate in fully regio- and chemoselective manner to obtain **87**.

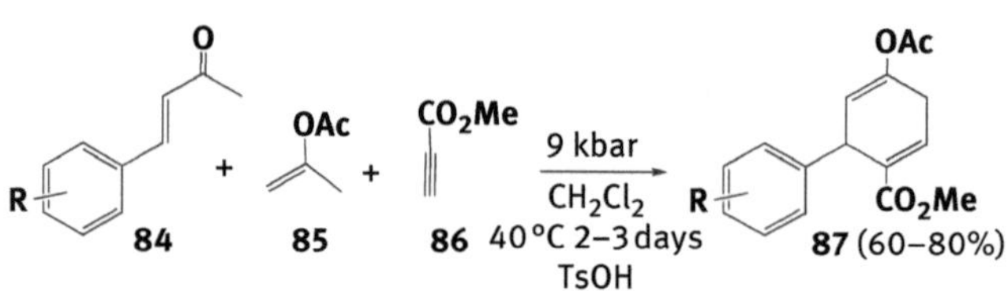

Scheme 8.18: Reaction of arylbutenones with isopropenyl acetate and methyl propiolate.

References

[1] Van Berkom, LWA., Kuster, GJT., & Scheeren, HW. High pressure: a promising tool for multicomponent reactions. Mol Divers 2003, 6, 271–281.

[2] Matsumoto, K. Synthesis under high pressure: Mannich reaction of ketones and esters with dichloromethane and secondary amines. Angew Chem Int Ed Engl 1982, 21, 922.

[3] Matsumoto, K., Uchida, T., Hashimoto, S., Yonezawa, Y., Iida, H., Kakehi, A., & Otani, S. Mannich reactions of indoles with dichloromethane and secondary amines under high pressure. Heterocycles 1993, 36, 2215–2219.

[4] Jenner, G. Effect of high pressure on sterically congested Passerini reactions, Tetrahedron Lett 2002, 43, 1235–1238.

[5] Yamada, T., Yanagi, T., Omote, Y., Miyazawa, T., Kuwata, S., Sugiura, M., & Matsumoto, K. Four-component condensation (Ugi Reaction) at high pressure: novel synthesis of peptides containing very bulky α,α-disubstituted glycines. J Chem Soc Chem Comm 1990, 1640–1641.

[6] van Berkom LWA., Kuster, GJ., Kalmoua, F., & Scheeren, JW. Novel domino (4+2)/(4+2)/(3+2) cycloaddition reaction leading to highly functionalised polycyclic nitroso acetals. Tetrahedron Lett 2003, 44, 5091–5093.

[7] Blanco-Ania, D., Aben, RWM., van Berkom LWA., Scheeren, HW., & Rutjes, FPJT. Synthesis of steroidal D-ring-fused pyrrolidines of dehydroepiandosterone. Eur J Org Chem 2017, 3729–3737.

[8] Gjt, Kuster., Rhj, Steeghs., & Scheeren, HW. Novel five/five- and six/five-membered bicyclic nitroso acetals from high-pressure-promoted cyclisation reactions of p-methoxybenzyl vinyl ether, 1-nitro-2-heteroaryl ethenes, and mono- and di-substituted olefins. Eur J Org Chem 2001, 553–560.

[9] Kuster, GJ., & Scheeren, HW. High pressure promoted tandem [4+2]/[3+2] cycloadditions on the solid phase. Tetrahedron Lett 1998, 39, 3613–3616.

[10] Kuster, GJ., & Scheeren, HW. The preparation of resin-bound nitroalkenes and some applications in high pressure promoted cycloadditions. Tetrahedron Lett 2000, 41, 515–519.

[11] Chataigner, I., Serge, R., & Piettre, SR. Multicomponent domino [4+2]/[3+2] cycloadditions of nitroheteroaromatics: an efficient synthesis of fused nitrogenated polycycles. Org Lett 2007, 9, 4159–4162.

[12] Matsumoto, K., Kim, JC., Hayashi, N., & Jenner, G. High pressure mediated three component Strecker synthesis of α-aminonitriles from ketones, aromatic amines and trimethylsilyl cyanide. Tetrahedron Lett 2002, 43, 9167–9169.

[13] Matsumoto, K., Kim, JC., Iida, H., Hamana, H., Kumamoto, K., Kotsuki, H., & Jenner, G. Multicomponent Strecker reaction under high pressure. Helv Chim Acta 2005, 88, 1734–1753.

[14] Jenner, G., Ben Salem, R., Kim, JC., & Matsumoto, K. Effect of pressure on the Strecker synthesis of hindered α-aminonitriles from ketones and aromatic amines. Tetrahedron Lett 2003, 44, 447–449.

[15] Kumamoto, K., Iida, H., Hamana, H., Kotsuki, H., & Matsumoto, K. Are multicomponent Strecker reactions of diketones with diamines under high pressure amenable to heterocyclic synthesis? Heterocycles 2005, 66, 675–681.

[16] Has-Becker, S., Bodmann, K., Kreuder, R., Santoni, G., Rein, T., & Reiser, O. High-pressure induced domino-Horner-Wadsworth-Emmons (HWE)-Michael reactions. Synlett 2001, 1395–1398.

[17] Minuti, L., Piazzolla, F., & Temperini, A. High pressure promoted multicomponent and metal-free synthesis of polyfunctionalized biaryls. Eur J Org Chem 2017, 5370–5377.

9 Supramolecular chemistry

A review on high-pressure synthesis of molecules that were employed afterward in supramolecular studies is given by Toda and Matsumoto [1]. Although some supramolecular processes are reported to be carried out under high pressure (host–guest processes, and rotaxane formation), this area is still not fully explored.

The use of high pressure to force encapsulation of small molecules in an open-cage fullerene derivative was reported by Kurotobi and Murata [2]. The full encapsulation of water molecule inside **1** was achieved when this process was carried out by applying pressure of 9 kbar and simultaneous heating at 120 °C (Scheme 9.1). It was found that the high pressure is important for the process since in identical conditions under 5 kbar only 40% encapsulation was obtained. Authors have some evidence that the encapsulation likely takes place by the elimination of water and the formation of tetraketone **2** with larger opening, which is smoothly hydrated back to **1**, thus reducing the possibility of the encapsulated water to escape.

Similar water encapsulation process was achieved for C_{70} via an analogous insertion mechanism [3]. The solution of diol **3** in toluene with some water added was subjected to high pressure (8.5–9 kbar) for 40 h at 120 °C. Under high-pressure/high-temperature conditions, 88% insertion of water was detected by ^{1}H NMR spectroscopy (Scheme 9.2). Water molecule was not efficiently encapsulated in tetrone **4**, because the opening was not sufficiently large. Just trace amounts of H_2O were encapsulated at 120 °C under 9 kbar [4].

Interesting results were obtained by Murata and coworkers [5] for tetraketone **5** (C_{60} fullerene **2** with enlarged orifice by insertion of sulfur atom). Complexation of **5** and methanol was conducted under 8 kbar at 150 °C for 24 h in chlorobenzene, to obtain a mixture of **CH_3OH@5, N_2@5** and **H_2O@5** in 35:57:3 ratio (Scheme 9.3). Complex **N_2@5** arises from nitrogen gas dissolved in chlorobenzene and indicates that high pressure could also effectively promote the encapsulation of gaseous molecules. Encapsulation ratio of 60% of methanol was achieved at 9 kbar after 50 h. Pressurization at 8 kbar for 24 h with ethanol gave no complexation, while traces of acetonitrile complex **MeCN@5** were detected. These results suggest that either cage opening of **5** is not wide enough or internal volume is not sufficiently large to house these molecules.

Standard procedure for the introduction of guest molecules into hemicarcerands [6] is prolonged heating in high boiling point solvents such as diphenyl ether. In such way, the encarcerated template molecule is removed and guest is incarcerated into an empty container. There are reports that guest could be encarcerated by the application of high pressure as well. Supramolecular guest encapsulation into hemicarcerand **6** at elevated pressure was achieved by Klärner and coworkers [7]. It was

https://doi.org/10.1515/9783110556841-009

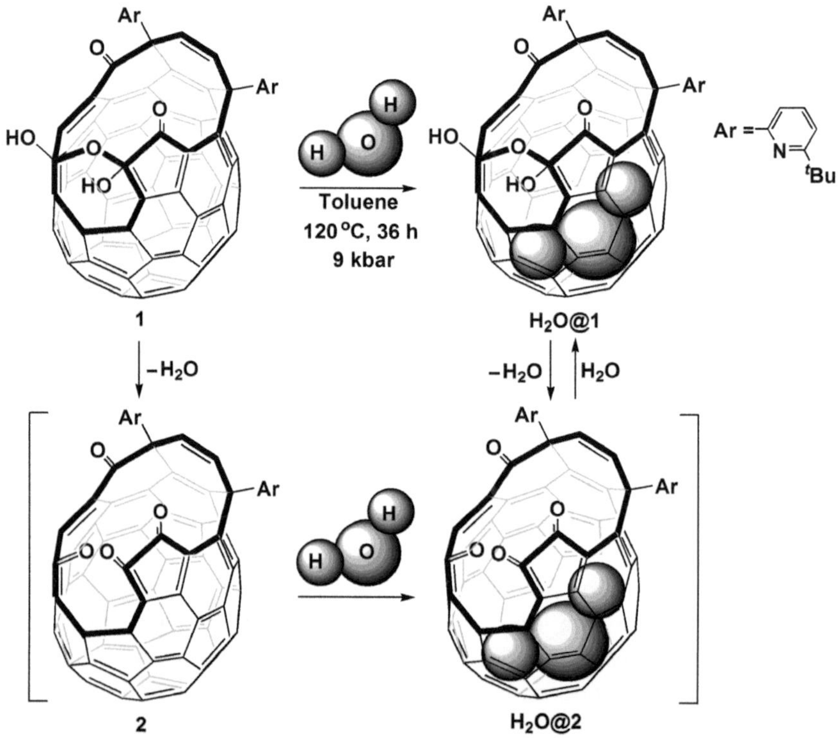

Scheme 9.1: Encapsulation of small molecules in an open-cage fullerene.

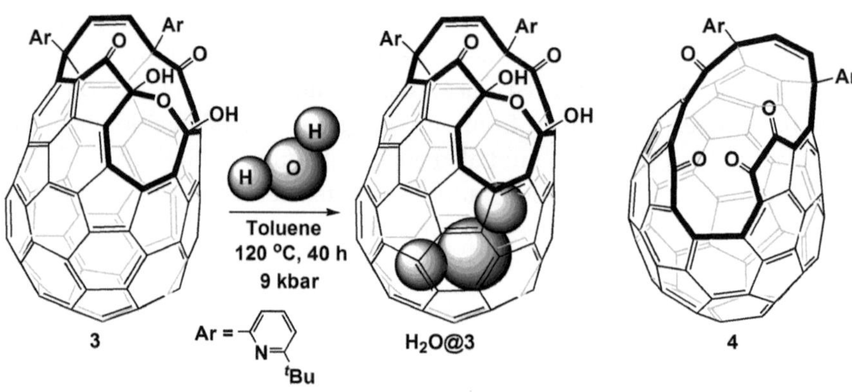

Scheme 9.2: Encapsulation of water in open-cage C_{70}.

found that under 11.5 kbar and heating at 110 °C for 3 days in acetonitrile, the hemicarceplex with two acetonitrile guests **$(CH_3CN)_2$@6** does not release acetonitrile and a new complex is formed, which was assigned to contain three acetonitrile molecules, which reverts back to **$(CH_3CN)_2$@6** within 1 h at atmospheric pressure and at 55 °C (Scheme 9.4).

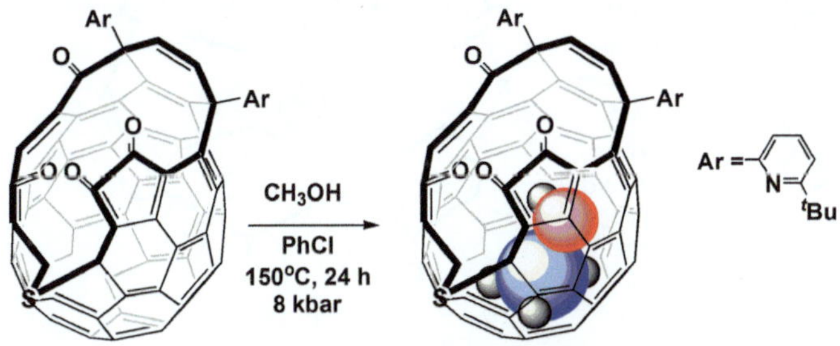

5

CH_3OH@5/N_2@5/H_2O@5/5
35 / 57 / 5 / 3

Scheme 9.3: Encapsulation of methanol in an open-cage fullerene.

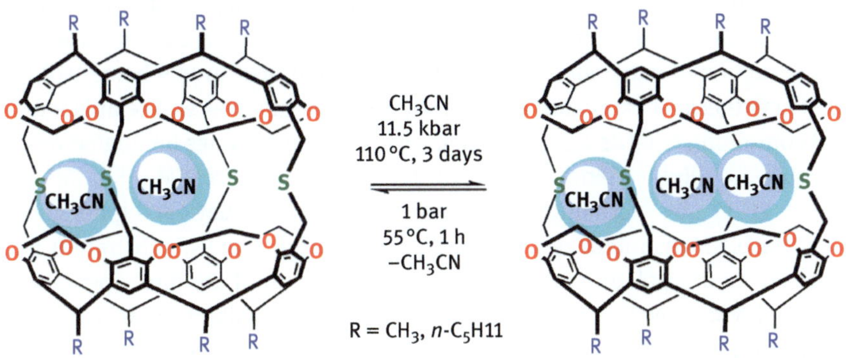

$(CH_3CN)_2$@6 **$(CH_3CN)_3$@6**

Scheme 9.4: Encapsulation of acetonitrile into hemicarcerand.

Different host–guest behavior was observed in the hemicarceplex *N,N*-dimethylacetamide **(DMA)@7**. When **DMA@7** was subjected to pressure of 14 kbar at room temperature in CH_2Cl_2, decomplexation of DMA occurred, followed by encapsulation of dichloromethane (Scheme 9.5) [8]. This result shows that the change of physical conditions (application of high pressure) alters the behavior of the hemicarceplex, and the guest exchange in hemicarceplexes could be achieved even at room temperature. Reversibility of the host–guest process was demonstrated by encapsulation of DMA in "empty" **CH_2Cl_2@7**; however, this process was less efficient.

Some other solvent molecules could be used to extrude DMA: acetonitrile (6 kbar, 80 °C, 18 h) and CH_2Br_2 (6 kbar, RT, 4 days, 65% extrusion). Pressurization of **DMA@7** in acetone (6 kbar, RT, 4 days) and ethyl iodide (6 kbar, RT, 7 days) led to 95% and 65% replacement of DMA, respectively. Molecular dynamics calculations

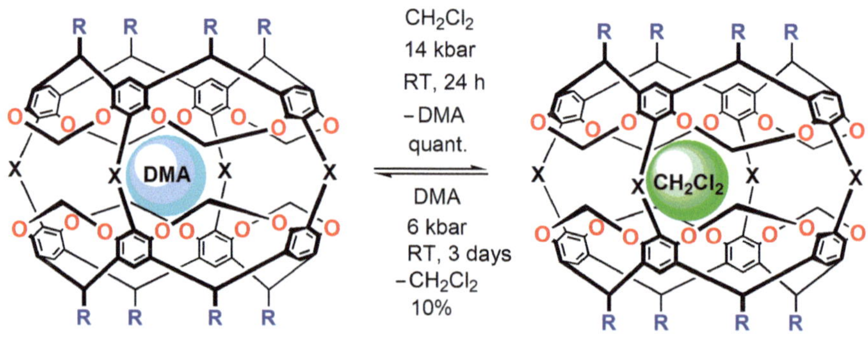

Scheme 9.5: Guest exchange in hemicarcerand.

indicate that in this guest exchange process under high pressure, the DCM molecule expels DMA from the hemicarcerand **7** in a single step in a "billiard-ball" fashion.

The construction of catenanes was facilitated by high pressure (Scheme 9.6) [9]. The template synthesis of catenanes reported by Stoddart involves as a template a [3] catenane **8** containing two tris-1,5-naphtho-57-crown-15 macrocycles templated by cyclobis(paraquat-4,4′-biphenylene). At 12 kbar, templation produces [4]-, [5]-, [6]- and [7]catenanes **11–15** by formation of new cyclobis(paraquat-4,4′-biphenylene) tetracations via bis-Menshutkin reaction of **9** and **10.** After 6 days, substrate **8** was fully consumed. In this self-assembly process, the formation of the linear pentacatenane is greatly favored over its branched isomer. When templation was carried out at ambient pressure, catenanes **11** and **12** were obtained after 14 days in 51% and 18% yields, respectively. Stoddart and coworkers have also reported that the yields of [2] catenane self-assembled from **9, 10** and BPP34C10 significantly raise from 18% to 42% at 15 kbar [10].

Stoddart et al. have also employed high pressure in the preparation of rotaxanes possessing central 9,10- or a 2,6-dioxyanthracene units [11]. Preparation of [2]rotaxane **18**•$4PF_6$ was achieved at 12 kbar by interlocking of dumbbell compound **17** with 1,4-bis(4,4′-pyridylpyridinium)xylene bis(hexafluorophosphate) **9**•$4PF_6$ and 1,4-bis(bromomethyl)benzene **10** (Scheme 9.7). In this process, a double Menshutkin pyridine alkylation reaction afforded the second rotaxane ring. In addition to **17**•$4PF_6$, small amounts of a second product, [3]rotaxane **19**•$8PF_6$ was isolated, which incorporates two bipyridinium-based tetracationic components. Identical reaction conditions were used for the preparation of [2] rotaxane **21**•$4PF_6$ possessing a 2,6-dioxyanthracene unit starting from **20, 9**•$4PF_6$ to **10** (Scheme 9.8).

An acceleration of the slipping process in the formation of pseudorotaxanes **26** and **27** by applying high pressure was noted by Tokunaga et al. [12].

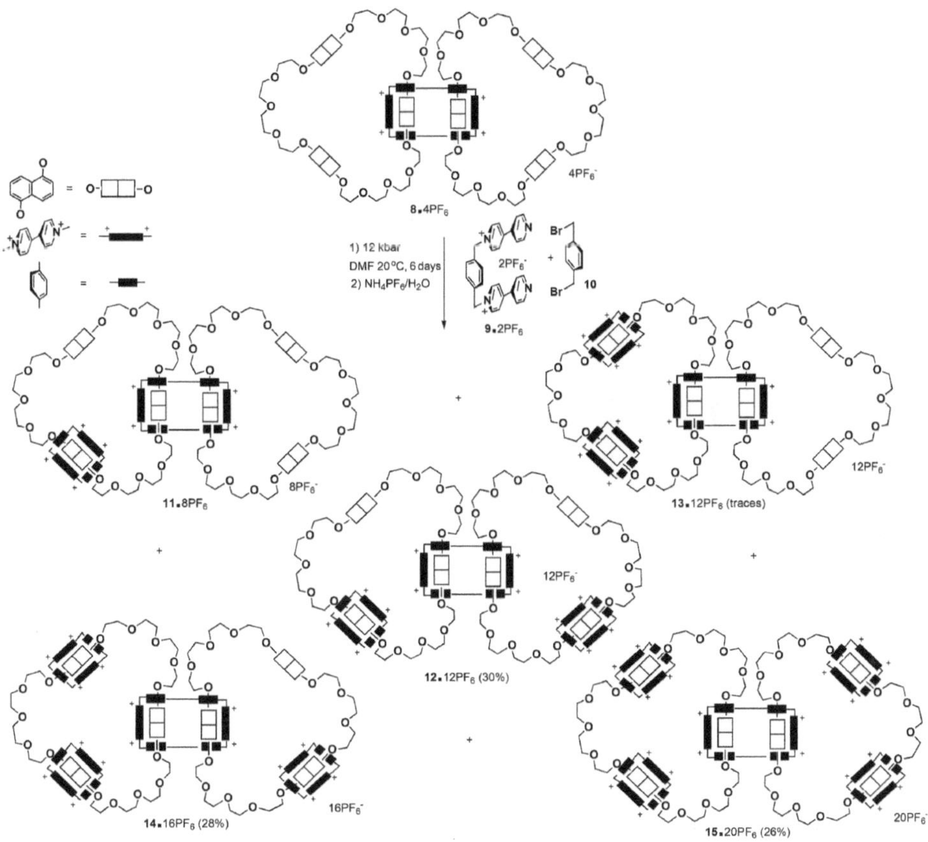

Scheme 9.6: The construction of catenanes under high pressure.

Supramolecular assembly of dibenzo[24] crown-8/bis(cyclohexylmethyl)ammonium salt **22** (DB24C8) and tetrabenzo[24] crown-8/dibenzylammonium salt **23** (DB24C8) with bis(cyclohexylmethyl) ammonium salts **24** and **25** was achieved under the pressures up to 5 kbar at room temperature (Scheme 9.9). At atmospheric pressure, conversions were lower than 10%. Application of higher temperatures (40 °C or 50 °C) also accelerates of the slipping reaction.

The rate constants for the formation and decomplexation of [2]pseudorotaxanes (Scheme 9.6) are affected by high pressure (3–5 kbar) and the polarity of solvent [13]. Activation volumes in nonpolar solvent $CDCl_3/CD_3CN$ are moderately negative, and polar DMSO-d_6/$CDCl_3$ is slightly negative. Activation volumes for rotaxane formation are larger than for the dissociation of rotaxanes in polar solvent. Obtained values for $\Delta V^{\neq}$ are not as large (negative) as those of typical reactions in which the molecularity decreases, which usually display large negative activation volumes for transition states during the formation of a covalent bond. Although formation of

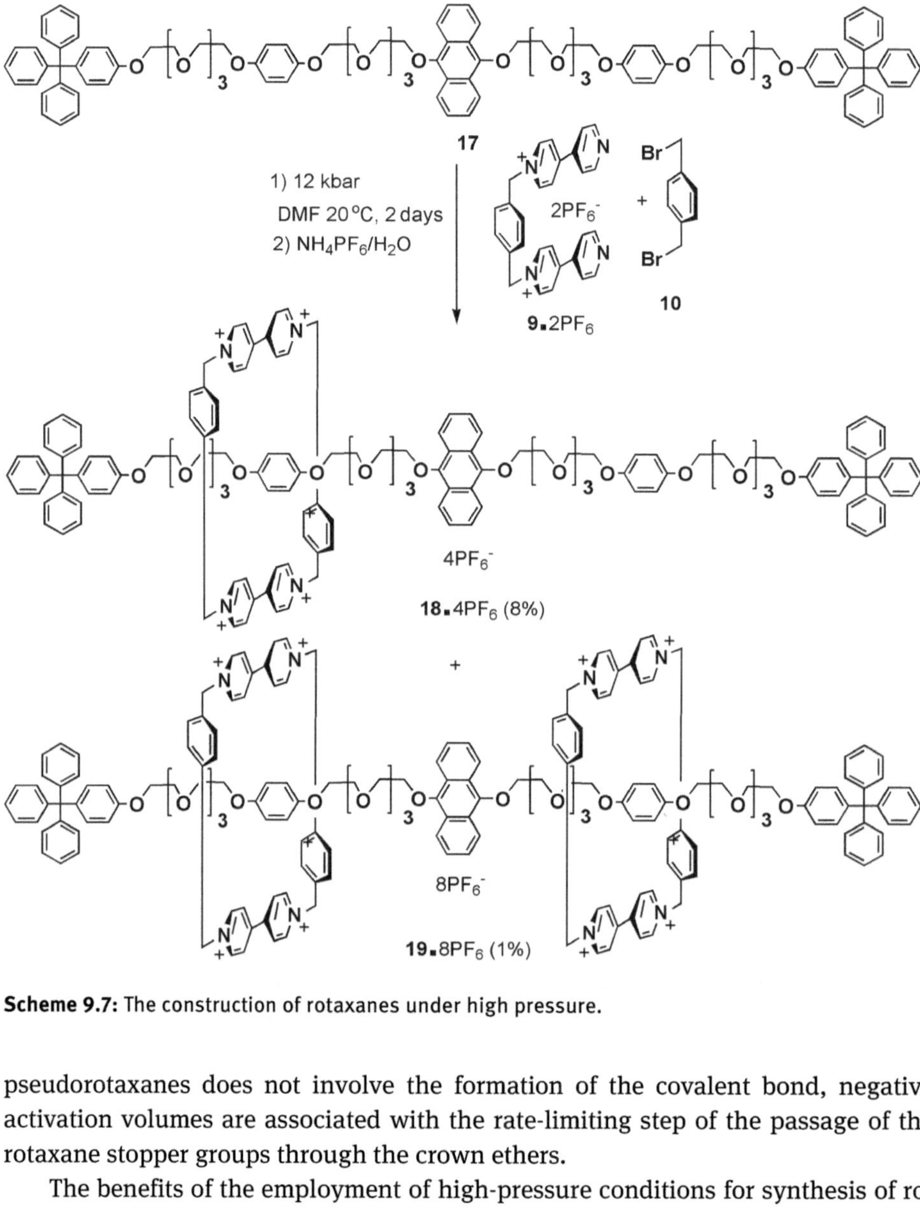

Scheme 9.7: The construction of rotaxanes under high pressure.

pseudorotaxanes does not involve the formation of the covalent bond, negative activation volumes are associated with the rate-limiting step of the passage of the rotaxane stopper groups through the crown ethers.

The benefits of the employment of high-pressure conditions for synthesis of rotaxanes by formation of C–N covalent bonds in a template-directed reaction are further illustrated in the following examples. Jeppesen et al. have prepared [2]rotaxane **29•4PF$_6$** from dumbbell component **28**, the dicationic precursor **9•2PF$_6$** and the dibromide **10** in 21% yield at 10 kbar (Scheme 9.10) [14]. Applied conditions are highly beneficial, as in classical conditions yield was only 8%. In an analogous manner, the clipping procedure was carried out in a reaction whereby the dumbbell **30**, **9•2PF$_6$**, and the dibromide **10** were subjected to 10 kbar pressure at room

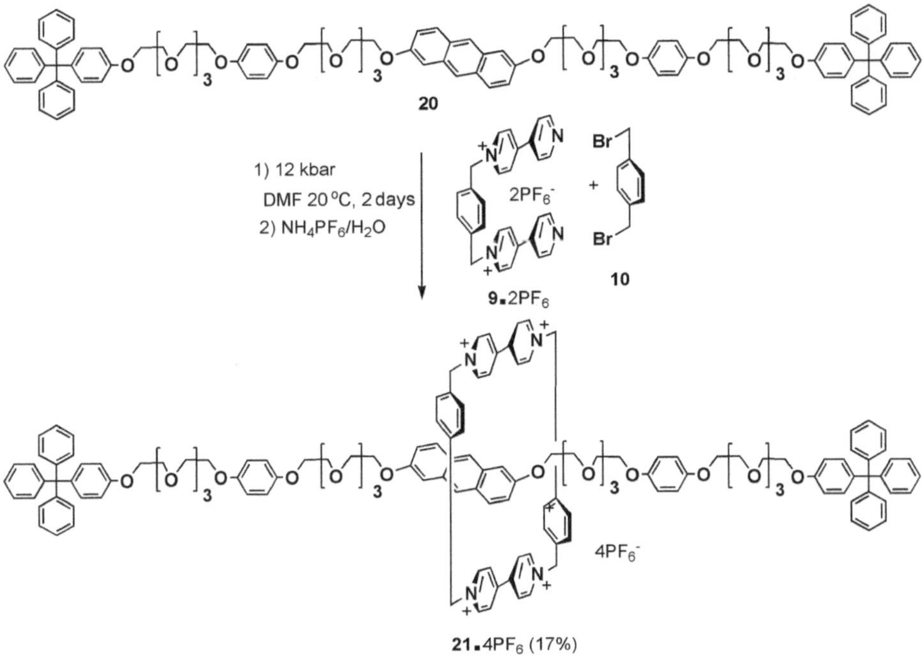

Scheme 9.8: Synthesis of rotaxanes under high pressure.

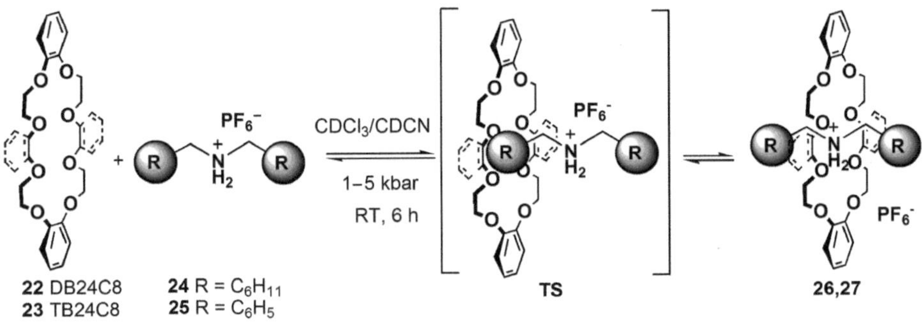

Scheme 9.9: Formation of pseudorotaxanes under high pressure.

temperature (Scheme 9.11). In this reaction, the [2]rotaxane **31•4PF$_6$** was isolated in 47% yield, in comparison to 15% yield obtained at atmospheric pressure after 10 days. Finally, the [2]rotaxane **33•4PF$_6$** was obtained under high pressure in 41% yield (Scheme 9.12) [15].

Pressure was used by Jeppesen and coworkers as a physical parameter to achieve kinetic control over the ratio of isomers produced in the synthesis of a [2] rotaxane (Scheme 9.13) [16]. Besides influencing the yield in the synthesis of

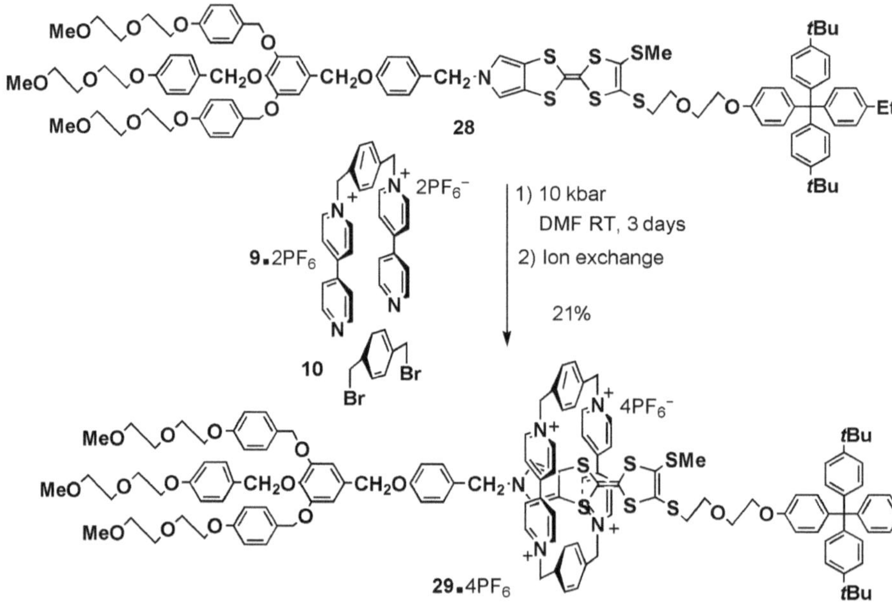

Scheme 9.10: Assembly of [2]rotaxane under high pressure.

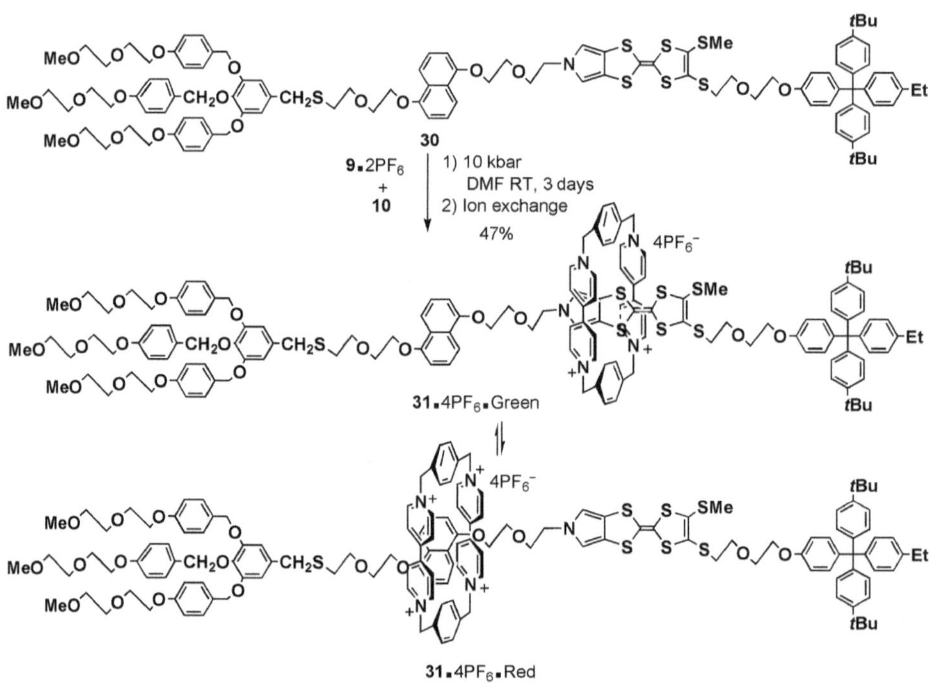

Scheme 9.11: Assembly of TTF [2]rotaxane under high pressure.

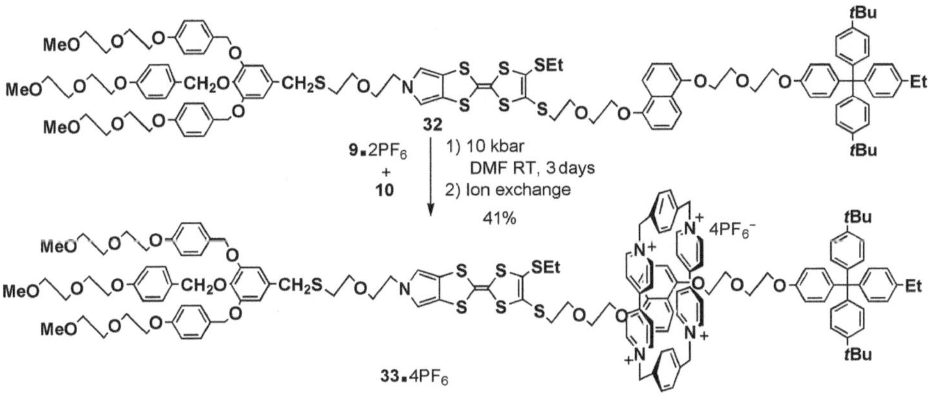

Scheme 9.12: Assembly of TTF [2]rotaxane under high pressure.

a two-station [2]rotaxane **35** pressure also determines the product distribution between the two possible coconformational isomers. Synthesis of the [2]rotaxane translational isomers **35•4PF₆**•HQ and **35•4PF₆**•MPTTF from a dumbbell containing monopyrrolotetrathiafulvalene (MPTTF) and hydroquinone (HQ) moieties was achieved at atmospheric and high pressure. Ratio of HQ:PTTF isomers depends on the applied pressure and the amount of PTTF isomer increases with the increase of applied pressure: 73:27 (1 bar), 53:47 (5 kbar), 42:58 (10 kbar) and 26:74 (15 kbar). The ratio obtained at atmospheric pressure is actually inverted at 15 kbar, thus the pressure could be used to control the distribution between the two possible co-conformational isomers. When mixture of two model dumbbell compounds containing only one, either HQ or MPTTF binding site were used in synthesis afforded [2]rotaxanes **35** and **36** in 24% yield and 39:61 ratio at 10 kbar, whereas synthesis at 1 bar gave lower yield (7%) with the favorable formation of **36** (**36:37** ratio 85:15). Results indicate that the formation of **35•4PF₆**•HQ is thermodynamically more stable, whereas the activation volume for the formation of **35•4PF₆**•MTPF is more negative; thus, the ratio is under kinetic control at high pressure.

The effect of applied pressure on the rotational dynamics of encapsulated guests was studied by Raymond in the flexible supramolecular host $[Ga_4L_6]^{12-}$ **38** (L = 1,5-bis (2,3-dihydroxybenzamido)naphthalene) (Scheme 9.14) [17]. The high-pressure NMR study of the encapsulated cationic ortho-substituted benzyl phosphonium guests **39–42** under high-pressure conditions (up to 1.5 kbar) in solution showed that the applied external pressure reduces either the host cavity size or its flexibility. These geometrical changes of host impose a more restricted motional dynamics for encapsulated guest molecules. Their free energy barriers for Ar–CH_2 bond rotation decrease in organic solvents CD_3OD and DMF-d_7 approximately by a factor of 2 as the pressure is increased to 1.5 kbar. There was no observed influence of pressure on

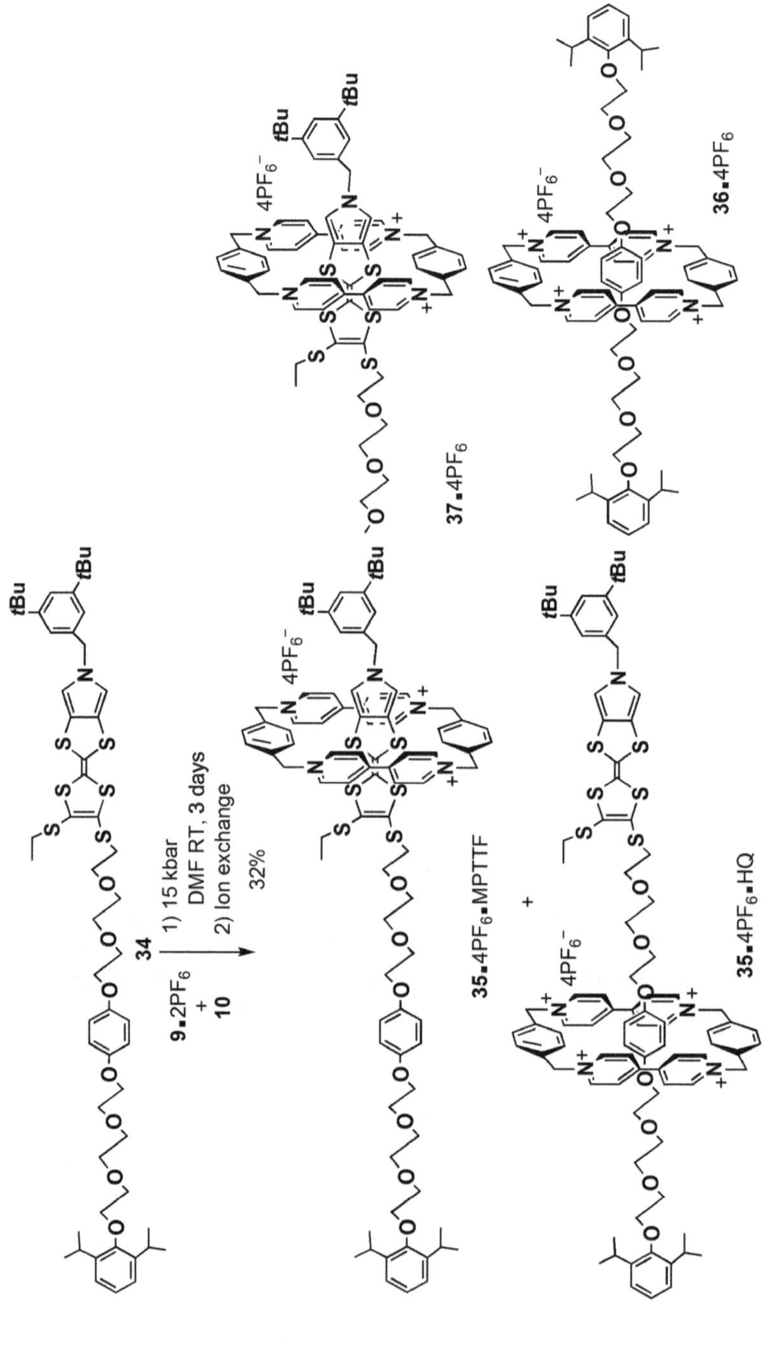

Scheme 9.13: Synthesis of a [2]rotaxanes.

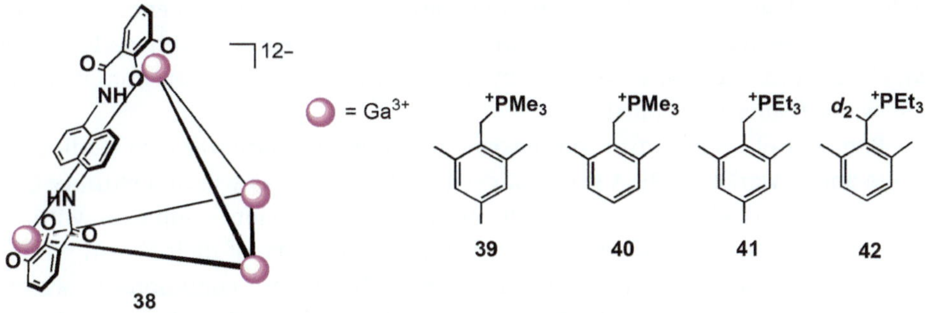

Scheme 9.14: Encapsulation of guests in the supramolecular host $[Ga_4L_6]^{12-}$.

rotational dynamics on encapsulated guest molecules in D_2O. It was demonstrated that the slowing down the Ar-CH_2 bond rotation could be effected by the increase of external pressure, or change from aqueous to organic solvents. However, it could not be established that a single and specific physical property of the solvent is responsible for the observed rate changes at high pressure.

By employment of UV spectrophotometry through a high-pressure window vessel with optical cell, le Noble et al. had found that the pressure affects the rate of transacylation reactions of esters **43** and **44** in the presence of β-cyclodextrin **45** (Scheme 9.15) [18]. Applications of pressures up to 1.8 kbar revealed that the volume change of the complexation step of **43** is negative, whereas the activation volume is slightly positive. Since the cyclodextrin cavity is spacious enough to contain six water molecules, their replacement with **43** leads to significant reduction of the overall volume indicating good fit of ester within the cavity. The volume decrease is associated with lower extent of hydrogen bonding of water molecules than in the bulk solvent. It is necessary for substrate **43** to partially withdraw in the transition

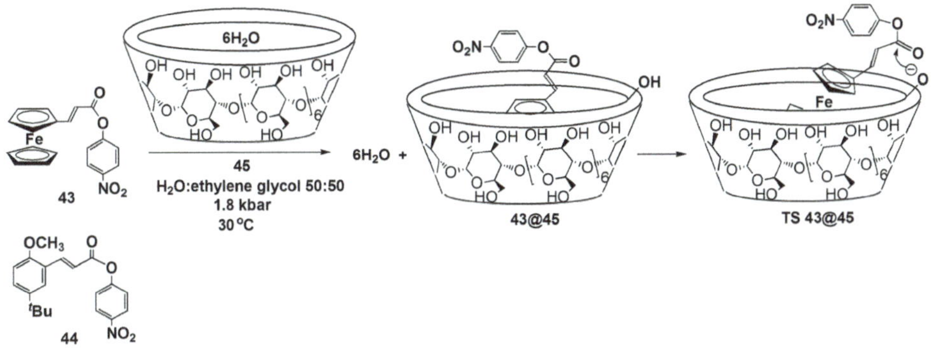

Scheme 9.15: Transacylation reactions of esters in the presence of β-cyclodextrin.

state to less bound TS to release *p*-nitrophenoxide, without the reenter of water. For guest **44**, two reactions pathways are operating: the reaction with cyclodextrin which is a fast one that is retarded by pressure, and solvolysis reaction which is a slow one that is accelerated.

Behavior of cyclodextrin complexes under high pressure was also studied by Inoue and coworkers [19]. A series of cyclodextrins such as γ-cyclodextrin (γ-CD) **47** was used as chiral hosts to form ternary 1:2 inclusion complexes with 2-anthracenecarboxylic acid **46** (Scheme 9.16). The outcome of their [4 + 4] photodimerizations was markedly affected under high-pressure conditions in water. The effect on the stereoselectivity is explained by the formation of more compact complexes, which also dramatically accelerate photodimerization reaction. Under high pressure, deeper/tighter inclusion is favored, which has an effect on host–guest interactions associated with the chiral recognition behavior of CD hosts. The irradiations under high hydrostatic pressure were carried out in a high-pressure vessel through sapphire windows. It was found that the product chirality of the asymmetric photodimerizations can be manipulated by change of the pressure applied. In the case when the differential activation volume ($\Delta\Delta V^{\ddagger}$) is equal to zero, the enantiomeric or diastereomeric pair will be formed. The change of applied pressure is the effective mean for change of the product distribution and chirality. Formation of dimer **48** is enhanced and its yield becomes dominant, on the expense of the other cyclodimers. The *anti/syn* ratios of head-to -head and head-to-tail photodimers obtained at high pressure are lower than those obtained at atmospheric pressure, indicating that $\Delta\Delta V^{\ddagger}$ for the complex precursor pairs to the *syn* isomers are smaller than those for *anti*. These results demonstrate that precursor complexes differ in volume and are sensitive to pressure.

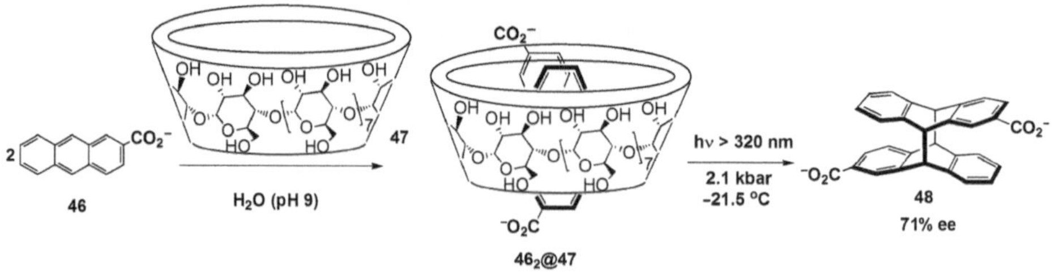

Scheme 9.16: [4 + 4] photodimerizations of inclusion complex of 2-anthracenecarboxylic acid and cyclodextrins.

Detailed pressure-dependent kinetic measurements of the host–guest interactions of α-CD and series of phenylazobenzenesulfonate guests were carried out by Merbach at pressures up to 2 kbar employing high-pressure stopped-flow spectrometer [20]. The complexation of guests takes place by a two-step process involving initial fast step that gives the intermediate G•α-CD*, followed by a rearrangement to form the final complex G•α-CD as a slower step:

$$\mathbf{S} + \boldsymbol{\alpha}\textbf{-CD} \underset{\substack{k_{1,r} \\ \text{fast}}}{\overset{k_{1,f}}{\rightleftharpoons}} \mathbf{S.\alpha\text{-}CD^{\star}} \underset{\substack{k_{2,r} \\ \text{slow}}}{\overset{k_{2,f}}{\rightleftharpoons}} \mathbf{S.\alpha\text{-}CD} \tag{9.1}$$

Negative reaction volume $\Delta V_1^{o} \approx -11$ to -4 cm^3 mol^{-1} between the ground states and the corresponding intermediate states G•α-CD* is observed in the first step. Contraction of volume can be explained by partial desolvation upon inclusion of the guest molecule, thus enabling their full hydrogen bonding with the bulk water molecules. In the second slow step, the sulfonate group is released from the trap by the primary hydroxy groups of the α-CD and is now solvated by the bulk water molecules. In the final state, the guest is thus further inserted in the host cavity. Polar interactions between the tail of the dye and the α-CD larger rim could induce conformational changes of the host cavity facilitating a slipping of the head of the dye out of the α-CD. The positive reaction volumes in the second step $\Delta V_2^{o} \approx +3$ to $+17$ cm^3 mol^{-1} reflect the breaking of the primary hydroxy group trap around the guest heads allowing an expansion of the host cavity. At the same time, the large α-CD rim is accessible to the tail of the guest molecules. Thus, a deeper encapsulation in the host induces a larger expansion of the α-CD cavity. The second transition state is associated with volume contraction, relative to the intermediate state S•α-CD*. The formation of hydrogen bonds between the guest tail and the secondary OH groups of the α-CD and the interactions of the primary OH groups of the host with the sulfonate head of the guest at the transition state determine a contraction of the cavity $\Delta V_2^{\ddagger} \approx -2.5$ to -16 cm^3 mol^{-1}.

The pressure effects in host–guest complexations with CDs have been also investigated by fluorescence and EPR spectroscopy. Fluorescence spectroscopy was applied in Turro's work to evidence of formation of complexes of CDs and 1,3-bischromophoric propanes in aqueous solution at pressures up to 2.6 kbar [21]. Sueishi et al. published a high-pressure study in which EPR detection was employed to identify complexation of CDs with radical nitroxide probes in water at 0.7 kbar [22].

The pressure dependence on the association constant K for host–guest complexes of molecular tweezers **49** and **50** with aromatic guests **51**–**53** was established by Klärner and coworkers (Scheme 9.17) [23]. The effect is relatively small in comparison to temperature dependence.

Upon the applied pressure up to 2 kbar in the case of **51@49** complex, it was established that the association constant values change from 20.0 to 25.8 L mol^{-1}. For studied complexes, only small variations of reaction volumes around zero in $CDCl_3$ and C_6D_6 were obtained (ΔV from + 1.5 to -3 cm^3 mol^{-1}). These results suggest

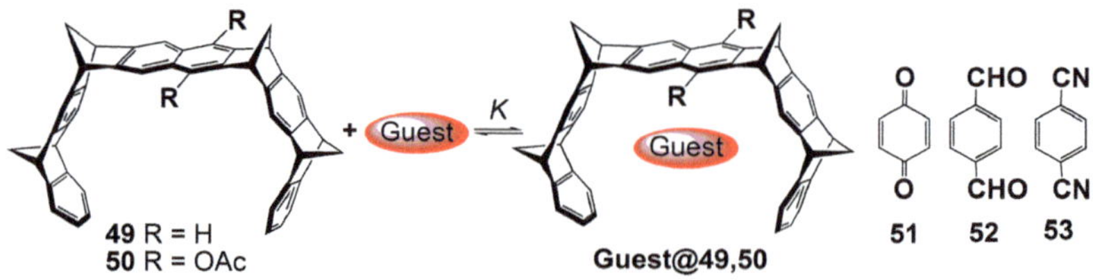

Scheme 9.17: Host–guest complexation of molecular tweezers and aromatic guests.

that the volume decrease as a result of complexation of the guest molecule within the cavity of tweezer is almost compensated by the volume increase resulted from the desolvation of host and guest during the complexation process.

References

[1] Toda, M., & Matsumoto, K. Constructions of supramolecules under high pressure. Rev High Press Sci Technol 1994, 3, 360–367.

[2] Kurotobi, K., & Murata, Y. A single molecule of water encapsulated in fullerene C_{60}. Science 2011, 333, 613–616.

[3] Zhang, R., Murata, M., Aharen, T., Wakamiya, A., Shimoaka, T., Hasegawa, T., & Murata, Y. Synthesis of a distinct water dimer inside fullerene C_{70}. Nature Chem 2016, 8, 435–441.

[4] Zhang, R., Futagoishi, T., Murata, M., Wakamiya, A., & Murata, Y. Synthesis and structure of an open-cage thiafullerene $C_{69}S$: reactivity differences of an open-cage C_{70} tetraketone relative to its C_{60} analogue. J Am Chem Soc 2014, 136, 8193–8196.

[5] Futagoishi, T., Murata, M., Wakamiya, A., & Murata, Y. Encapsulation and dynamic behavior of methanol and formaldehyde inside open-cage C_{60} derivatives. Angew Chem Int Ed 2017, 56, 1–6.

[6] Margetić, D. Carcerands and Hemicarcerands. In: J L. Atwood, ed. Comprehensive Supramolecular Chemistry II, vol. 6, Elsevier, Oxford, 2017, 115–137.

[7] Ruloff, R., Saudan, C., Merbach, AE., & Klärner, F-G. Future Perspectives: Application of High Pressure in Supramolecular Chemistry, In: High Pressure Chemistry, Synthetic, Mechanistic, and Supercritical Applications, R van Eldik & Klärner F-G, eds. Wiley-VCH, Weinheim, 2002, 348–368.

[8] Vazdar, M., Mitchell, AS., Warrener, RN., & Margetić, D. High pressure promoted exchange of guests from hemicarceplexes. Tetrahedron 2015, 71, 550–553.

[9] Amabilino, DB., Ashton, PR., Balzani, V., Boyd, SE., Credi, A., Young Lee, J., Menzer, S., Stoddart, JF., Venturi, M., & Williams, DJ. Oligocatenanes made to order. J Am Chem Soc 1998, 120, 4295–4307.

[10] Brown, CL., Philp, D., & Stoddart, JF. The self-assembly of a [2]catenane. Synlett 1991, 459–461.

[11] Ballardini, R., Balzani, V., Dehaen, W., Dell'Erba, AE., Raymo, FM., Stoddart, JF., & Venturi, M. Molecular meccano, 56. anthracene-containing [2]rotaxanes: synthesis, spectroscopic, and electrochemical properties. Eur J Org Chem 2000, 591–602.

[12] Tokunaga, Y., Wakamatsu, N., Ohbayashi, A., Akasaka, K., Saeki, S., Hisada, K., Goda, T., & Shimomura, Y. The effect of pressure on pseudorotaxane formation by using the slipping method. Tetrahedron Lett 2006, 47, 2679–2682.

[13] Tokunaga, Y., Wakamatsu, N., Ohiwa, N., Kimizuka, O., Ohbayashi, A., Akasaka, K., Saeki, S., Hisada, K., Goda, T., & Shimomura, Y. Effect of pressure on [2]pseudorotaxane formation and decomplexation and their corresponding activation volumes. J Org Chem 2010, 75, 4950–4956.

[14] Jeppesen, JO., Nielsen, KA., Perkins, J., Vignon, SA., Fabio, AD., Ballardini, R., Gandolfi, MT., Venturi, M., Balzani, V., Becher, J., & Stoddart, JF. Amphiphilic bistable rotaxanes. Chem Eur J 2003, 9, 2982–3007.

[15] Choi, JW., Flood, AH., Steuerman, DW., Nygaard, S., Braunschweig, AB., Moonen, NNP., Laursen, BW., Luo, Y., Delonno, E., Peters, AJ., Jeppesen, JO., Xu, K., Stoddart, JF., & Heath, JR. Ground-state equilibrium thermodynamics and switching kinetics of bistable [2]rotaxanes switched in solution, polymer gels, and molecular electronic devices. Chem Eur J 2006, 12, 261–279.

[16] Sørensen, A., Andersen, SS., Amar, H., Flood, AH., & Jeppesen, JO. Pressure effects in the synthesis of isomeric rotaxanes. Chem Commun 2013, 49, 5936–5938.

[17] Mugridge, JS., Zahl, A., van Eldik, R., Bergman, RG., & Raymond, KN. Solvent and pressure effects on the motions of encapsulated guests: tuning the flexibility of a supramolecular host. J Am Chem Soc 2013, 135, 4299–4306.

[18] le Noble, WJ., Srivastava, S., Breslow, R., & Trainor, G. Effect of pressure on two cyclodextrin-promoted ester hydrolyses. J Am Chem Soc 1983, 105, 2745–2748.

[19] Yang, C., Nakamura, A., Fukuhara, G., Origane, Y., Mori, T., Wada, T., & Inoue, Y.. Pressure and temperature-controlled enantiodifferentiating [4+4] photocyclodimerization of 2-anthracenecarboxylate mediated by secondary face- and skeleton-modified γ-cyclodextrins. J Org Chem 2006, 71, 3126–3136.

[20] Abou-Hamdan, A., Bugnon, P., Saudan, C., Lye, PG., & Merbach, AE. High-pressure studies as a novel approach in determining inclusion mechanisms: thermodynamics and kinetics of the host-guest interactions for R-cyclodextrin complexes. J Am Chem Soc 2000, 122, 592–602.

[21] Turro, NJ., Okubo, T., & Weed, GC. Enhancement of intramolecular excimer formation of 1,3-bichromophoric propanes via application of high pressure and via complexation via complexation with cyclodextrins. Protection from oxygen quenching. Photochem Photobiol 1982, 35, 325–329.

[22] Sueishi, Y., Tobisako, H., & Kotake, Y. Distinctive pressure effects on the association equilibrium in cyclodextrin group-inclusion complex as studied with electron paramagnetic resonance. J Phys Chem B 2004, 108, 12623–12627.

[23] Ruloff, R., Seelbach, UP., Merbach, AE., & Klärner, F-G. Molecular tweezers as synthetic receptors: the effect of pressure and temperature on the formation of host-guest complexes. J Phys Org Chem 2002, 15, 189–196.

10 Miscellaneous reactions

Some of the chemical processes that were not described in the previous chapters are listed in this chapter. These mostly include conformational changes such as *trans/cis* isomerization and hydrogen transfer.

10.1 Conformational changes

Plieninger and coworkers have shown that the ring-opening process of dimethyl-l,2-diphenyl-cyclobutene-*cis*-3,4-dicarboxylate **1** to (*E,Z*)-dimethyl-3,4-diphenylmuconate **2** is affected by pressure (Scheme 10.1) [1]. By the application of a pressure of 9.5 kbar, reaction is accelerated 1.4 times. Interestingly, the (*Z,Z*) dimethyl-3,4-diphenyl-muconate **3** was also obtained. The formation of this product by the thermal electrocyclic ring-opening process (**1**↔**2**) is forbidden by the Woodward–Hoffmann rules and the mechanism of its origin was sought. Kinetic measurements revealed the rate constants that indicate that product **3** is formed by an *E*–*Z* isomerization of **2** (**2**↔**3** process). Rate acceleration of this reaction by a factor of 11 is observed at 9.5 kbar, whereas the forthcoming ring closure process (**3**↔**4**) is very slow.

In the follow-up study, Plieninger and coworkers have obtained the activation volumes for the ring opening of methyl and dimethyl derivatives of cyclobutene **1**. The activation volume for conrotatory ring opening of the monomethyl derivative **6** (**6**↔**8** process) is negative, $\Delta V^{\neq} = -12.7$ cm^3 mol^{-1} (70 °C, *n*-pentane) (Scheme 10.2) [2]. The ring opening of the structurally related, dimethyl-3,4-dimethyl-l,2-diphenyl-cyclobutene-*cis*-3,4-dicarboxylate **7**, possessing additional methyl group at 3,4-positions is also pressure accelerated, affording dimethyl-(*E,Z*)-2,5-dimethyl-3,4-diphenyl muconate **10**. Activation volume for **7**↔**10** process is determined to be $\Delta V^{\neq} = -6.9$ cm^3 mol^{-1}, (70 °C, *n*-pentane). In contrast to previous experiments with **1**, *cis*–*trans* isomerization of **8** and **10** to obtain related (*Z,Z*)-muconates **9** and **11** does not take place under high-pressure conditions.

Kinetic measurements of the excited state *s-cis*↔*s-trans* isomerization rate of 2-(2-propenyl)anthracene **12** to **13** were carried out by Hara et al. in a high-pressure optical cell equipped with sapphire window in *n*-alkanes (pentane, hexane, octane and decane) (Scheme 10.3) [3]. The steady-state and picosecond time-resolved fluorescence spectroscopy measurements revealed that there is a viscosity dependence

https://doi.org/10.1515/9783110556841-010

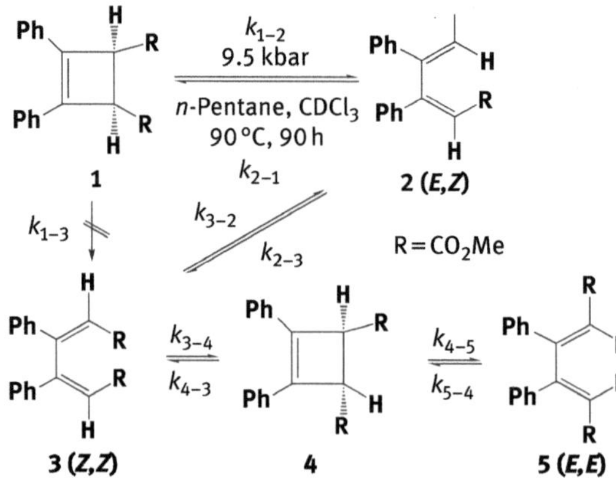

Scheme 10.1: Ring opening of dimethyl-l,2-diphenyl-cyclobutene-*cis*-3,4-dicarboxylate.

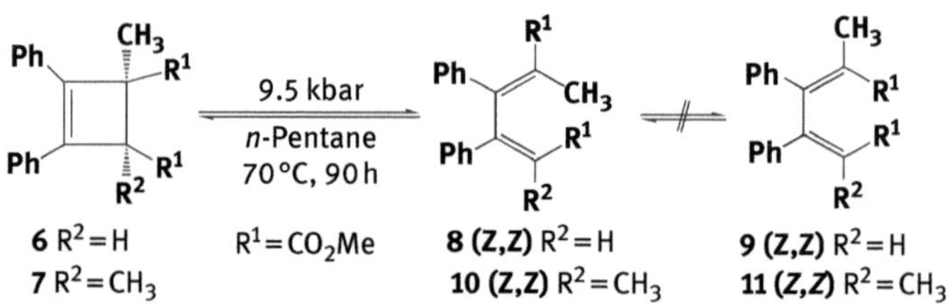

Scheme 10.2: Ring opening of methyl and dimethyl derivatives of cyclobutene.

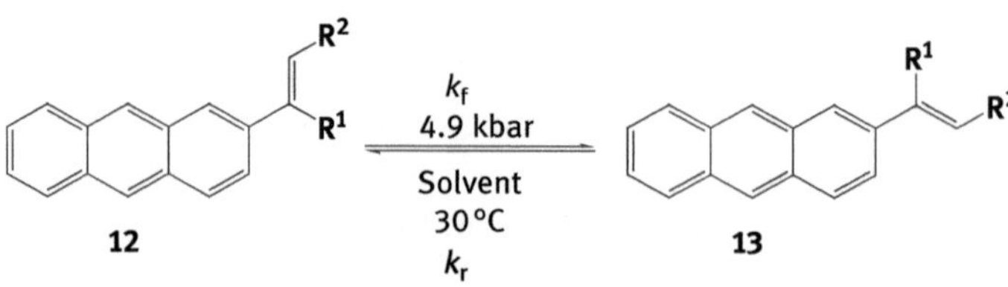

Scheme 10.3: *s-cis*⟷*s-trans* isomerization of 2-(2-propenyl)anthracene.

of the isomerization rate constant, which slightly decrease with the increased pressure (and increased solvent viscosity).

The solution of *α*-hexachlorocyclohexane ***α*-14** subjected to high pressure at 140 °C in HMPT (hexamethylphosphortriamide) isomerizes to ***γ*-14** in 30% yield (Scheme 10.4) [4]. Due to the HMPT base, the isomerization is in the competition with dehydrochlorination which is about 70% yield toward the positionally isomeric trichlorobenzenes. In the same reaction conditions, *γ*-HCH isomer (lindane ***γ*-14**) is converted to *α*-isomer ***α*-14** in only 1% yield, indicating that under high

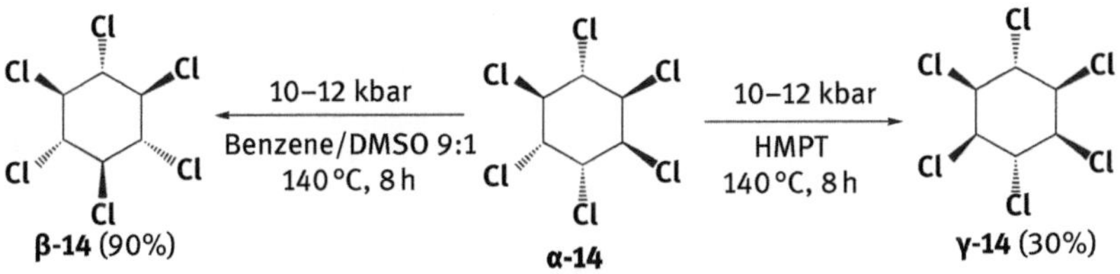

Scheme 10.4: Isomerization of α-hexachlorocyclohexane.

pressure, the equilibrium is in favor of lindane. An occurrence of other isomers was not observed. In the benzene/DMSO solution, isomerization of ***α*-14** goes surprisingly to ***β*-14** in 90% yield, without the formation of other isomers. The influence of the difference in polarity of aprotic solvent on the course of a reaction was noted. In benzene or the mixture with other aprotic solvents or at atmospheric pressure, no isomerization was observed.

The rate of racemization of (–)2′-methyl-2-nitro-biphenyl-6-carboxylic acid **15** and its methyl ester **16** is increased under pressure of 9 kbar by the factor of 5–10 (Scheme 10.5) [5]. The activation volumes were estimated for acid and methyl ester ($\Delta V^{\neq}$ = –28.3 and –32.0 $cm^3\ mol^{-1}$), respectively. The unusually high negative value for the acid is the consequence of the strong increase in racemization speed between 1 bar and 1 kbar. High pressure often helps to overcome steric hindrance and a similar effect seems to be strong here. The strong influence of the pressure is therefore especially noteworthy because the volume of starting compound and end product does not differ ($\Delta V_r = 0$), while in most reactions with high negative activation volume ΔV is very negative.

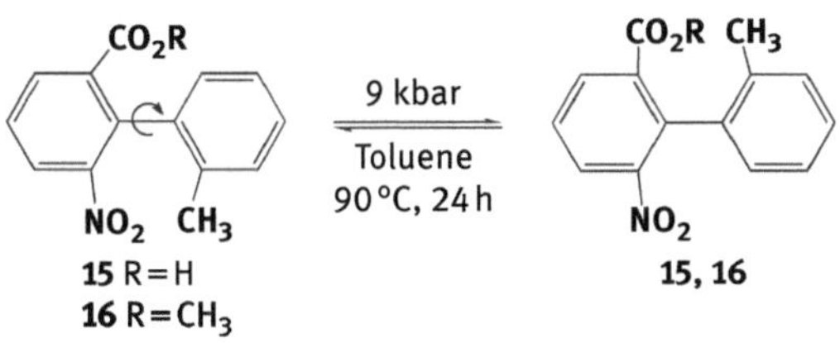

Scheme 10.5: Racemization of (–)2′-methyl-2-nitro-biphenyl-6-carboxylic acid.

The measured picosecond fluorescence lifetime as a function of pressure revealed the solvent viscosity effects on the twisted intramolecular charge-transfer state (TICT) in the excited state of 4,4′-diaminodiphenyl sulfone **17** (DAPS) (Scheme 10.6) [6]. Three different alcohols were employed (1-propanol, 1-butanol and 1-pentanol), and possible conformational transformation for the TICT formation were depicted in Scheme 10.6. Solvent relaxation control was established at lower pressure kinetics of TICT formation. With the increase of the pressure, the reaction mechanism changes

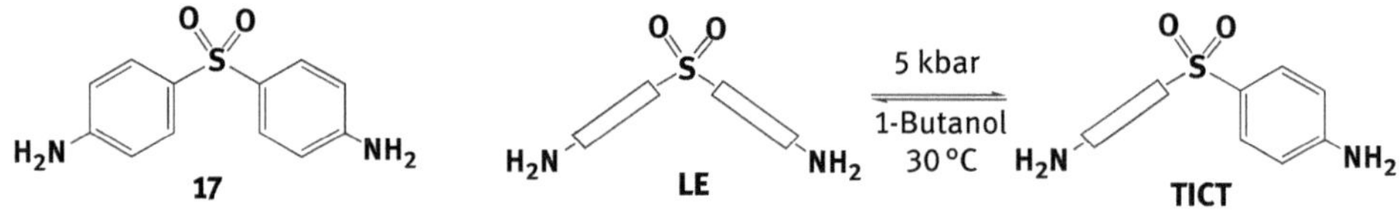

Scheme 10.6: Conformational transformation of 4,4′-diaminodiphenyl sulfone.

to high viscosity regime. Here the reaction proceeds through the unrelaxed path which is not dependent on the solvent coordinate on the free energy surface of the S_1 excited state. This behavior is called pressure-tuning effect of solvent viscosity.

Inoue et al. have found that the product chirality can be controlled, and in some cases actually switched, by changing the pressure from 1 bar to 4 kbar in the photosensitized enantiodifferentiating isomerization process of cyclooctene (Scheme 10.7) [7]. The enantiodifferentiating geometrical *Z*–*E* photoisomerization of cyclooctene **19*ZZ***, which is sensitized by optically active sensitizers (aromatic esters, benzene(poly) carboxylates), produces the antipodal (*E*)-cyclooctenes, (*S*)-(−)- and (*R*)-(+)-enantiomers, depending on the applied sensitizers. This enantiodifferentiating photoisomerization of **19*ZZ*** being significantly affected by pressure resulting in inversion of product chirality was also noted in the previous study [8]. The differential activation volumes $\Delta\Delta V^{\neq}_{S\text{-}R}$ ($=\Delta V^{\neq}_{S} - \Delta V^{\neq}_{R}$) vary from +3.50 to −5.56 $cm^3\ mol^{-1}$, and strongly depend on the sensitizer employed. The application of pressure as a tool in the control of the weak interactions which determine the stereochemical outcome in the excited state presumably involves a structurally well-defined exciplex intermediate, where the transfer of chiral information in the excited state takes place, that is, a photochemical induction of molecular chirality. For the isomerization of **18Z** carried out in supercritical carbon dioxide, the absolute $\Delta\Delta V^{\neq}_{S\text{-}R}$ values are greater than those obtained in pentane (+45 to −27.5 $cm^3\ mol^{-1}$) indicating that the transition-state structure and/or enantiodifferentiation mechanism involved differ [9]. In the case of (*Z*,*Z*)-cycloocta-1,5-diene **19ZZ**, the product ee did not display a simple pressure dependence [10]. Instead of

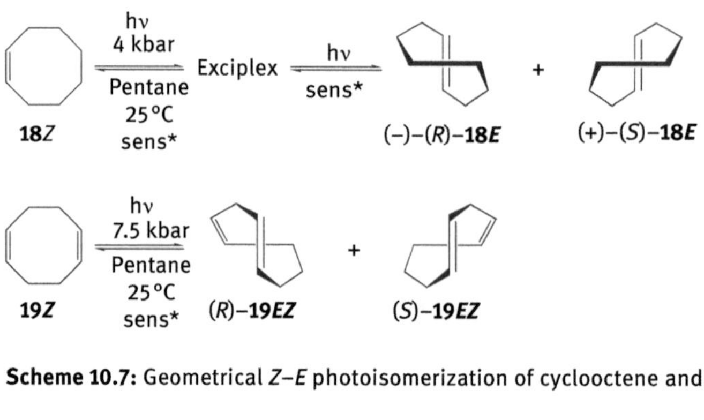

Scheme 10.7: Geometrical *Z*–*E* photoisomerization of cyclooctene and cycloocta-1,5-diene.

linear function, pressure dependence features different relationship in different pressure ranges, indicating that the $\Delta\Delta V^{\neq}$ and enantiodifferentiation mechanism diverge in each pressure range. This behavior might be caused by the change in the exciplex conformation with the pressure (more compact conformer).

Girard has investigated the DNA conformational changes upon pressurization. The solid-state crystal of the d(GGTATACC) oligonucleotide was subjected to pressures up to 20 kbar at 22 °C [11]. It was found that the geometry of base pairs is well preserved under compression up to at least 14 kbar. A continuous and gentle shortening of the average base-pair spacing upon increased hydrostatic pressure was noted. At normal pressure, the average base-pair distance is 3.34 Å at which shortens to 3.07 Å at 20 kbar. It was found that the double-helix base-paired architecture behaves as a molecular spring and that the nucleic acids have a capacity to endure high pressure and keep an almost invariable geometry of base pairs. In contrast, Takahashi and Sugimoto demonstrated that the volumetric characteristics of G-quadruplex DNA are affected by the application of pressure [12]. In the case of the thrombin binding aptamer, 5′-GGTTGGTGTGGTTGG-3′ in solution, the increase in pressure from 1 bar to 4 kbar (at room temperature) induces the unfolding of the G-quadruplex DNA structure. The structural characteristics are also affected by molecular crowding agents (ethylene glycol, PEG200 and PEG4000) influencing the denaturation under high pressure.

To evaluate the effects of high pressure on the structure, single-stranded complementary oligonucleotides (DNA, GCCCGCATATAT), (RNA, AUAUAUGCGCGC), (RNA, GGCCGGUUAAUU) and (RNA, AAUUAACCGGCC) were submitted to 6 kbar for 18 h [13]. The high pressure affected the heteroduplex and homoduplex structures, and both DNA–RNA and RNA–RNA duplexes change their conformation. On the other hand, the heteroduplex did not change its conformation.

Furthermore, it has been shown that at 6 kbar, poly(dGdC) changes its conformation from B-DNA to Z-DNA. The role of water at high pressure conformational changes on the transition has been interpreted in terms of the different water activity of B- and Z-forms of DNA or economy of hydration [14].

High-pressure effects on the conformation of proteins were employed in the formation of a molecular template. Gamoh and Kotsuki reported that the bovine serum albumin-adsorbed aminopropylsilica gel made a template with 4-cholesten-3-one in aqueous ethanol under high-pressure conditions (2–4 kbar, 4–12 h) [15].

10.2 Sigmatropic shifts

The study of high-pressure interconversion of previtamin D_3 **20** into vitamin D_3 **21** in three different solvents (benzene/toluene 3:8, ethanol and ethanol/water 9:1) was carried out by Dauben et al. to ascertain the reaction mechanism (Scheme 10.8) [16]. Assumed concerted [1,7] sigmatropic hydrogen shift process was corroborated by

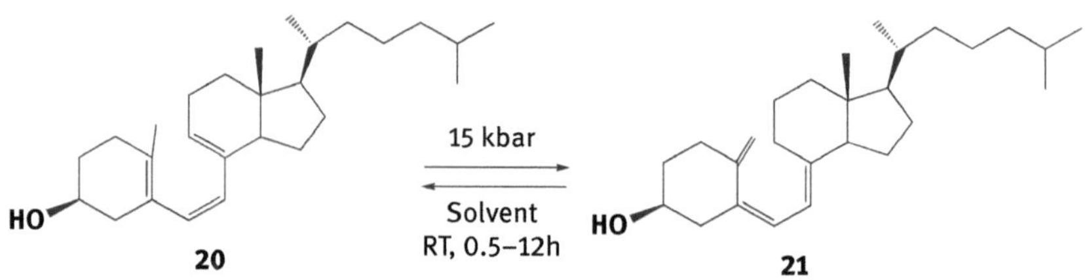

20

21

Scheme 10.8: Interconversion of previtamin D_3 into vitamin D_3.

experimental findings. There was evident rate increase by the application of 15 kbar pressure and serious increase in the amount of vitamin D_3 (82% in 11.5 h), in comparison to 80% in 13.3 days at atmospheric pressure (in benzene/toluene). Essentially no solvent dependence on rates was noted, at 1 bar or 15 kbar, ruling out an ionic mechanism. The values of activation volumes obtained (forward process $\Delta V^{\neq}{}_1 = -5.45$, -5.70 and -5.15 cm^3 mol^{-1}, backward process $\Delta V^{\neq}{}_2 = -5.66$, -6.11 and -5.60 cm^3 mol^{-1}, in benzene/toluene, ethanol and ethanol/water, respectively) indicate a concerted [1,7] sigmatropic hydrogen shift mechanism, since a stepwise diradical mechanism would have a positive $\Delta V^{\neq}$.

The experimental findings on the [1,5] sigmatropic shifts under high-pressure conditions show a range of activation volumes ($\Delta V^{\neq}$) that are dependent on the reaction mechanism and they vary from +10 to −30 cm^3 mol^{-1}. The stepwise diradical mechanism features a positive $\Delta V^{\neq}$ and the concerted mechanism typically has negative values (from −4 to −30 cm^3 mol^{-1}), with the larger ones possessing partial diradical character.

A clear distinction between two possible mechanisms of [1,5] sigmatropic shift reaction (concerted or biradical) was demonstrated by le Noble. Whereas the thermal [1,5] shift of 2-alkoxypyridine *N*-oxide **22** to *N*-alkoxy-2-pyridone **23** when R is benzyl it proceeds via a concerted process, but when R is benzhydryl it proceeds via diradicals at atmospheric pressure (Scheme 10.9) [17]. There was no solvent dependency on the reaction rates. Activation parameters could be used as a criterion for concertedness of the processes. Under high pressure, the activation volume for the concerted process (R = benzyl) is highly negative ($\Delta V^{\neq} = -30$ cm^3 mol^{-1}), whereas the R = benzhydryl proceeding via the stepwise mechanism has a positive value ($\Delta V^{\neq} = +10$ cm^3 mol^{-1}). For the benzyl substrate, rate acceleration by the increase of pressure is noted, whereas for benzhydryl substrate, rate retardation was found. These results could be rationalized by the assumption that the concerted reactions primarily involve a new bond formation, which is associated with a contraction of volume. Stepwise diradical process, on the other hands, is characterized by the original formation of radicals by bond cleavage which is characterized by the expansion of volume. By such, the concerted reactions are accelerated, whereas stepwise processes are retarded.

High-pressure kinetics of the symmetry-allowed sigmatropic shift reactions of 5-(trimethylsilyl)cyclopentadiene **24**, 5-formylpentamethylcyclopentadiene **25** and bullvalene **26** published by le Noble and coworkers showed a wide variety

22 R = Bn, Benzhydryl 23

Scheme 10.9: [1,5] H-shift of 2-alkoxypyridine *N*-oxide.

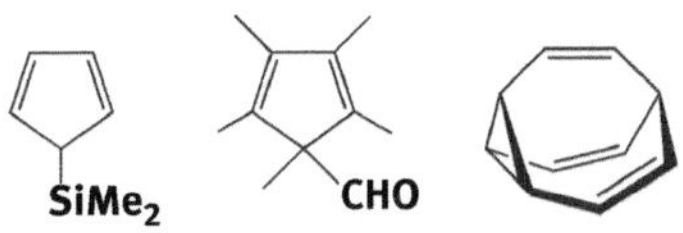

24 25 26

Scheme 10.10: Substrates for sigmatropic shift kinetic studies.

of mechanisms (Scheme 10.10) [18]. The hydrogen migration at 1.5 kbar in **24** has a very large contraction of volume in transition state ($\Delta V^{\neq}$ = −26.5 cm^3 mol^{-1} in $CDCl_3$ at 30 °C), suggesting that substantial charge separation is occurring. This result of zwitterionic transition state is well supported by large solvent effects on the reaction rate (in benzene $\Delta V^{\neq}$ = −12.5 cm^3 mol^{-1}). Much less charge separation could be deduced for formyl shift in **25** and degenerate shift in bullvalene **26** on the basis of their $\Delta V^{\neq}$ values (−4 and −0.5 cm^3 mol^{-1}, respectively, in benzene/freon and CS_2). In the case of bullvalene, the bond cleavage and bond formation processes in the very fast, degenerate Cope rearrangement proceed via a loose biradical transition state.

The [1,5] sigmatropic shift in the cycloheptatriene system, 3,3-diphenyl-3, 3a-dihydrocyclohepta[b]furan-2-one **27** to 3,3-diphenyl-3,6-dihydrocyclohepta[b]furan-2-one **28** showed some interesting features (Scheme 10.11) [19]. Since this transformation does not display sensitivity to solvent polarity that rules out ionic routes, the mechanism (concerted or biradical) was established by determination of high-pressure reaction kinetics. For this transformation, reaction volume was determined to be +3 cm^3 mol^{-1} and activation volume is −2.2 cm^3 mol^{-1}. This small $\Delta V^{\neq}$ value (in comparison to other 1,5-sigmatropic reactions) in conjunction with small pressure effects indicates that transition state might be biradicaloid in character. In the cycloheptatriene skeleton, the proton moves over the longer distance and shows stronger biradical character in the transition state.

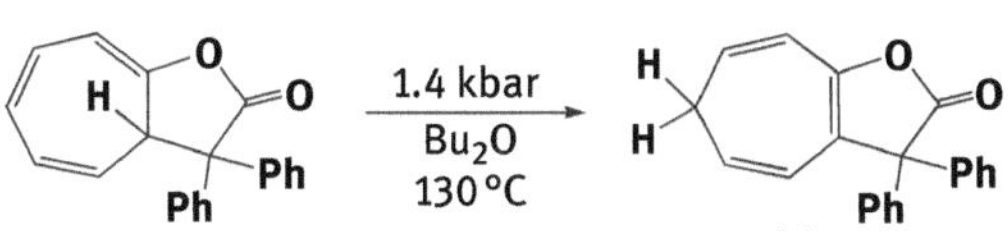

27 28

Scheme 10.11: [1,5] Sigmatropic shift in cycloheptatriene.

Klärner has found that the sigmatropic 1,5-hydrogen shift competes with an intramolecular Diels–Alder reaction in the thermal rearrangement of *Z*-1,3,8-nonatriene *Z*-**29** (Scheme 10.12) [20]. The hydrogen shift is the major reaction at ambient pressure and the intramolecular Diels–Alder reaction is the dominating process taking place at high pressure. The selectivity change induced by pressure shows that the number of new rings that are being formed in the transition state is important for the pressure-induced mechanistic effect. For the *Z*-**29**↔*cis*-**31** process, activation volume was established to be highly negative ($\Delta V^{\neq} = -24.8\ cm^3\ mol^{-1}$, 150 °C, *n*-hexane).

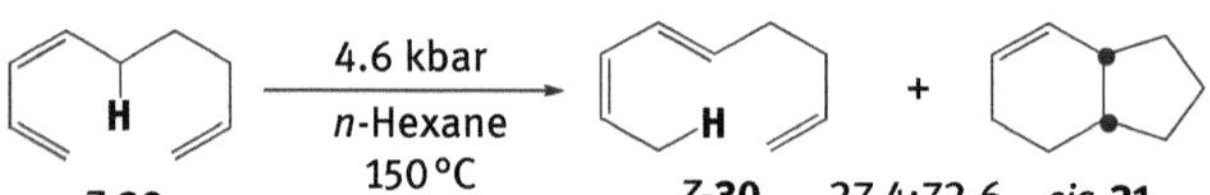

Z-**29** *Z*-**30** 27.4:72.6 *cis*-**31**
1 bar 68.8:31.2

Scheme 10.12: Rearrangement of *Z*-1,3,8-nonatriene.

10.3 Valence isomerization

The aromatization of benzene valence isomers under high pressure indicated that the aromatization of hexamethyl (Dewar benzene) **32** is accelerated by pressure (also found by Plieninger and coworkers) [21], whereas there was almost no effect on the aromatization process of the parent Dewar benzene **33**. (Scheme 10.13) [22]. Kinetic studies by le Noble (from ambient pressure to 2.15 kbar) revealed that the activation volume for hexamethylbenzene **32** is $\Delta V^{\neq} = -12\ cm^3\ mol^{-1}$ and the reaction volume is $-22\ cm^3\ mol^{-1}$ (140 °C, bromobenzene). For the aromatization of **33**, the small positive activation volume ($\Delta V^{\neq} = +5\ cm^3\ mol^{-1}$) in pyridine (42 °C) was established. For the early part of the aromatization of **33** is accelerated by pressure, albeit by a very small factor. The profitable packing of the methyl groups in the case of **32** was postulated to be the major difference among two reactions.

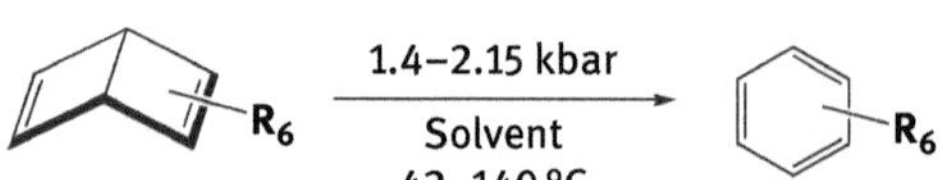

32 R = Me
33 R = H

34 R = Me (1.38 kbar, 40°C, bromobenzene)
35 R = H (2.15 kbar, 42°C, pyridine)

Scheme 10.13: Valence isomerization of Dewar benzene.

10.4 Hydrogen transfer

A hydrogen atom transfer mechanism leading to a pair of radicals in the rate-determining step is well supported by very negative volumes of activation for the reactions of hydroarenes with 2,3-dichloro-5,6-dicyano-1,4-quinone (DDQ) **36** or *o*-chloranil **38** producing the corresponding arenes **39** (Scheme 10.14) [23]. High-pressure measurements (1–3,500 bar) gave $\Delta V^{\neq}$ values for these two reagents in *tert*-butyl methyl ether (MTBE) solvent: $\Delta V^{\neq} = -13$ to -25 and in MeCN/AcOEt (1/1): -15 to -29 cm^3 mol^{-1}. On the other hand, a moderate dependence on solvent polarity was established. The stark pressure dependence of the kinetic deuterium isotope effect (k_H/k_D) for the reaction of 9,10-dihydroanthracene-D_4 with **37** was noted. The $k_H/k_D = 10.8$ at atmospheric pressure decreases to $k_H/k_D = 5.0$ at 3 kbar in MTBE. Similarly, volume of activation is much more negative going from nondeuterated to deuterated 9,10-dihydroanthracene $\Delta V^{\neq} = -25.9$ to -38.6 cm^3 mol^{-1}, respectively. The observed dependency was attributed to a tunneling in the hydrogen transfer. The change in the DDQ reagent to thymoquinone in the reaction of tetralin causes differences in k_H/k_D, which points to their mechanisms: stepwise hydrogen atom transfer and the pericyclic hydrogen transfer, respectively.

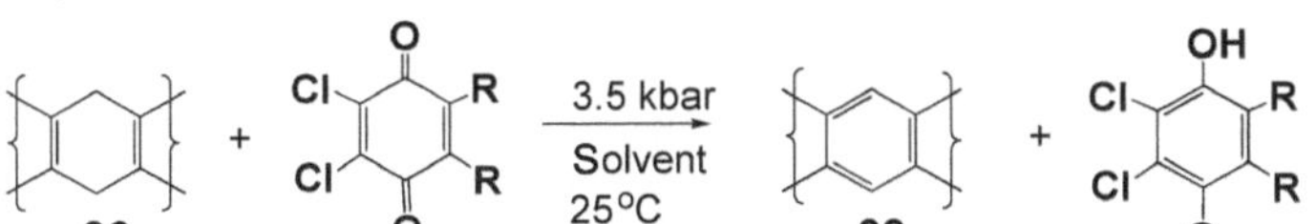

Scheme 10.14: Hydrogen atom transfer in hydroarene reactions with DDQ.

Similar $\Delta V^{\neq}$ values and the effect of pressure on the k_H/k_D were obtained by Isaacs et al. for the hydrogen transfer reaction of DDQ and leuco crystal violet [24]. In variance, the decomposition of diphenyldiazomethane **42** by benzoic acid **45** (Scheme 10.15) which occurs by a slow proton transfer has almost identical activation volumes for two isotopes: $\Delta V^{\neq}(H) = -13$ and $\Delta V^{\neq}(D) = -13$ cm^3 mol^{-1}, and the k_H/k_D isotope effects: 4.5/4.7 at 1 bar/2 kbar, respectively. Large negative $\Delta V^{\neq}$ values are mainly due to the electrostriction of solvent molecules brought about by the development of charges in transition state. Obtained $\Delta V^{\neq}$ values are very similar to -15 cm^3 mol^{-1} (1.67 kbar), which was measured for the proton transfer reaction between 4-nitrophenylnitromethane and 1,1′,3,3′-tetramethylguanidine [25]. In addition, small k_H/k_D were found for the series of hydrogen transfer reactions from phenols to radical [26]. Similar values of $\Delta V^{\neq} = -13 \pm 3$ cm^3 mol^{-1}, and lack of effect of pressure on the k_H/k_D values were found up to 2 kbar. These findings indicate that transition state and reactants are not influenced by a large extent by pressure with regard to their basic properties.

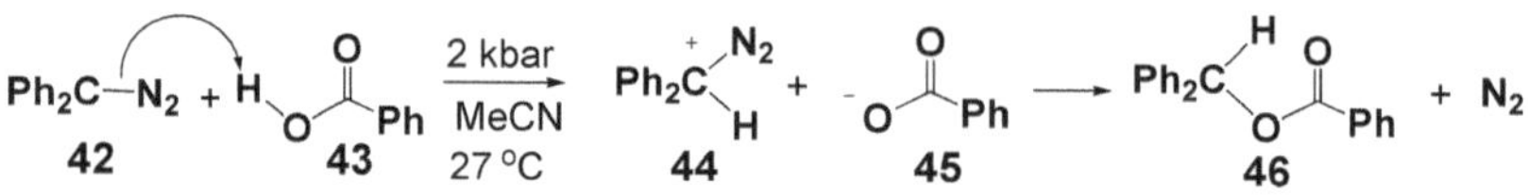

Scheme 10.15: Mechanism of decomposition of diphenyldiazomethane.

Hydrogen transfer equilibrium was also not greatly affected by pressure as judged from changes in the proton affinity of the methyl orange species induced by pressure up to 1 kbar [27]. The acid/base equilibrium of methyl orange in aqueous solution (20 °C) showed just a minor pressure facilitation of the deprotonation of methyl orange, with pK values ranging from 3.505 at atmospheric pressure to 3.445 at 1 kbar. In addition to small effect of increased pressure on the acid–base properties of methyl orange, just a small negative change in $\Delta V^{\neq}$ was established (from −6.9 to $-1.7\ cm^3\ mol^{-1}$, at 1 bar and 1 kbar, respectively).

References

[1] Mündnich, R., Plieninger, H., & Vogler, H. Hochdruckversuche VIII. Die Thermische Isomerisierung von 1,2-Diphenyl-cyclobuten-(1)-*cis*-3,4-dicarbonsäure-dimethyl-ester zu 3,4-Diphenylmuconsäure-estern. Tetrahedron 1977, 33, 2661–2665.

[2] Mündnich, R., & Plieninger, H. Hochdruckversuche IX. Die Thermische Isomerisierung von 1,2-Diphenyl-3-methyl-cyclobuten-(1)-*cis*-3,4-dicarbonsäuredimethylester und 3,4-Dimethyl -1,2-diphenyl-cyclobuten-(1)-*cis*-3,4-dicarbonsäuredimethylester. Tetrahedron 1978, 34, 887–889.

[3] Hara, K., Ito, N., & Kajimoto, O. High-pressure studies of dynamic solvent effects on large amplitude isomerization: 2-(2-propenyl)anthracene. J Phys Chem A 1997, 101, 2240–2244.

[4] Roemer-Mähler, J., Bieniek, D., & Korte, F. Hochdruckreaktionen V. Isomerisierung von *α*-Hexachlorcyclohexan zu *γ*- und *β*-Isomeren unter Hohen Drucken. Tetrahedron Lett 1972, 13, 4807–4808.

[5] Plieninger, H., & Schnelle, H-O. Über den Einfluss Hohen Druckes aud die Racemisierungsgescwhindigkeit der (-)2′-Methyl-2-nitrobiphenyl-6-carbonsäure und Ihres Methylesters. Tetrahedron 1977, 33, 1197–1199.

[6] Hara, K., Bulgarevich, DS., & Kajimoto, O. Pressure tuning of solvent viscosity for the formation of twisted intramolecular charge-transfer state in 4,4′-diaminodiphenyl sulfone in alcohol solution. J Chem Phys 1996, 104, 9431–9436.

[7] Inoue, Y., Wada, T., Asaoka, S., Sato, H., & Pete, J-P. Photochirogenesis: multidimensional control of asymmetric photochemistry. Chem Commun 2000, 251–259.

[8] Inoue, Y., Matsushima, E., & Wada, T. Pressure and temperature control of product chirality in asymmetric photochemistry. enantiodifferentiating photoisomerization of cyclooctene sensitized by chiral benzenepolycarboxylates. J Am Chem Soc 1998, 120, 10687–10696.

[9] Saita, R., Kaneda, M., Wada, T., Katoh, A., & Inoue, Y. First asymmetric photosensitization in supercritical fluid. exceptionally high pressure/density dependence of optical yield in photosensitized enantiodifferentiating isomerization of cyclooctene. Chem Lett 2002, 31, 860–861.

[10] Kaneda, M., Asaoka, S., Ikeda, H., Mori, T., Wada, T., & Inoue, Y. Discontinuous pressure effect upon enantiodifferentiating photosensitized isomerization of cyclooctene Chem Commun 2002, 1272–1273.

[11] Girard, E., Prangé, T., Dhaussy, A-C., Migianu-Griffoni, E., Lecouvey, M., Chervin, J-C., Mezouar, M., Kahn, R., & Four, R. Adaptation of the base-paired double-helix molecular architecture to extreme pressure. Nucl Acids Res 2007, 35, 4800–4808.

[12] Takahashi, S., & Sugimoto, N. Effect of pressure on the stability of G-quadruplex DNA: thermodynamics under crowding conditions. Angew Chem Int Ed 2013, 52, 13774–13778.

[13] Krzyzaniak, A., Fürste, JP., Erdmann, VA., Salanski, P., Jurczak, J., & Barciszewski, J. High pressure effects on conformation of homo- and heteroduplexes of nucleic acids. Biochimie 1996, 78, 862–867.

[14] Barciszewski, J., Jurczak, J., Porowski, S., Specht, T., & Erdmann, VA. The role of water structure in conformational changes of nucleic acids in ambient and high-pressure conditions. Eur J Biochem 1999, 260, 293–307.

[15] Gamoh, K., & Kotsuki, H. Molecular template formation on serum albumin under a high-pressure condition. Bunseki Kagaku 2000, 49, 551–556.

[16] Dauben, WH., Kowalczyk, BA., & Funhoff, DJH. Organic reactions at high pressure. Interconversion of previtamin D_3 and vitamin D_3.Tetrahedron Lett 1988, 29, 3021–3024.

[17] le Noble, WJ., & Daka, M R. Effect of pressure on concerted and stepwise sigmatropic shifts. J Am Chem Soc 1978, 100, 5961–5962.

[18] Schulman, EM., Merbach, AE., Turin, M., Wedinger, R., & le Noble, WJ. Effect of pressure on the rates of sigmatropic shifts. volume as an indication of a wide mechanistic range. J Am Chem Soc 1983, 105, 3988–3991.

[19] Sugiyama, S., & Takeshita, H. The pressure effect on the 1,5-sigmatropic hydrogen shift of tropylidene derivative, 3,3-diphenyl-3,3a-dihydrocyclohepta[b]furan-2-one. Chem Lett 1986, 15, 1203–1204.

[20] Diedrich, MK., Hochstrate, D., F-G, Klärner., & Zimny, B. The effect of pressure on pericyclic rearrangements (intramolecular Diels-Alder reaction, Cope rearrangement, and electrocyclic reaction). High Press Res 1994, 13, 7–11.

[21] Mündnich, R., & Plieninger, H. Hochdruckversuche VI. Über die Umlagerung von Hexamethyldewarbenzol in Hexamethylbenzol. Tetrahedron 1976, 32, 2335–2337.

[22] le Noble, WJ., Brower, KR., Brower, C., & Chang, S. Pressure effects on the rates of aromatization of hexamethyl (Dewar benzene) and Dewar benzene. Volume as a factor in crowded molecules. J Am Chem Soc 1982, 104, 3150–3152.

[23] Wurche, F., Sicking, W., Sustmann, R., F-G, Klärner., & Rüchardt, C. The effect of pressure on hydrogen transfer reactions with quinones. Chem Eur J 2004, 10, 2707–2721.

[24] Isaacs, NS., Javaid, K., & Rannala, E. Reactions at high pressure. Part 5. The effect of pressure on some primary kinetic isotope effects. J Chem Soc Perkin Trans 2 1978, 0, 709–711.

[25] Hubbard, CD., Wilson, CJ., & Caldin, EF. Activation volumes and volumes of reaction for the proton-transfer reaction between 4-nitrophenylnitromethane and 1,1′,3,3′-tetramethylguanidine in some aprotic solvents. J Am Chem Soc 1976, 98, 1870–1875.

[26] Palmer, DA., & Kelm, H. Pressure effects on the rates of some hydrogen/deuterium transfer reactions between 2,2-Diphenyl-1-picrylhydrazyl and substituted phenols. Austral J Chem 1977, 30, 1229–1239.

[27] Suleimenov, OM., & Boily, J-F. Pressure dependence of the acid/base equilibrium of methyl orange in aqueous solutions to 1,000 bar at 20 °C. J Solut Chem 2006, 35, 541–549.

Author Index

https://doi.org/10.1515/9783110556841-011

Subject Index